THREE-DIMENSIONAL ELECTROMAGNETICS

Proceedings of the Second International Symposium

Methods in Geochemistry and Geophysics
(Volumes 1–28 are out of print)

1. A.S. RITCHIE — CHROMATOGRAPHY IN GEOLOGY
2. R. BOWEN — PALEOTEMPERATURE ANALYSIS
3. D.S. PARASNIS — MINING GEOPHYSICS
4. I. ADLER — X-RAY EMISSION SPECTROGRAPHY IN GEOLOGY
5. THE LORD ENERGLYN AND L. BREADLY — ANALYTICAL GEOCHEMISTRY
6. A.J. EASTON — CHEMICAL ANALYSIS OF SILICATE ROCKS
7. E.E. ANGINO AND G.K. BILLINGS — ATOMIC ABSORPTION SPECTROMETRY IN GEOLOGY
8. A. VOLBORTH — ELEMENTAL ANALYSIS IN GEOCHEMISTRY, A: MAJOR ELEMENTS
9. P.K. BHATTACHARYA AND H.P. PATRA — DIRECT CURRENT GEOELECTRIC SOUNDING
10. J.A.S. ADAMS AND P. GASPARINI — GAMMA-RAY SPECTROMETRY OF ROCKS
11. W. ERNST — GEOCHEMICAL FACIES ANALYSIS
12. P.V. SHARMA — GEOPHYSICAL METHODS IN GEOLOGY
13. C.H. CHEN (Editor) — COMPUTER-AIDED SEISMIC ANALYSIS AND DISCRIMINATION
14A. O. KOEFOED — GEOSOUNDING PRINCIPLES, 1. RESISTIVITY SOUNDING MEASUREMENTS
14B. H.P. PATRA AND K. MALLICK — GEOSOUNDING PRINCIPLES, 2. TIME-VARYING GEOELECTRIC SOUNDINGS
15. A.A. KAUFMAN AND G.V. KELLER — THE MAGNETOTELLURIC SOUNDING METHOD
16. A.A. KAUFMAN AND G.V. KELLER — FREQUENCY AND TRANSIENT SOUNDINGS
17. C.H. CHEN (Editor) — SEISMIC SIGNAL ANALYSIS AND DISCRIMINATION
18. J.E. WHITE — UNDERGROUND SOUND APPLICATION OF SEISMIC WAVES
19. M.N. BERDICHEVSKY AND M.S. ZHDANOV — ADVANCED THEORY OF DEEP GEOMAGNETIC SOUNDINGS
20A. A.A. KAUFMAN AND G.V. KELLER — INDUCTIVE MINING PROSPECTING, PART I: THEORY
21. A.W. WYLIE — NUCLEAR ASSAYING OF MINING BOREHOLES - AN INTRODUCTION
22. C.H. CHEN (Editor) — SEISMIC SIGNAL ANALYSIS AND DISCRIMINATION III
23. R.P. PHILP — FOSSIL FUEL BIOMARKERS
24. R.B. JOHNS (Editor) — BIOLOGICAL MARKERS IN THE SEDIMENTARY RECORD
25. J.C. D' ARNAUD GERKENS — FOUNDATION OF EXPLORATION GEOPHYSICS
26. P. TYGEL AND P. HUBRAL — TRANSIENT WAVES IN LAYERED MEDIA
27. A.A. KAUFMAN AND GV. KELLER — INDUCTION LOGGING
28. J.G. NEGI AND P.D. SARAF — ANISOTROPY IN GEOELECTROMAGNETISM
29. V.P. DIMRI — DECONVOLUTION AND INVERSE THEORY - APPLICATION TO GEOPHYSICAL PROBLEMS
30. K.-M. STRACK — EXPLORATION WITH DEEP TRANSIENT ELECTROMAGNETICS
31. M.S. ZHDANOV AND G.V. KELLER — THE GEOELECTRICAL METHODS IN GEOPHYSICAL EXPLORATION
32. A.A. KAUFMAN AND A.L. LEVSHIN — ACOUSTIC AND ELASTIC WAVE FIELDS IN GEOPHYSICS, I
33. A.A. KAUFMAN AND P.A. EATON — THE THEORY OF INDUCTIVE PROSPECTING
34. A.A. KAUFMAN AND P. HOEKSTRA — ELECTROMAGNETIC SOUNDINGS
35. M.S. ZHDANOV AND P.E. WANNAMAKER — THREE-DIMENSIONAL ELECTROMAGNETICS

Methods in Geochemistry and Geophysics, 35

THREE-DIMENSIONAL ELECTROMAGNETICS

Proceedings of the Second International Symposium

Edited by

Michael S. ZHDANOV

and

Philip E. WANNAMAKER
University of Utah
Salt Lake City
UTAH, U.S.A.

2002
ELSEVIER

Amsterdam - Boston - London - New York - Oxford - Paris
San Diego - San Francisco - Singapore - Sydney - Tokyo

ELSEVIER SCIENCE B.V.
Sara Burgerhartstraat 25
P.O. Box 211, 1000 AE Amsterdam, The Netherlands

First edition 2002

Library of Congress Cataloging in Publication Data
A catalog record from the Library of Congress has been applied for.

British Library Cataloguing in Publication Data

Zhdanov, M. S. (Mikhail Semenovich)
Three-dimensional electromagnetics : proceedings of the Second International Symposium. - (Methods in geochemistry and geophysics ; 35)
1.Electric prospecting - Congresses
I.Title II.Wannamaker, P. E.
622.1'54

ISBN: 0 444 50429 X
ISSN: 0076-6895

∞ The paper used in this publication meets the requirements of ANSI/NISO Z39.48-1992 (Permanence of Paper).
Printed in The Netherlands.

PREFACE

The Second International Symposium on Three-Dimensional Electromagnetics was held at the University of Utah, Salt Lake City, in 1999. This symposium honors the memory of Gerald W. Hohmann, late Professor of Geophysics at the University of Utah, and follows the highly successful First Symposium held in Ridgefield, CT, organized in 1995 by Schlumberger–Doll Research (M. Oristaglio and B. Spies, Eds.). Nearly 25 years ago, Jerry published a seminal paper on EM scattering from 3-D structures in the earth due to diverse sources using the integral equation method [1]. This achievement and the considerable body of subsequent work largely inspired by it, have shown how unique the behavior of fully 3-D structures can be. Very sadly, Jerry died May 23, 1992, at the age of 51 from complications of a cancer. His career history is described further in the citation for Honorary Membership in the Society of Exploration Geophysicists [2].

The G.W. Hohmann Memorial Trust for Teaching and Research in Applied Electrical Geophysics was established to be a lasting and active memorial to Jerry's achievements as scientist and teacher. This symposium was held on behalf of the Trust and, through meeting proceeds and donations, and matching funds from the SEG, we have created several student scholarships and a career achievement award from the Society [3].

There were three full days of presentations at this Symposium, with 32 oral and 56 poster papers spread over six half-day sessions: three sessions on 3D EM modeling, one session on 3D EM inversion and two sessions on 3D EM in practice. The sheer number of works, their diversity, and the truly international character of the efforts attested to the vigor with which the problems of the field are pursued today, despite adverse cyclicity in the resource industry or the inscrutable politics of national support mechanisms.

The authors of the papers submitted to the Symposium have been encouraged to produce full papers for submission to a book, published as a companion to that of the First Symposium [4]. Sixteen articles, representing the work of 36 authors from different academic, government, and industrial research labs, form the volume that we present to the electromagnetic induction community.

The papers in the volume were grouped in three parts:

Part I 3-D EM modeling,
Part II 3-D EM inversion,
Part III 3-D EM in practice.

The presented papers cover a wide range of topics in forward modeling and inversion based on new fast approximate approaches and new efficient solutions by integral equation, finite difference and finite elements techniques. If 1980s were the decade of rapid development in 3D seismics, the 1990s became the decade of growing interest of practical geophysicists in 3D EM modeling and inversion methods. It is well worth comparing the papers here with those of the proceedings of the First Symposium to understand what fundamentally new ideas have arisen and what concepts have become established into frequent practice. However, the general message is clear – 3-D modeling and inversion is a reality, and not an illusion.

Finally, we have many to thank for the production of this volume. Our first acknowledgement is to the authors themselves, for their contribution to this manuscript. The members of the Trust, especially Michael Oristaglio, are thanked for encouraging the Second Symposium and for it's hosting at the University of Utah. We are thankful to Perry Eaton from Newmont Gold, and John Weaver from the University of Victoria, BC, the Symposium technical co-chairs, for their help with selecting papers and organizing the program. We are indebted to Nikolay Golubev from the Center for Electromagnetic Modeling and Inversion, University of Utah, for his help with the correction of the papers and producing the final document.

Profits from sales of this book will be donated to the G.W. Hohmann Trust for it's stated purposes of teaching and research in electromagnetic geophysics.

MICHAEL ZHDANOV
PHIL WANNAMAKER

University of Utah
Salt Lake City, Utah, USA
25 October, 2001

[1] Hohmann, Gerald W., 1975, Three-dimensional induced polarization and electromagnetic modeling: Geophysics, 40, 309–324.

[2] Honorary Membership Citation for Gerald W. Hohmann, by Stanley H. Ward and Misac Nabighian, Sixty-second Annual International Meeting, Society of Exploration Geophysicists, Presidential Session, October, 1992.

[3] Inman, J., 1998, Hohmann Trust awards first fellowship, scholarship: The Leading Edge, October, 1368.

[4] Three-dimensional electromagnetics, M. Oristaglio and B. Spies (Eds.), 1999, Geophys. Devel. Ser., no. 7, Soc. Explor. Geophys. Tulsa, 724 pp.

CONTENTS

Part II. 3-D EM INVERSION

Part I
3-D EM Modeling

Chapter 1

AN INTEGRAL EQUATION SOLUTION TO THE GEOPHYSICAL ELECTROMAGNETIC FORWARD-MODELLING PROBLEM

Colin G. Farquharson and Douglas W. Oldenburg

UBC – Geophysical Inversion Facility, Department of Earth and Ocean Sciences, University of British Columbia, Vancouver, Canada

Abstract: We investigate the use of edge element basis vectors in an integral equation solution for three-dimensional geophysical electromagnetic modelling. Expansion of the total electric field within the region of anomalous conductivity in terms of these basis vectors gives a bilinear variation of each component of the field within a cell in the two directions perpendicular to the component (and so a divergence-free but not curl-free field within a cell), and continuity of the tangential electric field between two cells. In addition, we use a form of the electric field integral equation that explicitly involves the charge densities on cell faces associated with any discontinuity of the normal component of the current density. The two types of integrals in the integral equation — the volume integration of the scattering current within each cell, and the surface integration of the charge density on the faces of each cell — are computed using Gaussian quadrature. The system of equations to be solved is constructed using the Galerkin approach. In this preliminary study, we consider the simple case of a homogeneous halfspace as the background model. Comparisons with results from the literature and other codes have been promising. We include here two examples: one for a grounded electric line source at low frequency (3 Hz) on the surface of a halfspace ($\sigma = 0.02$ S/m) in which a more conductive vertical prism ($\sigma = 0.2$ S/m) is buried, and one for a magnetic dipole source-receiver combination over a conductive cube ($\sigma = 100$ S/m) in a resistive ($\sigma = 10^{-4}$ S/m) background.

1. INTRODUCTION

Traditional integral equation formulations for electromagnetic modelling in geophysics (e.g., Hohmann, 1975), which use pulse basis functions to represent the electric field, fail for large conductivity contrasts. Newman and Hohmann (1988) provide a means of remedying this problem by grouping together pulse basis functions to explicitly form current loops. We have implemented, as we shall describe, the more sophisticated edge element basis functions in the anticipation that the inaccuracies of the traditional approach can be avoided. Edge element basis functions have desirable properties: they are divergence free but not curl free, and give a solution for which the tangential electric field is, by construction, continuous across cell faces. They have been successfully used in finite element solutions to electromagnetic forward modelling (e.g., Jin, 1993). The use of similarly sophisticated, although different, basis functions (linear variation of each electric field component in all three dimensions, and imposition of continuity of tangential field components and normal current density) has been presented by Slob and van den Berg (1999).

We start by deriving the form of the integral equation upon which our numerical solution is based, a form that explicitly involves the charge densities on cell faces across which there is a difference in conductivity, and thus directly reduces at zero frequency to the integral equation solution for the direct-current (DC) resistivity forward-modelling problem.

2. THE INTEGRAL EQUATION

Consider Maxwell's two curl equations in the frequency domain (assuming a time dependence of $e^{-i\omega t}$, and the quasi-static approximation):

$$\nabla \times \mathbf{E} = i\omega\mu\mathbf{H}, \tag{1.1}$$

$$\nabla \times \mathbf{H} = \sigma\mathbf{E} + \mathbf{J}^{\mathrm{I}}, \tag{1.2}$$

and the statement of conservation of charge:

$$\nabla \cdot \left(\sigma\mathbf{E}\right) = -\nabla \cdot \mathbf{J}^{\mathrm{I}}, \tag{1.3}$$

where $\mathbf{E}(\mathbf{r},\omega)$ and $\mathbf{H}(\mathbf{r},\omega)$ are the total electric and magnetic fields, $\mu = \mu_0$ is the magnetic permeability of free space, $\sigma = \sigma(\mathbf{r})$ is the electrical conductivity of our model Earth, and $\mathbf{J}^{\mathrm{I}}$ is the impressed electric current density. Taking the curl of Equation (1.1), using Equation (1.2) to eliminate $\mathbf{H}$, and exploiting the vector identity $\nabla \times \nabla \times \mathbf{E} = -\nabla^2\mathbf{E} + \nabla(\nabla \cdot \mathbf{E})$, gives

$$-\nabla^2\mathbf{E} + \nabla(\nabla \cdot \mathbf{E}) - i\omega\mu\sigma\mathbf{E} = i\omega\mu\mathbf{J}^{\mathrm{I}}. \tag{1.4}$$

Consider now a background conductivity $\sigma_{\mathrm{b}}(\mathbf{r})$, and the electric and magnetic fields $\mathbf{E}_{\mathrm{b}}$ and $\mathbf{H}_{\mathrm{b}}$ that exist in this background model for the impressed current density $\mathbf{J}^{\mathrm{I}}$. The two Maxwell's curl equations for this scenario are

$$\nabla \times \mathbf{E}_{\mathrm{b}} = i\omega\mu\mathbf{H}_{\mathrm{b}}, \tag{1.5}$$

$$\nabla \times \mathbf{H}_{\mathrm{b}} = \sigma_{\mathrm{b}}\mathbf{E}_{\mathrm{b}} + \mathbf{J}^{\mathrm{I}}, \tag{1.6}$$

and the statement of conservation of charge is

$$\nabla \cdot \left(\sigma_{\mathrm{b}}\mathbf{E}_{\mathrm{b}}\right) = -\nabla \cdot \mathbf{J}^{\mathrm{I}}. \tag{1.7}$$

Performing the same manipulations as for the total fields gives

$$-\nabla^2\mathbf{E}_{\mathrm{b}} + \nabla(\nabla \cdot \mathbf{E}_{\mathrm{b}}) - i\omega\mu\sigma_{\mathrm{b}}\mathbf{E}_{\mathrm{b}} = i\omega\mu\mathbf{J}^{\mathrm{I}}. \tag{1.8}$$

Now consider the total electric field in the model with $\sigma = \sigma(\mathbf{r})$ to be the sum of the background electric field and a 'secondary' part: $\mathbf{E} = \mathbf{E}_{\mathrm{b}} + \mathbf{E}_{\mathrm{s}}$. Also consider $\sigma(\mathbf{r}) = \sigma_{\mathrm{b}}(\mathbf{r}) + \Delta\sigma(\mathbf{r})$. Substituting these two expressions into Equation (1.4) gives

$$-\nabla^2(\mathbf{E}_{\mathrm{b}} + \mathbf{E}_{\mathrm{s}}) + \nabla(\nabla \cdot (\mathbf{E}_{\mathrm{b}} + \mathbf{E}_{\mathrm{s}})) - i\omega\mu(\sigma_{\mathrm{b}} + \Delta\sigma)(\mathbf{E}_{\mathrm{b}} + \mathbf{E}_{\mathrm{s}}) = i\omega\mu\mathbf{J}^{\mathrm{I}}. \tag{1.9}$$

Expanding the terms in parentheses, and using Equation (1.8) to cancel many of the resulting terms, gives

$$-\nabla^2\mathbf{E}_s + \nabla(\nabla\cdot\mathbf{E}_s) - i\omega\mu\sigma_b\mathbf{E}_s = i\omega\mu\Delta\sigma\mathbf{E}. \tag{1.10}$$

Likewise substituting $\mathbf{E} = \mathbf{E}_b + \mathbf{E}_s$ and $\sigma = \sigma_b + \Delta\sigma$ into Equation (1.3), and using Equation (1.7) to eliminate $\mathbf{J}^I$, gives

$$\nabla\cdot\mathbf{E}_s = -\frac{1}{\sigma_b}\nabla\sigma_b\cdot\mathbf{E}_s - \frac{1}{\sigma_b}\nabla\cdot(\Delta\sigma\,\mathbf{E}). \tag{1.11}$$

We now assume that our background model is a homogeneous halfspace, meaning that $\nabla\sigma_b$ is non-zero only on the Earth–air interface (at $z = 0$). Rewriting Equation (1.11) using more symbolic terms that avoid derivatives of conductivity discontinuities (both at the Earth–air interface and between cells once we discretise the anomalous region) gives

$$\nabla\cdot\mathbf{E}_s = \nabla\cdot\mathbf{E}_s\big|_{z=0} + \nabla\cdot\mathbf{E}_s\big|_{V_a}, \tag{1.12}$$

where the first term on the right-hand side represents the contribution from the Earth–air interface, and the second term represents the contribution from the region of anomalous conductivity (which occupies the volume V_a). The differential equation for the secondary electric field (Equation (1.10)) therefore becomes (cf. Hohmann, 1987)

$$\nabla^2\mathbf{E}_s - \nabla(\nabla\cdot\mathbf{E}_s|_{z=0}) + i\omega\mu\sigma_b\mathbf{E}_s = -i\omega\mu\Delta\sigma\mathbf{E} + \nabla(\nabla\cdot\mathbf{E}_s|_{V_a}). \tag{1.13}$$

Rewriting this equation using notation that simplifies the right-hand side:

$$\nabla^2\mathbf{E}_s - \nabla(\nabla\cdot\mathbf{E}_s|_{z=0}) + i\omega\mu\sigma_b\mathbf{E}_s = \mathbf{Q}, \tag{1.14}$$

where $\mathbf{Q} = -i\omega\mu\Delta\sigma\mathbf{E} + \nabla(\nabla\cdot\mathbf{E}_s|_{V_a})$. Consider now a vector Green's function $\mathbf{g}^k$ such that

$$\nabla^2\mathbf{g}^k(\mathbf{r};\mathbf{r}') - \nabla(\nabla\cdot\mathbf{g}^k(\mathbf{r};\mathbf{r}')|_{z=0}) + i\omega\mu\sigma_b\mathbf{g}^k(\mathbf{r};\mathbf{r}') = -\delta(\mathbf{r}-\mathbf{r}')\hat{\mathbf{u}}_k, \tag{1.15}$$

where $\hat{\mathbf{u}}_k$ is the unit vector in the kth direction. All components of $\mathbf{g}^k \to 0$ as $|\mathbf{r}| \to \infty$. What conditions $\mathbf{g}^k$ satisfies on the Earth–air interface will be discussed in Section 7. Taking the scalar product of $\mathbf{g}^k(\mathbf{r};\mathbf{r}')$ and Equation (1.14), the scalar product of $\mathbf{E}_s$ and Equation (1.15), subtracting the second resulting equation from the first, and integrating over all space, gives

$$\begin{aligned}\int_V \Big\{&\mathbf{g}^k\cdot[\nabla'^2\mathbf{E}_s - \nabla'(\nabla'\cdot\mathbf{E}_s|_0) + i\omega\mu\sigma_b\mathbf{E}_s]\\ &- \mathbf{E}_s\cdot[\nabla'^2\mathbf{g}^k - \nabla'(\nabla'\cdot\mathbf{g}^k|_0) + i\omega\mu\sigma_b\mathbf{g}^k]\Big\}\,\mathrm{d}v'\\ = \int_V &\Big\{\mathbf{g}^k\cdot\mathbf{Q} + \mathbf{E}_s\cdot\delta(\mathbf{r}-\mathbf{r}')\hat{\mathbf{u}}_k\Big\}\,\mathrm{d}v'.\end{aligned} \tag{1.16}$$

Simplifying both the right- and left-hand sides gives

$$\begin{aligned}\int_V \Big\{\mathbf{g}^k\cdot\nabla'^2\mathbf{E}_s - \mathbf{E}_s\cdot\nabla'^2\mathbf{g}^k\Big\}\,\mathrm{d}v' - \int_V \Big\{\mathbf{g}^k\cdot\nabla'(\nabla'\cdot\mathbf{E}_s|_0) - \mathbf{E}_s\cdot\nabla'(\nabla'\cdot\mathbf{g}^k|_0)\Big\}\,\mathrm{d}v'\\ = \int_V \mathbf{g}^k\cdot\mathbf{Q}\,\mathrm{d}v' + E_k^s,\end{aligned} \tag{1.17}$$

where E_k^{s} is the kth component of the secondary electric field. Rewriting the first integral on the left-hand side of the above equation in terms of the components of $\mathbf{E}_{\mathrm{s}}$ and $\mathbf{g}^k$ gives

$$\int_V \{\mathbf{g}^k \cdot \nabla^{2\prime}\mathbf{E}_{\mathrm{s}} - \mathbf{E}_{\mathrm{s}} \cdot \nabla^{2\prime}\mathbf{g}^k\}\mathrm{d}v' = \int_V \{g_x^k \nabla^{2\prime} E_x^{\mathrm{s}} - E_x^{\mathrm{s}} \nabla^{2\prime} g_x^k\}\mathrm{d}v'$$

$$+ \int_V \{g_y^k \nabla^{2\prime} E_y^{\mathrm{s}} - E_y^{\mathrm{s}} \nabla^{2\prime} g_y^k\}\mathrm{d}v' + \int_V \{g_z^k \nabla^{2\prime} E_z^{\mathrm{s}} - E_z^{\mathrm{s}} \nabla^{2\prime} g_z^k\}\mathrm{d}v', \tag{1.18}$$

$$= \int_{\partial V} \{g_x^k \nabla' E_x^{\mathrm{s}} - E_x^{\mathrm{s}} \nabla' g_x^k\} \cdot \hat{\mathbf{n}}\,\mathrm{d}s' + \int_{\partial V} \{g_y^k \nabla' E_y^{\mathrm{s}} - E_y^{\mathrm{s}} \nabla' g_y^k\} \cdot \hat{\mathbf{n}}\,\mathrm{d}s'$$

$$+ \int_{\partial V} \{g_z^k \nabla' E_z^{\mathrm{s}} - E_z^{\mathrm{s}} \nabla' g_z^k\} \cdot \hat{\mathbf{n}}\,\mathrm{d}s', \tag{1.19}$$

using the second scalar Green's theorem (Tai, 1993). But these three surface integrals vanish because all components of $\mathbf{E}_{\mathrm{s}}$ and $\mathbf{g}^k$ are zero on the boundary ∂V (at infinity). Integrating by parts the second integral of the left-hand side of Equation (1.17) gives

$$\int_V \{\mathbf{g}^k \cdot \nabla'(\nabla' \cdot \mathbf{E}_{\mathrm{s}}|_0) - \mathbf{E}_{\mathrm{s}} \cdot \nabla'(\nabla' \cdot \mathbf{g}^k|_0)\}\mathrm{d}v'$$

$$= \int_V \{\nabla' \cdot \mathbf{g}^k\, \nabla' \cdot \mathbf{E}_{\mathrm{s}}|_0 - \nabla' \cdot \mathbf{E}_{\mathrm{s}}\, \nabla' \cdot \mathbf{g}^k|_0\}\mathrm{d}v', \tag{1.20}$$

$$= \int_{-\infty}^{\infty}\int_{-\infty}^{\infty} \{\nabla' \cdot \mathbf{g}^k|_{z=0}\, \nabla' \cdot \mathbf{E}_{\mathrm{s}}|_{z=0} - \nabla' \cdot \mathbf{E}_{\mathrm{s}}|_{z=0}\, \nabla' \cdot \mathbf{g}^k|_{z=0}\}\mathrm{d}x'\,\mathrm{d}y', \tag{1.21}$$

which is equal to zero. Hence, from Equation (1.17), and reinstating $\mathbf{Q}$,

$$E_k^{\mathrm{s}} = i\omega\mu \int_V \mathbf{g}^k \cdot \mathbf{E}\,\Delta\sigma\,\mathrm{d}v' - \int_V \mathbf{g}^k \cdot \nabla'(\nabla' \cdot \mathbf{E}_{\mathrm{s}}|_{V_{\mathrm{a}}})\,\mathrm{d}v'. \tag{1.22}$$

Integrating the second term on the right-hand side by parts, and using the fact that $\mathbf{g}^k$ is zero on the boundary, gives

$$E_k^{\mathrm{s}} = i\omega\mu \int_V \mathbf{g}^k \cdot \mathbf{E}\,\Delta\sigma\,\mathrm{d}v' + \int_V \nabla' \cdot \mathbf{g}^k\, \nabla' \cdot \mathbf{E}_{\mathrm{s}}|_{V_{\mathrm{a}}}\,\mathrm{d}v'. \tag{1.23}$$

Considering the three components of the secondary electric field, and adding the background electric field to both sides, gives the vector integral equation:

$$\mathbf{E} = \mathbf{E}_{\mathrm{b}} + i\omega\mu \int_V \underline{\mathbf{G}}^{(1)} \cdot \mathbf{E}\,\Delta\sigma\,\mathrm{d}v' + \int_V \mathbf{G}^{(2)}\, \nabla' \cdot \mathbf{E}_{\mathrm{s}}|_{V_{\mathrm{a}}}\,\mathrm{d}v', \tag{1.24}$$

where

$$\underline{\mathbf{G}}^{(1)} = \begin{pmatrix} g_x^x & g_y^x & g_z^x \\ g_x^y & g_y^y & g_z^y \\ g_x^z & g_y^z & g_z^z \end{pmatrix} \quad \text{and} \quad \mathbf{G}^{(2)} = \begin{pmatrix} \nabla' \cdot \mathbf{g}^x \\ \nabla' \cdot \mathbf{g}^y \\ \nabla' \cdot \mathbf{g}^z \end{pmatrix}. \tag{1.25}$$

This integral equation as written reduces at zero frequency to the integral equation for the DC resistivity problem, and we ensure this correspondence remains intact between our numerical implementation of Equation (1.24) and numerical implementations for the DC problem (e.g., Snyder, 1976).

The integral equation for the magnetic field is obtained from Equation (1.24) by $\mathbf{H} = \nabla \times \mathbf{E}/i\omega\mu$:

$$\mathbf{H} = \mathbf{H}_\mathrm{b} + \int_V \underline{\mathbf{G}}^{(3)} \cdot \mathbf{E}\ \Delta\sigma\ \mathrm{d}v', \tag{1.26}$$

where each column of $\underline{\mathbf{G}}^{(3)}$ is given by $1/(i\omega\mu)\nabla\times$ of each column of $\underline{\mathbf{G}}^{(1)}$. The second integral in Equation (1.24) does not contribute to the magnetic field.

3. NUMERICAL SOLUTION — GALERKIN APPROACH

The numerical solution of the integral equation is obtained using the method of weighted residuals. The only contributions to both integrals in Equation (1.24) are from the anomalous region. Since our background model is a homogeneous halfspace, $\nabla \cdot \mathbf{E}_\mathrm{b}$ is zero within this region, and we can therefore replace $\nabla \cdot \mathbf{E}_\mathrm{s}$ in Equation (1.24) by $\nabla \cdot \mathbf{E}$, giving, after re-ordering the terms and explicitly limiting the integrations to the anomalous region V_a,

$$\mathbf{E} - i\omega\mu \int_{V_\mathrm{a}} \underline{\mathbf{G}}^{(1)} \cdot \mathbf{E}\,\Delta\sigma\,\mathrm{d}v' - \int_{V_\mathrm{a}} \mathbf{G}^{(2)}\,\nabla' \cdot \mathbf{E}\ \mathrm{d}v' = \mathbf{E}_\mathrm{b}, \tag{1.27}$$

which is a linear operator acting on the total electric field:

$$\mathbf{L}\big[\mathbf{E}\big] = \mathbf{E}_\mathrm{b}. \tag{1.28}$$

The total electric field is expressed as a linear combination of basis vectors:

$$\mathbf{E}(\mathbf{r}) \approx \sum_{j=1}^{N} c_j \mathbf{v}_j(\mathbf{r}). \tag{1.29}$$

Substituting this representation into Equation (1.28) gives

$$\sum_{j=1}^{N} c_j \mathbf{L}\big[\mathbf{v}_j\big] = \mathbf{E}_\mathrm{b} + \mathbf{R}, \tag{1.30}$$

where the residual $\mathbf{R}$ represents the error introduced by the approximation in Equation (1.29). Taking the inner product of Equation (1.30) with each of a set of weight functions, $\mathbf{w}_i(\mathbf{r}), i = 1,\ldots,M$, and requiring that the residual $\mathbf{R}$ be orthogonal to these weight functions, that is, $\langle \mathbf{w}_i, \mathbf{R}\rangle = 0$ for all i, results in the matrix equation

$$\underline{\mathbf{A}}\,\mathbf{x} = \mathbf{b}, \tag{1.31}$$

where $A_{ij} = \langle \mathbf{w}_i, \mathbf{L}[\mathbf{v}_j]\rangle$, $x_i = c_i$, and $b_i = \langle \mathbf{w}_i, \mathbf{E}_\mathrm{b}\rangle$. The inner product is defined as $\langle \mathbf{w}, \mathbf{v}\rangle = \int_V \mathbf{w} \cdot \mathbf{v}\,\mathrm{d}v'$. We adopt the Galerkin approach in which the basis vectors are used as the weight functions.

4. EDGE ELEMENT BASIS VECTORS

It is assumed that the region of anomalous conductivity can be represented by a grid of cuboidal cells within each of which the conductivity is uniform. Within a cell, we approximate the total electric field by a linear combination of twelve basis vectors, four directed in the x-direction, four directed in the y-direction, and four in the z-direction. The four basis vectors directed in the x-direction vary bilinearly with y and z, the four in the y-direction vary with x and z, and those in the z-direction vary with x and y. The total electric field within each cell is therefore divergence free by construction. It can also have a non-zero curl.

Explicitly, the x-component of the electric field in an example cell (see Figure 1a) is given by (e.g., Jin, 1993):

$$E_x(\mathbf{r})\hat{\mathbf{u}}_x = c_1\mathbf{v}_1(\mathbf{r}) + c_2\mathbf{v}_2(\mathbf{r}) + c_3\mathbf{v}_3(\mathbf{r}) + c_4\mathbf{v}_4(\mathbf{r}), \tag{1.32}$$

$$\begin{aligned} &= c_1\{(y - y_c + l_y)(z - z_c + l_z)/4l_y l_z\}\hat{\mathbf{u}}_x \\ &\quad + c_2\{(y_c + l_y - y)(z - z_c + l_z)/4l_y l_z\}\hat{\mathbf{u}}_x \\ &\quad + c_3\{(y - y_c + l_y)(z_c + l_z - z)/4l_y l_z\}\hat{\mathbf{u}}_x \\ &\quad + c_4\{(y_c + l_y - y)(z_c + l_z - z)/4l_y l_z\}\hat{\mathbf{u}}_x, \end{aligned} \tag{1.33}$$

where y_c and z_c are the y- and z-coordinates of the centre of the cell, and $2l_y$ and $2l_z$ are the extents of the cell in the y- and z-directions. With the above definitions of these four basis vectors, $v_1 = 1$ and $v_2 = v_3 = v_4 = 0$ on the edge of the cell with $y = y_c + l_y$ and $z = z_c + l_z$, $v_2 = 1$ and $v_1 = v_3 = v_4 = 0$ on the edge with $y = y_c - l_y$ and $z = z_c + l_z$, and likewise for the other two edges. The x-component of the electric field is thus effectively equal to c_1, c_2, c_3 and c_4, respectively, on the x-directed edges of the cell (see Figure 1a).

In an analogous manner to Equation (1.33), the y-component of the electric field within this example cell is given by

$$\begin{aligned} E_y(\mathbf{r})\hat{\mathbf{u}}_y &= c_5\{(x - x_c + l_x)(z - z_c + l_z)/4l_x l_z\}\hat{\mathbf{u}}_y \\ &\quad + c_6\{(x_c + l_x - x)(z - z_c + l_z)/4l_x l_z\}\hat{\mathbf{u}}_y \\ &\quad + c_7\{(x - x_c + l_x)(z_c + l_z - z)/4l_x l_z\}\hat{\mathbf{u}}_y \\ &\quad + c_8\{(x_c + l_x - x)(z_c + l_z - z)/4l_x l_z\}\hat{\mathbf{u}}_y, \end{aligned} \tag{1.34}$$

and the z-component of the electric field by

$$\begin{aligned} E_z(\mathbf{r})\hat{\mathbf{u}}_z &= c_9\{(x - x_c + l_x)(y - y_c + l_y)/4l_x l_y\}\hat{\mathbf{u}}_z \\ &\quad + c_{10}\{(x_c + l_x - x)(y - y_c + l_y)/4l_x l_y\}\hat{\mathbf{u}}_z \\ &\quad + c_{11}\{(x - x_c + l_x)(y_c + l_y - y)/4l_x l_y\}\hat{\mathbf{u}}_z \\ &\quad + c_{12}\{(x_c + l_x - x)(y_c + l_y - y)/4l_x l_y\}\hat{\mathbf{u}}_z, \end{aligned} \tag{1.35}$$

where x_c is the x-coordinate of the centre of the cell and $2l_x$ is the extent of the cell in the x-direction.

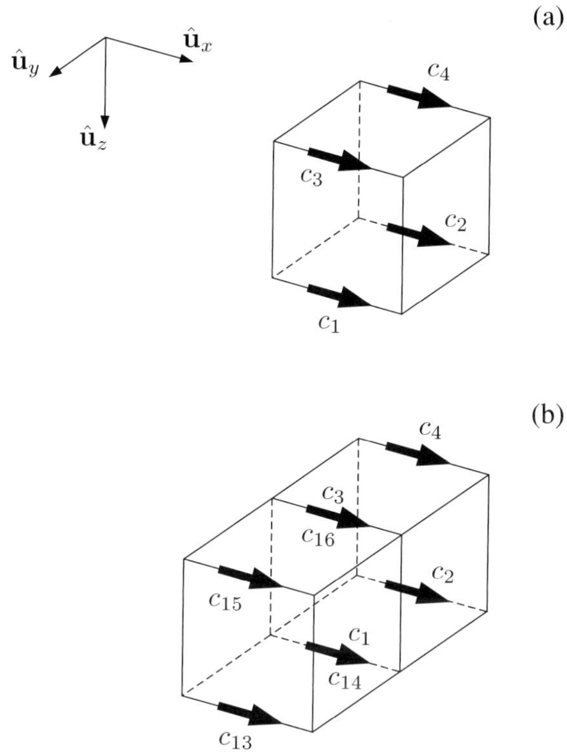

Figure 1. (a) An example cell. $c_1 \ldots c_4$ are the coefficients of the four basis vectors that approximate the x-component of the total electric field within this cell. (b) The example cell and its neighbour in the positive y-direction. $c_{13} \ldots c_{16}$ are the coefficients of the corresponding basis vectors in this neighbouring cell.

When two cells share a face, the basis function in the one cell that is equal to 1 on one of the shared edges and the basis function in the other cell that is equal to 1 on the same edge are treated as a single basis vector, meaning their coefficients are then equal. Consider, for example, the two cells in Figure 1b which share the face for which $y = y_c^{(1)} + l_y^{(1)} = y_c^{(2)} - l_y^{(2)}$, where the superscripts indicate the number of the cell. Suppose the basis vectors for $j = 1, \ldots, 12$ apply to cell 1 and those for $j = 13, \ldots, 24$ apply to cell 2. This means that both v_1 and v_{14} are equal to 1 on the edge for which $y = y_c^{(1)} + l_y^{(1)} = y_c^{(2)} - l_y^{(2)}$ and $z = z_c^{(1)} + l_z^{(1)} = z_c^{(2)} + l_z^{(2)}$, and both v_3 and v_{16} are equal to 1 on the edge for which $y = y_c^{(1)} + l_y^{(1)} = y_c^{(2)} - l_y^{(2)}$ and $z = z_c^{(1)} - l_z^{(1)} = z_c^{(2)} - l_z^{(2)}$. These four basis vectors are therefore considered as just two distinct basis vectors: $(v_1 + v_{14})$ and $(v_3 + v_{16})$ with coefficients $\tilde{c}_1 = c_1 = c_{14}$ and $\tilde{c}_3 = c_3 = c_{16}$. This tying together of basis vectors associated with shared cell edges is implemented by summing the corresponding columns and rows of the system of equations in Equation (1.31). The consequence is that the tangential component of the total electric field is continuous, by construction, across any interface between cells.

5. THE GREEN'S FUNCTIONS — PART I

Our background model is a homogeneous halfspace. However, to simplify the computations required when constructing Equation (1.31), it is assumed that the anomalous region is sufficiently far from the Earth–air interface that, when both $\mathbf{r}$ and $\mathbf{r}'$ are within the anomalous region, the contributions to the Green's functions $\underline{\mathbf{G}}^{(1)}$ and $\mathbf{G}^{(2)}$ from the Earth–air interface can be ignored. Hence, $\underline{\mathbf{G}}^{(1)}$ is a diagonal tensor whose elements are all equal to the wholespace Green's function:

$$g_x^x = g_y^y = g_z^z = g^w(\mathbf{r};\mathbf{r}') = \frac{1}{4\pi}\frac{e^{ik_b|\mathbf{r}-\mathbf{r}'|}}{|\mathbf{r}-\mathbf{r}'|}, \tag{1.36}$$

where $k_b^2 = i\omega\mu\sigma_b$, and

$$\mathbf{G}^{(2)} = \begin{pmatrix} \partial g^w/\partial x' \\ \partial g^w/\partial y' \\ \partial g^w/\partial z' \end{pmatrix} = \nabla' g^w. \tag{1.37}$$

6. EVALUATION OF THE INTEGRALS

The volume integrals associated with the inner products $\langle \mathbf{w}_i, \mathbf{L}[\mathbf{v}_j]\rangle$ and $\langle \mathbf{w}_i, \mathbf{E}_b\rangle$ in Equation (1.31) are computed using Gaussian quadrature. The two types of integrals associated with $\underline{\mathbf{L}}[\mathbf{v}_j]$ are computed as follows. The first integral, that is,

$$\mathbf{L}^{(1)}\left[\mathbf{v}_j\right] = i\omega\mu \int_{V_a} \underline{\mathbf{G}}^{(1)}(\mathbf{r};\mathbf{r}')\cdot\mathbf{v}_j(\mathbf{r}')\,\Delta\sigma(\mathbf{r}')\,\mathrm{d}v', \tag{1.38}$$

is evaluated using Gaussian quadrature when $\mathbf{r}$ is outside the cell over which the integration is taking place, and using the trapezoidal rule (with typically $10\times10\times10$ evaluation points) when $\mathbf{r}$ is within the cell so that $\mathbf{r}'$ does not correspond to a Gaussian quadrature node for the inner product integrations. The elements of $\underline{\mathbf{G}}^{(1)}$ are singular at $\mathbf{r}'=\mathbf{r}$ because of their $1/|\mathbf{r}-\mathbf{r}'|$ dependence. However, the singularity does not contribute to the integral. To see this, split the integral in Equation (1.38) into one part for a sphere of radius ϵ centred on $\mathbf{r}'=\mathbf{r}$, and one part over the rest of the cell in which $\mathbf{v}_j$ is non-zero. For small ϵ, the basis vector can be considered as being constant within the sphere, and the exponential term in the components of $\underline{\mathbf{G}}^{(1)}$ can be considered to be constant and equal to its value for $|\mathbf{r}-\mathbf{r}'|=0$, which is 1. The integration over the sphere becomes:

$$i\omega\mu \int_{V_\epsilon} \underline{\mathbf{G}}^{(1)}(\mathbf{r};\mathbf{r}')\cdot\mathbf{v}_j(\mathbf{r}')\,\Delta\sigma(\mathbf{r}')\,\mathrm{d}v' \approx i\omega\mu\,\Delta\sigma_J \frac{1}{4\pi}\underline{\mathbf{I}}\cdot\mathbf{v}_j(\mathbf{r}) \int_{\zeta=0}^{\epsilon} \frac{1}{\zeta}4\pi\zeta^2\,\mathrm{d}\zeta, \tag{1.39}$$

where $\Delta\sigma_J$ is the anomalous conductivity in the cell, $\underline{\mathbf{I}}$ is the identity tensor, and $\zeta = |\mathbf{r}-\mathbf{r}'|$. This integral vanishes as $\epsilon\to 0$.

The second integral in $\mathbf{L}\left[\mathbf{v}_j\right]$ is

$$\mathbf{L}^{(2)}\left[\mathbf{v}_J\right] = \int_{V_J} \mathbf{G}^{(2)}(\mathbf{r};\mathbf{r}')\nabla\cdot\mathbf{v}_j(\mathbf{r}')\,\mathrm{d}v', \tag{1.40}$$

where V_J is the volume of the cell in which $\mathbf{v}_j$ is non-zero. $\nabla\cdot\mathbf{E}$ is non-zero only on the interfaces between cells, where it is equal to the surface charge density (scaled by the permittivity of free space ε_0). Hence (e.g., Li and Oldenburg, 1991),

$$\nabla\cdot\mathbf{v}_j = \left(\frac{\sigma_J}{\sigma_n} - 1\right)\mathbf{v}_j\cdot\hat{\mathbf{n}} \tag{1.41}$$

where σ_J is the conductivity in the cell, σ_n is the conductivity of the neighbouring cell (or of the background if there is no neighbour), and $\hat{\mathbf{n}}$ is the normal to the face of the cell. The integral in Equation (1.40) therefore becomes

$$\mathbf{L}^{(2)}[\mathbf{v}_j] = \int_{\partial V_J} \mathbf{G}^{(2)}(\mathbf{r};\mathbf{r}')\left(\frac{\sigma_j}{\sigma_n} - 1\right)\mathbf{v}_j(\mathbf{r}')\cdot\hat{\mathbf{n}}\,\mathrm{d}s'. \tag{1.42}$$

The above integrand is singular at $\mathbf{r}'=\mathbf{r}$. This singularity gives rise to a contribution of $2\pi\,\tau(\mathbf{r})/\varepsilon_0$, where τ is the surface charge density (Snyder, 1976). Thus,

$$\mathbf{L}^{(2)}[\mathbf{v}_j] = \int_{\partial V_J,\,\mathbf{r}'\neq\mathbf{r}} \mathbf{G}^{(2)}(\mathbf{r};\mathbf{r}')\left(\frac{\sigma_j}{\sigma_n} - 1\right)\mathbf{v}_j(\mathbf{r}')\cdot\hat{\mathbf{n}}\,\mathrm{d}s' - 2\pi\,\tau(\mathbf{r})/\varepsilon_0. \tag{1.43}$$

However, because the second term above exists only on the faces of the cell, it does not contribute to the volume integration over the cell in the calculation of the inner product $\langle\mathbf{w}_i,\mathbf{L}[\mathbf{v}_j]\rangle$, and so can be ignored. The surface integral in Equation (1.43) is evaluated using Gaussian quadrature.

7. THE GREEN'S FUNCTIONS — PART II

Once the electric field within the anomalous region is known, the field anywhere can be computed using the simple re-arrangement of Equation (1.27):

$$\mathbf{E} = \mathbf{E}_b + i\omega\mu\int_{V_a} \underline{\mathbf{G}}^{(1)}\cdot\mathbf{E}\,\Delta\sigma\,\mathrm{d}v' + \int_{V_a} \mathbf{G}^{(2)}\,\nabla'\cdot\mathbf{E}\,\mathrm{d}v'. \tag{1.44}$$

The form of $\mathbf{g}^k$, and hence that of $\underline{\mathbf{G}}^{(1)}$ and $\mathbf{G}^{(2)}$, for the homogeneous halfspace is now required. The most straightforward way to see what $\mathbf{g}^x$, $\mathbf{g}^y$ and $\mathbf{g}^z$ should be is to recognise that they are equivalent to the Schelkunoff **A**-potentials described in Ward and Hohmann (1987) for x-, y- and z-directed electric current density dipole sources. As such, $\mathbf{g}^x$ has only x- and z-components, $\mathbf{g}^y$ only y- and z-components, and $\mathbf{g}^z$ only a z-component. Following equations (1.198)–(1.201) of Ward and Hohmann, the components of $\mathbf{g}^x$ satisfy

$$\frac{1}{\sigma_b}\left(\frac{\partial g_x^x}{\partial x} + \frac{\partial g_z^x}{\partial z}\right)\bigg|_{z=0^+} = 0, \tag{1.45}$$

$$g_z^x|_{z=0^+} = g_z^x|_{z=0^-}, \tag{1.46}$$

$$g_x^x|_{z=0^+} = g_x^x|_{z=0^-}, \tag{1.47}$$

$$\frac{\partial g_x^x}{\partial z}\bigg|_{z=0^+} = \frac{\partial g_x^x}{\partial z}\bigg|_{z=0^-}, \tag{1.48}$$

on the surface of the halfspace ($z = 0$, z positive down). The components of $\mathbf{g}^y$ satisfy the same set of conditions with x replaced by y. Following equations (1.182) and (1.183) of Ward and Hohmann, the component of $\mathbf{g}^z$ satisfies

$$\frac{1}{\sigma_b}\frac{\partial g_z^z}{\partial z}\bigg|_{z=0^+} = 0, \tag{1.49}$$

$$g_z^z|_{z=0^+} = g_z^z|_{z=0^-}. \tag{1.50}$$

Alternatively, to determine what conditions must be satisfied by $\mathbf{g}^k$ on the Earth–air interface, consider in component form the second and third terms on the right-hand side of Equation (1.44), that is, the secondary electric field (cf. Equation (1.23)):

$$E_k^s = i\omega\mu \int_{V_a} \mathbf{g}^k \cdot \mathbf{E}\,\Delta\sigma\, dv' + \int_{V_a} \nabla' \cdot \mathbf{g}^k\, \nabla' \cdot \mathbf{E}_s\, dv'. \tag{1.51}$$

We can rewrite the second integral using Equation (1.11):

$$E_k^s = i\omega\mu \int_{V_a} \mathbf{g}^k \cdot \mathbf{E}\,\Delta\sigma\, dv' - \int_{V_a} \nabla' \cdot \mathbf{g}^k \frac{1}{\sigma_b} \nabla' \cdot \left(\Delta\sigma\, \mathbf{E}\right) dv'. \tag{1.52}$$

Integrating the second integral by parts gives

$$E_k^s = i\omega\mu \int_{V_a} \mathbf{g}^k \cdot \mathbf{E}\,\Delta\sigma\, dv' + \frac{1}{\sigma_b}\int_{V_a} \nabla'\left(\nabla' \cdot \mathbf{g}^k\right) \cdot \mathbf{E}\,\Delta\sigma\; dv', \tag{1.53}$$

$$= \int_{V_a} \left\{ i\omega\mu\, \mathbf{g}^k + \frac{1}{\sigma_b}\nabla'\left(\nabla' \cdot \mathbf{g}^k\right)\right\} \cdot \mathbf{E}\,\Delta\sigma\; dv', \tag{1.54}$$

where the expression in the braces is the more familiar form of the electric field Green's tensor for a uniform background. Across the Earth-air interface, the tangential components of the secondary E- and H-fields are continuous. In order for our computed values of E_x^s to be continuous, each component of $\left\{i\omega\mu\, \mathbf{g}^x + \nabla'\left(\nabla' \cdot \mathbf{g}^x\right)/\sigma_b\right\}$ must be continuous, and for E_y^s to be continuous, each component of $\left\{i\omega\mu\, \mathbf{g}^y + \nabla'\left(\nabla' \cdot \mathbf{g}^y\right)/\sigma_b\right\}$ must be continuous. Similarly, for H_x^s and H_y^s to be continuous, each component of $\left\{-\partial_z \mathbf{g}^y + \partial_y \mathbf{g}^z\right\}$ and $\left\{\partial_z \mathbf{g}^x - \partial_x \mathbf{g}^z\right\}$ must be continuous. Finally, E_z^s is zero on the Earth–air interface. To satisfy this, each component of $\left\{i\omega\mu\, \mathbf{g}^z + \nabla'\left(\nabla' \cdot \mathbf{g}^z\right)/\sigma_b\right\}$ must be zero at $z = 0$. These conditions reduce to the same as those listed in Equations (1.45)–(1.50).

With the above interface conditions, and the defining differential equation (Equation (1.15)) and primary solution (Equation (1.36)), the components of $\mathbf{g}^k$ are determined using standard techniques (see, for example, Ward and Hohmann, 1987):

$$g_x^x(\mathbf{r};\mathbf{r}') = g^w(\mathbf{r};\mathbf{r}') + \frac{1}{2\pi}\int\limits_{\lambda=0}^{\infty} \frac{u-\lambda}{2u(u+\lambda)} \mathrm{e}^{-u(z+z')}\lambda\, J_0(\lambda\rho)\,\mathrm{d}\lambda, \quad z \geq 0, \tag{1.55}$$

$$= \frac{1}{2\pi}\int\limits_{\lambda=0}^{\infty} \frac{1}{(u+\lambda)} \mathrm{e}^{-uz'}\mathrm{e}^{\lambda z}\lambda\, J_0(\lambda\rho)\,\mathrm{d}\lambda, \quad z < 0, \tag{1.56}$$

$$g_z^x(\mathbf{r};\mathbf{r}') = -\frac{1}{2\pi}\frac{(x-x')}{\rho}\int\limits_{\lambda=0}^{\infty} \frac{1}{\lambda(u+\lambda)} \mathrm{e}^{-u(z+z')}\lambda^2\, J_1(\lambda\rho)\,\mathrm{d}\lambda, \quad z \geq 0, \tag{1.57}$$

$$= -\frac{1}{2\pi}\frac{(x-x')}{\rho}\int\limits_{\lambda=0}^{\infty} \frac{1}{\lambda(u+\lambda)} \mathrm{e}^{-uz'}\mathrm{e}^{\lambda z}\lambda^2\, J_1(\lambda\rho)\,\mathrm{d}\lambda, \quad z < 0, \tag{1.58}$$

$$g_z^z(\mathbf{r};\mathbf{r}') = g^w(\mathbf{r};\mathbf{r}') - \frac{1}{2\pi}\int\limits_{\lambda=0}^{\infty} \frac{1}{2u} \mathrm{e}^{-u(z+z')}\lambda\, J_0(\lambda\rho)\,\mathrm{d}\lambda, \quad z \geq 0, \tag{1.59}$$

$$= 0, \quad z < 0, \tag{1.60}$$

where $u = \sqrt{\lambda^2 - k_b^2}$, $\rho = \sqrt{(x-x')^2 + (y-y')^2}$, and J_0 and J_1 are the Bessel functions of the first kind of orders 0 and 1. The above Hankel transforms are computed using the digital filtering routine of Anderson (1982).

Given the halfspace Green's functions, and the total electric field within the anomalous region, the electric field is calculated wherever required in the halfspace using Equation (1.44), and the magnetic field is calculated using Equation (1.26). The integrals are computed using Gaussian quadrature and the forms given in Equations (1.38) and (1.42).

8. EXAMPLES

Here we present two examples: one a comparison with results from a DC resistivity modelling program since the form of our integral equation solution was strongly influenced by those for the DC problem, and one for an airborne electromagnetic transmitter–receiver geometry over a conductivity contrast of $100 : 10^{-4}$.

The model for the first example comprises a vertical prism of 0.2 S/m in a halfspace of 0.02 S/m, as shown in Figure 2. The centre of the prism is 1000 m from the furthest end of the 100 m long grounded electric line source. The top surface of the prism is 100 m below the surface of the halfspace, and its extents in the x-, y- and z-directions are 120, 200 and 400 m, respectively. The total electric field along a profile over the centre of the prism was computed, and the values are shown by the crosses in the top panel of Figure 3. A frequency of 3 Hz was used. The prism was divided into $5 \times 5 \times 5$ cells. The volume integrations were carried out using $2 \times 2 \times 2$ nodes, and the surface integrals using 5×5 nodes. The electric field without the prism present, that is, the background

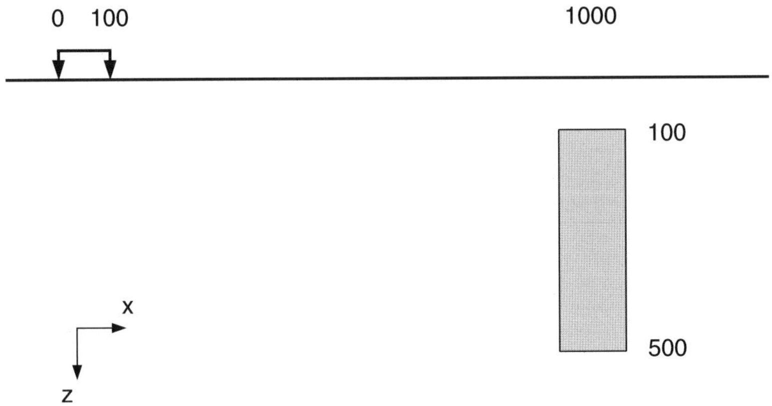

Figure 2. The geometry of the first example presented in this paper. The halfspace background has a conductivity of 0.02 S/m. The prism has a conductivity of 0.2 S/m, and has extents of 120 and 200 m in the *x*- and *y*-directions, respectively, and is centred below the *x*-axis.

field, is also shown in Figure 3. The real part of the secondary field is displayed in the bottom panel in Figure 3. The electric field for this model was also computed using the DC resistivity modelling program 'DCIP3D' (Li et al., 1999), and is indicated by the squares in Figure 3. The real part of the electric field within the prism as computed using the integral equation solution is shown in Figure 4. The left panel in this figure shows the horizontal component of the total field over the uppermost layer of quadrature nodes, and the panel on the right shows the component in the x–z-plane for the cells down through the centre of the prism.

The second example is for a conductive cube in a resistive halfspace, as shown in Figure 5. The conductivity of the cube was 100 S/m, and that of the halfspace was 10^{-4} S/m. The vertical component of the magnetic field was computed 5 m from a unit vertical magnetic dipole source for a range of locations of the source–receiver pair over the cube (see Figure 5). The frequency was 900 Hz. The computed values of the secondary magnetic field are shown by the circles in Figure 6. The cube was divided into $5 \times 5 \times 5$ cells, the volume integrations were carried out using $3 \times 3 \times 3$ nodes, and the surface integrals using 15×15 nodes. The computed total electric field at the uppermost plane of volume integration nodes for the source at $x = 2.5$ m, $y = 0$ m is shown in Figure 7. The field for a sphere of the same conductivity and volume as the cube but in free space (Ellis, 1995) is indicated by the lines in Figure 6. It is clear from this figure that the integral equation solution presented here is successful even for such a large conductivity contrast as the one in this example.

9. COMPUTATIONAL EFFICIENCY

The current implementation of our integral equation solution is slow. This is because of the large number of times the Green's functions and basis vectors are evaluated in

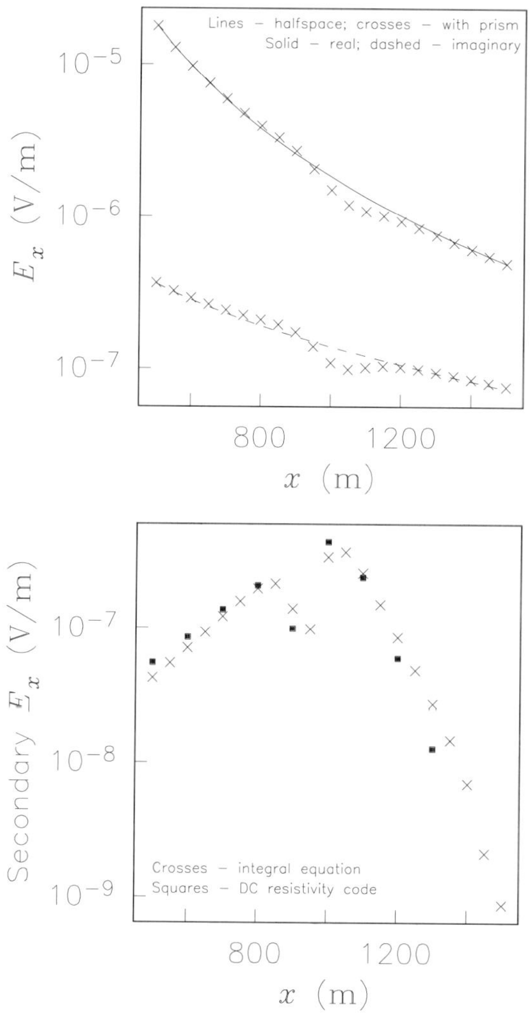

Figure 3. Top panel: the x-component of the background (lines) and total (crosses) electric fields on the surface of the halfspace over the top of the prism (which is centred at $x = 1000$ m). Bottom panel: the real part of the secondary field as computed by the integral equation code presented in this paper (crosses), and the secondary field computed using the DC resistivity forward modelling code DCIP3D (squares).

the Gaussian quadrature integrations. The situation is exacerbated by the second level of integrations for the inner products required in the Galerkin approach. As an illustration, the second example in the preceding section (for the discretisation of $5 \times 5 \times 5$ cells) took 3.5 h on a 733 MHz Pentium III computer using $2 \times 2 \times 2$ nodes for the volume integrations and 5×5 nodes for the surface integrals, and took 4¼ days using $3 \times 3 \times 3$ nodes for the volume integrations and 15×15 nodes for the surface integrals. (The results for these two cases were similar, but with noticeable improvement on the flanks

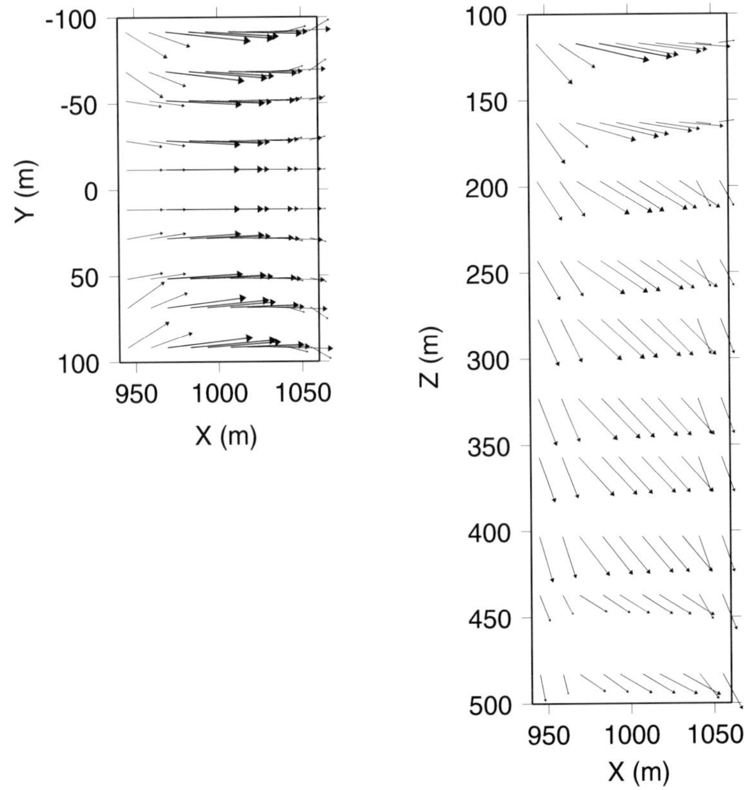

Figure 4. The real part of the computed total electric field within the prism. The prism was divided into 5 x 5 x 5 cells with 2 x 2 x 2 quadrature nodes within each cell. The left panel shows the field over the top-most plane of quadrature nodes, and the right panel shows the field in the cells down through the centre of the prism. The longest arrow corresponds to a field strength of 8.4 x 10^{-7} V/m.

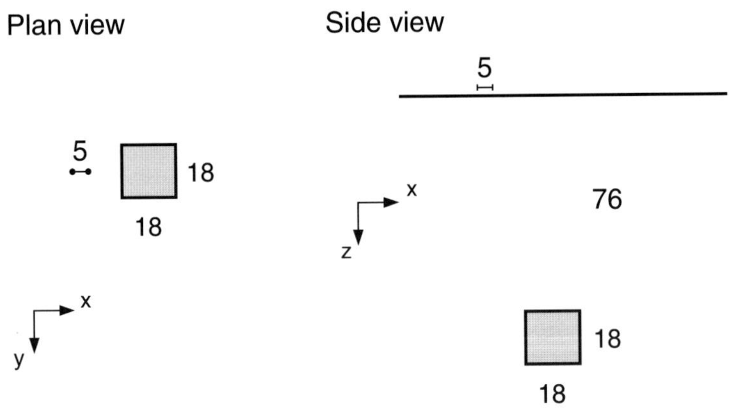

Figure 5. The geometry of the second example. A magnetic dipole source–receiver pair is considered. The cube has a conductivity of 100 S/m, and the halfspace a conductivity of 10^{-4} S/m. The origin is on the Earth–air interface directly above the centre of the cube. All dimensions are in metres.

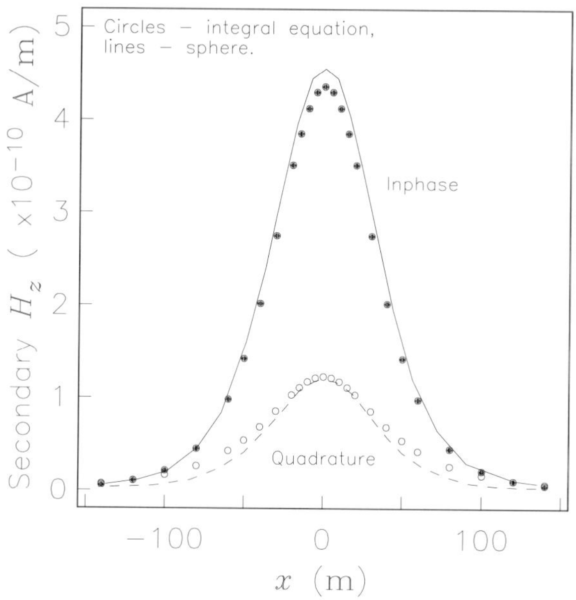

Figure 6. The vertical component of the secondary magnetic field for the second example. The circles show the values computed using our integral equation solution. The lines are for an equivalent sphere in free space. The abscissa is the location of the centre of the source–receiver pair.

of the responses shown in Figure 6 for the greater number of quadrature nodes.) For both these times, over 99% was taken up with the integrations needed to construct the matrix equation. The introduction of efficient ways to compute the essentially convolution-type integrals is therefore needed to make the technique viable for general usage. However, even the current formulation is valuable in that it can supply an independent check on results obtained from other numerical solutions of the geophysical electromagnetic forward-modelling problem.

10. CONCLUSIONS

We have implemented edge element basis functions in the numerical solution of the electric field integral equation. We have also ensured that the form of the integral equation upon which our solution is based, especially once the discretisation of the model has been introduced, reduces at zero frequency to that for the DC resistivity problem. We feel that the resulting treatment of charges, and hence current density, on the interfaces between cells of different conductivities, and between cells and the background, plays just as important a role as the use of the divergence-free but not curl-free edge element basis vectors. Tests to date have agreed well with published results and those from other algorithms, including those for models with large conductivity contrasts.

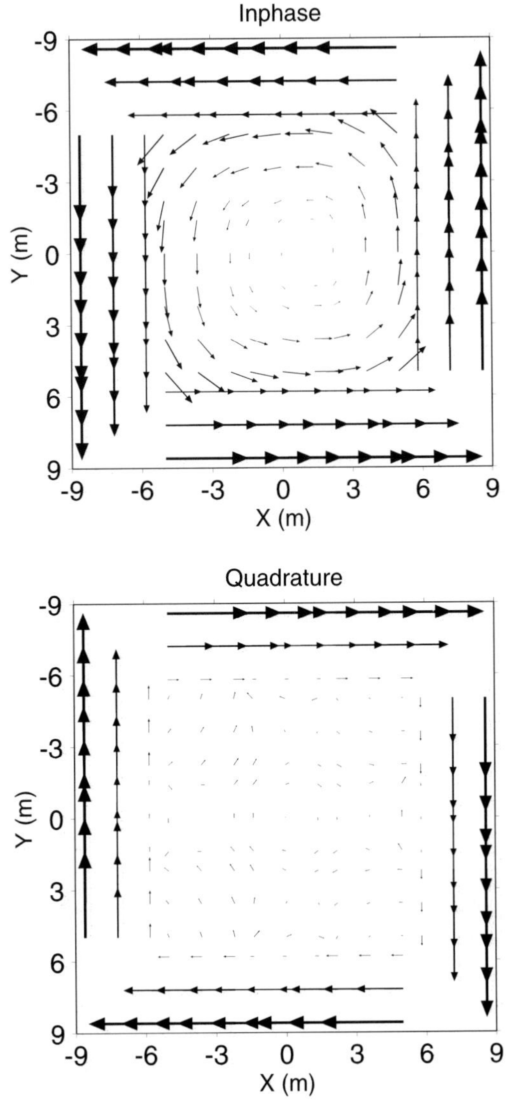

Figure 7. The computed total electric field over the top plane of quadrature nodes for the source at $x = 2.5$ m, $y = 0$ m. The cube was divided into 5 x 5 x 5 cells, with 3 x 3 x 3 quadrature nodes within each cell. The longest arrow corresponds to a field strength of 4×10^{-9} V/m.

ACKNOWLEDGEMENTS

This work was supported by NSERC, and both the Consortium for the Joint and Cooperative Inversion of Geophysical and Geological Data (the 'JACI' Consortium) and the Consortium for the 3D Inversion of DC Resistivity and IP Data (the 'INDI' Consortium). The following were participants: Placer Dome, BHP Minerals, Noranda

Exploration, Cominco Exploration, Falconbridge, INCO Exploration and Technical Services, Hudson Bay Exploration and Development, Kennecott Exploration Company, Newmont Gold Company, WMC Exploration, and CRA Exploration Party. We are grateful for their involvement. We would also like to thank Dmitry Avdeev, Gregory Newman and Phil Wannamaker for their thorough and constructive reviews which led to many improvements in this paper.

REFERENCES

Anderson, W.L., 1982. Fast Hankel transforms using related and lagged convolutions. ACM Trans. Math. Software, 8, 344–368.

Ellis, R.G., 1995. Airborne EM Forward Modeling in 3D. In: 1994 Annual Report of the JACI Consortium, University of British Columbia.

Hohmann, G.W., 1975. Three-dimensional induced-polarization and electromagnetic modeling. Geophysics, 40, 309–324.

Hohmann, G.W., 1987. Numerical modeling for electromagnetic methods of geophysics. In: M.N. Nabighian (Ed.), Electromagnetic Methods in Applied Geophysics. SEG, Tulsa, OK.

Jin, J., 1993. The Finite Element Method in Electromagnetics. John Wiley, New York.

Li, Y. and Oldenburg, D.W., 1991. Aspects of charge accumulation in d.c. resistivity experiments. Geophys. Prospect., 39, 803–826.

Li, Y., Oldenburg, D.W. and Shekhtman, R., 1999. DCIP3D: A Program Library for Forward Modelling and Inversion of DC Resistivity and Induced Polarization Data over 3D Structures. In: Report of the INDI Consortium, University of British Columbia.

Newman, G.A. and Hohmann, G.W., 1988. Transient electromagnetic responses of high-contrast prisms in a layered earth. Geophysics, 53, 691–706.

Slob, E.C. and van den Berg, P.M., 1999. Integral-equation method for modeling transient diffusive electromagnetic scattering. In: M. Oristaglio and B. Spies (Eds.), Three-Dimensional Electromagnetics. SEG, Tulsa, OK.

Snyder, D.D., 1976. A method for modeling the resistivity and IP response of two-dimensional bodies. Geophysics, 41, 997–1015.

Tai, C.-T., 1993. Dyadic Green Functions in Electromagnetic Theory. IEEE Press, New York.

Ward, S.H. and Hohmann, G.W., 1987. Electromagnetic theory for geophysical applications. In: M.N. Nabighian (Ed.), Electromagnetic Methods in Applied Geophysics. SEG, Tulsa, OK.

Chapter 2

COMPRESSION IN 3-D INTEGRAL EQUATION MODELING

Michael S. Zhdanov, Oleg Portniaguine and Gabor Hursan

University of Utah, Salt Lake City, UT 84112, USA

Abstract: The integral equations (IE) method is a powerful tool for forward electromagnetic (EM) modeling. However, due to a dense matrix arising from the IE formulation, practical application of the IE method is limited to modeling of relatively small bodies. The use of a compression technique can overcome this limitation. The compression transformation is formulated as a multiplication by a compression matrix. Using this matrix as a preconditioner to an integral equation, we convert the originally dense matrix of the problem to a sparse matrix, which reduces its size and speeds up computations. Thus, compression helps to overcome practical limitations imposed on the numerical size of the anomalous domain in IE modeling. With the compression, the flexibility of the IE method approaches that of finite-difference (FD) or finite-element (FE) methods, allowing modeling of large-scale conductivity variations.

---

1. INTRODUCTION

The integral equations (IE) method is a powerful tool for forward electromagnetic (EM) modeling. The basic principles of constructing integral equations in 3-D cases were outlined by Hohmann (1975) and Weidelt (1975). A comprehensive implementation of the IE methods was realized by Xiong (1992) in the SYSEM code. The main advantage of the IE method in comparison with the FD and FE methods is the fast and accurate simulation of compact 3-D bodies in a layered background (Hohmann, 1975; Weidelt, 1975; Wannamaker, 1991; Xiong, 1992). At the same time, the area of the FD and FE methods is modeling of EM fields in complex structures with large-scale conductivity variations. In principle, the IE method can handle these models; however, the tremendous demand on computer resources places practical limits on its use. This happens due to the large dense matrix arising from IE formulation.

Another advantage that the IE method has over the FD and FE methods is its greater suitability for inversion. IE formulation readily contains a sensitivity matrix, which can be recomputed at each inversion iteration at little expense. With FD, in contrast, this matrix has to be established anew on each iteration at a cost equal to the cost of the full forward simulation.

However, for inversion purposes, and for greater flexibility in forward simulations, the IE method has to overcome severe practical limitations imposed on the numerical size of the anomalous domain.

Several approximate methods have been developed recently in this direction. These are: localized nonlinear approximation (Habashi et al., 1993), quasi-linear approximation (Zhdanov and Fang, 1996a,b), quasi-linear series (Zhdanov and Fang, 1997), and quasi-analytical approximation (Zhdanov et al., 2000). At the same time, most of these methods produce approximate solutions, therefore we still seek to develop a rigorous IE forward modeling technique.

In this paper we develop a technique that makes the IE method a more flexible tool for simulations in complex geological structures. This technique is based on applying compression to the solution of IE. Compression is routinely applied in the telecommunications industry for data transmission (Klarke, 1995; Natravali, 1995; Ramstad et al., 1995; Henson et al., 1996; Losano and Laget, 1996; Bhaskaran and Konstantinides, 1997). To decrease the traffic through an information channel, the data are compressed at the transmitting end, passed through the channel in compact form, and decompressed to their original form at the receiving end.

A similar approach can be applied to numerical simulations. That is, original equations are transformed into a compressed domain and the solution is obtained there in a compact form (Portniaguine, 1999).

We introduce a compression matrix and explain the principles of compression with an *interpolation pyramid* in multiple dimensions. A compression matrix is then used as a preconditioner to the discretized IE. This approach is similar to the use of preconditioners for FD solutions to express multi-resolution and multigrid methods in numerical simulations (Bank and Xu, 1994; Manteuffel et al., 1994; McCormick, 1994; Ainsworth et al., 1997).

Using compression, we convert the original dense matrix of the forward problem to a sparse matrix. This reduces the memory required for storage and speeds up computations.

2. FORWARD MODELING WITH 3-D INTEGRAL EQUATIONS

Let us represent a 3-D distribution of conductivity $\tilde{\sigma}$ as a sum of background (normal) complex conductivity $\tilde{\sigma}_b$ and anomalous conductivity $\tilde{\sigma}$, which is non-zero only within local domain D.

We assume that magnetic permeability μ is constant everywhere and is equal to that of a free-space $\mu = 4\pi \times 10^{-7}$ H/m. This model is excited by a harmonic source of radian frequency ω. The complex conductivity includes the effect of displacement currents: $\tilde{\sigma} = \sigma - i\omega\varepsilon$, where σ and ε are electrical conductivity and dielectric permittivity.

The vectors of total electric $\mathbf{E}$ and magnetic $\mathbf{H}$ fields in this model can be presented as a sum of background (normal) and anomalous (scattered) fields:

$$\mathbf{E} = \mathbf{E}^{\mathrm{b}} + \mathbf{E}^{\mathrm{a}}, \qquad \mathbf{H} = \mathbf{H}^{\mathrm{b}} + \mathbf{H}^{\mathrm{a}}, \tag{2.1}$$

The background field $\mathbf{E}^{\mathrm{b}}$ is a field generated by the given sources in the background model $\tilde{\sigma}_b$, and the anomalous $\mathbf{E}^{\mathrm{a}}$ field is cased by the presence of anomalous conductivity $\Delta\tilde{\sigma}$.

The anomalous field is presented as an integral over the excess currents in the inhomogeneous domain D:

$$\mathbf{E}^{a}(\mathbf{r}_j) = \iiint_D \widehat{\mathbf{G}}_E(\mathbf{r}_j \mid \mathbf{r}) \cdot \mathbf{j}^{a}(\mathbf{r})\,\mathrm{d}v, \tag{2.2}$$

$$\mathbf{H}^{a}(\mathbf{r}_j) = \iiint_D \widehat{\mathbf{G}}_H(\mathbf{r}_j \mid \mathbf{r}) \cdot \mathbf{j}^{a}(\mathbf{r})\,\mathrm{d}v, \tag{2.3}$$

where $\widehat{\mathbf{G}}_E(\mathbf{r}_j \mid \mathbf{r})$ and $\widehat{\mathbf{G}}_H(\mathbf{r}_j \mid \mathbf{r})$ are the electric and magnetic Green's tensors defined for an unbounded conductive medium with the background conductivity $\tilde{\sigma}_b$. Excess current $\mathbf{j}^{a}(\mathbf{r})$ at the point $\mathbf{r}$ is determined by the equation

$$\mathbf{j}^{a}(\mathbf{r}) = \Delta\tilde{\sigma}(r)\left(\mathbf{E}^{b}(r) + \mathbf{E}^{a}(r)\right). \tag{2.4}$$

Expression (2.2) becomes an integral equation with respect to anomalous electric field $\mathbf{E}^{a}(\mathbf{r})$, if point $\mathbf{r}_j$ is inside D.

Inserting Equation (2.4) into (2.2) produces:

$$\mathbf{E}^{a}(\mathbf{r}_j) = \iiint_D \widehat{\mathbf{G}}_E(\mathbf{r}_j \mid \mathbf{r}) \cdot \Delta\tilde{\sigma}(r)\left(\mathbf{E}^{b}(r) + \mathbf{E}^{a}(r)\right)\mathrm{d}v. \tag{2.5}$$

Because the background field is known, it is convenient to rewrite (2.5) as:

$$\mathbf{E}^{a}(\mathbf{r}_j) = \iiint_D \widehat{\mathbf{G}}_E(\mathbf{r}_j \mid \mathbf{r}) \cdot \Delta\tilde{\sigma}(r)\mathbf{E}^{a}(r)\,\mathrm{d}v + \mathbf{E}^{B}(r_j), \tag{2.6}$$

where $\mathbf{E}^{B}(r_j)$ is the Born approximation at point $\mathbf{r_j}$:

$$\mathbf{E}^{B}(\mathbf{r}_j) = \iiint_D \widehat{\mathbf{G}}_E(\mathbf{r}_j \mid \mathbf{r}) \cdot \Delta\tilde{\sigma}(r)\mathbf{E}^{b}(r)\,\mathrm{d}v. \tag{2.7}$$

We express vector Equation (2.6) via individual scalar components. Let us denote:

$$\mathbf{E}^{a} = \left(E^{a}_{x}\ E^{a}_{y}\ E^{a}_{z}\right) = \left(E^{a}_{1}\ E^{a}_{2}\ E^{a}_{3}\right), \tag{2.8}$$

where lower index 1 denotes the x-component, 2 denotes the y component, and 3 denotes the z component.

Then, Expression (2.6) breaks into three equations:

$$E^{a}_{n}(\mathbf{r}_j) = \sum_{m=1}^{3}\left(\iiint_D G_{E_{nm}}(\mathbf{r}_j \mid \mathbf{r}) \cdot \Delta\tilde{\sigma}(r)E^{a}_{m}(r)\,\mathrm{d}v\right) + E^{B}_{n}(r_j), \tag{2.9}$$

where $G_{E_{nm}}$ ($n = 1,2,3$, and $m = 1,2,3$) are the components of the Green's tensor.

For the purpose of computer simulations we must discretize Equation (2.9). To accomplish this, we divide domain D into rectangular cells. Individual cells are denoted D_k, and the total number of cells is N_c. We assume that anomalous conductivity is constant within each individual cell, and the cell size is small enough to consider the anomalous electric field to be constant inside the cell (Xiong, 1992).

Under these assumptions the triple integral over the entire domain D (in Expression (2.9)) transforms into the sum of contributions from the individual cells:

$$E_n^{\mathrm{a}}(\mathbf{r_j}) = \sum_{m=1}^{3}\sum_{k=1}^{N_c}\left(\iiint_{D_k} G_{E_{nm}}(\mathbf{r}_j \mid \mathbf{r})\,\mathrm{d}v \cdot \Delta\tilde{\sigma}(\mathbf{r}_k)E_m^{\mathrm{a}}(\mathbf{r}_k)\right) + E_n^{\mathrm{B}}(\mathbf{r}_j), \tag{2.10}$$

where $n = 1,2,3$, $\mathbf{j} = 1,\ldots,N_c$; $\mathbf{r_j}$ is the central point of the cell with the index j, and $\mathbf{r_k}$ is the central point of the cell with the index k.

Equation (2.10) can be written using matrix notations as follows:

$$\mathbf{e} = \widehat{\mathbf{G}}\widehat{\mathbf{S}}\mathbf{e} + \mathbf{b}, \tag{2.11}$$

where $\widehat{\mathbf{G}}$ is the matrix of a size $3N_c \times 3N_c$, of the known Green's tensor integrated over elementary cell D_k, with the scalar components $G_{(l,\,i)}$:

$$G_{(j+(n-1)N_c,\,k+(m-1)N_c)} = \iiint_{D_k} G_{E_{nm}}(\mathbf{r}_j \mid \mathbf{r})\,\mathrm{d}v, \tag{2.12}$$

$\widehat{\mathbf{S}}$ is a sparse diagonal matrix, of a size $3N_c \times 3N_c$, with the diagonal elements $S_{(l,\,l)}$ equal to the known conductivities within each cell:

$$S_{(k+(m-1)N_c,\,k+(m-1)N_c)} = \Delta\tilde{\sigma}(r_k), \tag{2.13}$$

$\mathbf{b}$ is a vector of a length $3N_c$, containing three components of the Born approximation, with the scalar components $b_{(l)}$:

$$b_{(j+(n-1)N_c)} = E_n^{\mathrm{B}}(r_j). \tag{2.14}$$

and $\mathbf{e}$ is a vector of a length $3N_c$, containing three components of the unknown anomalous field E_n^{a} in every cell, with the scalar components $e_{(l)}$:

$$e_{(j+(n-1)N_c)} = E_n^{\mathrm{a}}(\mathbf{r_j}). \tag{2.15}$$

Thus, a forward electromagnetic modeling problem is reduced to numerically solving Equation (2.11) to find the unknown vector $\mathbf{e}$ representing a discretized anomalous electric field inside a domain D. One way of solving (2.11) is by using the block-relaxation method (Xiong, 1992). Here we discuss an alternative method, that is, solving (2.11) using compression.

The main problem with the integral equations method is that in general 3-D cases matrix $\widehat{\mathbf{G}}$ in Equation (2.11) may be very large. Assume that anomalous domain D is a rectangular prism. Each side of the prism is divided by N to produce rectangular prismatic cells. Then the number of cells is $N_c = N^3$. The number of scalar components in matrix $\widehat{\mathbf{G}}$ is $(3 * N_c^3)^2 = 9 * N^6$. We can see that this number grows as the sixth power of N. This growth is the main limiting factor of the integral equation method. If $N = 5$ the problem is small and readily solvable. Yet, for $N = 10$ the size of the problem becomes very large.

By using the compression technique, we may reduce the size of the problem.

3. COMPRESSION MATRIX

First, let us consider the compression with an interpolation pyramid for a 1-D function.

An interpolation pyramid consists of levels (grids), with each level twice as coarse as the previous one. The first and finest grid is that of an original discretized curve.

On each grid we distinguish odd nodes, or *reference nodes*, and even nodes, or *intermediate nodes*. Values at reference nodes can be used to predict (by interpolation) the values at the intermediate nodes. If the curve is smooth, then the difference between predicted and original values is small. That is the key point which enables compression. Take a curve uniformly discretized at $(2N+1)$ points, which we denote as a vector $\mathbf{v}_1$:

$$\mathbf{v}_1 = \{d_1, d_2, d_3, \ldots, d_{2N}, d_{2N+1}\}. \tag{2.16}$$

Consider the transformation of $\mathbf{v}_1$ to $\mathbf{v}_2$, where values at odd points $(1,3,\ldots,2N+1)$ are retained as they were, and values at even points $(2,4,\ldots,2N)$ are transformed into residuals between the predicted (by interpolation from the odd grid) and the original values. For example, points 1 and 3 are retained as they were, and point 2 is transformed as the half-sum of the values at points 1 and 3, minus the original value at point 2. Such transformation can be described by the matrix, which we denote as $\mathbf{W}_1$:

$$\mathbf{v}_2 = \widehat{\mathbf{W}}_1 \mathbf{v}_1. \tag{2.17}$$

Matrix $\widehat{\mathbf{W}}_1$ has the following structure:

$$\begin{bmatrix} 1 & 0 & 0 & \ldots & 0 & 0 & 0 & 0 \\ 1/2 & -1 & 1/2 & \ldots & 0 & 0 & 0 & 0 \\ 0 & 0 & 1 & \ldots & 0 & 0 & 0 & 0 \\ \ldots & \ldots & \ldots & \ldots & \ldots & \ldots & \ldots & \ldots \\ 0 & 0 & 0 & \ldots & 0 & 1/2 & -1 & 1/2 \\ 0 & 0 & 0 & \ldots & 0 & 0 & 0 & 1 \end{bmatrix}. \tag{2.18}$$

Note that matrix $\widehat{\mathbf{W}}_1$ is inverse to itself

$$\widehat{\mathbf{W}}_1 \widehat{\mathbf{W}}_1 = \widehat{\mathbf{I}}$$

which means it is positive definite.

The next level uses a grid that is twice as coarse. Now, the intermediate values have indices $(3,7,11,\ldots,2N-1)$. Again, they are predicted using reference values with indices $(1,5,9,13,\ldots,2N+1)$, and the originals are subtracted from the predictions. This transformation is described by the matrix $\widehat{\mathbf{W}}_2$:

$$\mathbf{v}_3 = \widehat{\mathbf{W}}_2 \mathbf{v}_2. \tag{2.19}$$

where matrix $\widehat{\mathbf{W}}_2$ has the following structure

$$\begin{bmatrix} 1 & 0 & 0 & 0 & 0 & \ldots & 0 & 0 & 0 & 0 & 0 \\ 0 & 1 & 0 & 0 & 0 & \ldots & 0 & 0 & 0 & 0 & 0 \\ 1/2 & 0 & -1 & 0 & 1/2 & \ldots & 0 & 0 & 0 & 0 & 0 \\ \ldots & \ldots & \ldots & \ldots & \ldots & \ldots & \ldots & \ldots & \ldots & \ldots & \ldots \\ 0 & 0 & 0 & 0 & 0 & \ldots & 1/2 & 0 & -1 & 0 & 1/2 \\ 0 & 0 & 0 & 0 & 0 & \ldots & 0 & 0 & 0 & 1 & 0 \\ 0 & 0 & 0 & 0 & 0 & \ldots & 0 & 0 & 0 & 0 & 1 \end{bmatrix}. \tag{2.20}$$

We repeat this process for L pyramid levels:

$$\mathbf{v}_L = \widehat{\mathbf{W}}_{L-1} \ldots \widehat{\mathbf{W}}_2 \widehat{\mathbf{W}}_1 \mathbf{v}_1,$$

and

$$\hat{\mathbf{v}}_L = \widehat{\mathbf{W}}_c \mathbf{v}_1, \tag{2.21}$$

where $\widehat{\mathbf{W}}_c$ is a *compression matrix*:

$$\widehat{\mathbf{W}}_c = \widehat{\mathbf{W}}_{L-1} \ldots \widehat{\mathbf{W}}_2 \widehat{\mathbf{W}}_1. \tag{2.22}$$

The resulting vector $\mathbf{v}_L$ contains only few meaningful values. The rest of the values are close to zero. Thus, vector $\mathbf{v}_L$ can be approximated by a sparse vector. The final stage of compression involves thresholding. We find all values in the vector $\mathbf{v}_L$ which are close to zero (say, less than 1 percent of vector maximum) and set them equal to zero:

$$\mathbf{v}'_L = \mathit{threshold}(\widehat{\mathbf{W}}_c \mathbf{v}_1, 0.01), \tag{2.23}$$

where vector $\mathbf{v}'_L$ is the resulting sparse vector. Operator *threshold* denotes threshold transformation.

4. COMPRESSION IN THREE DIMENSIONS

In multiple dimensions, compression with an interpolation pyramid is a series of subsequent 1-D linear interpolations on coarsening grids. Compressing a 3-D function, for example, we would first reduce values at even nodes in the x-direction, then reduce the values at even nodes at the y-direction, then in the z-direction. After that we would consider a coarse grid, and start the process from the x-direction, etc.

To illustrate how compression is done, let us consider the following example. We take a thin conductive plate, as shown in Figure 1. We divide this plate into 81 blocks $N_x = 9$, $N_y = 9$ and $N_z = 1$. Components of electrical Green's tensor for the cell in the middle of the domain (the influence of the middle cell on all other cells) are shown in Figure 2.

Consider the compression of one component of Green's tensor on this uniform 9×9 grid (shown in Figure 3). In three dimensions, compression with an interpolation pyramid is a series of 1-D linear interpolations on coarsening grids. Assume that the original function (one component of Green's tensor given on 3-D mesh $N_x \times N_y \times N_z$) is denoted as a vector $\mathbf{d}$. First, values at nodes with even x indices $i_x = 2, 4, 6, 8$ (denoted by stars in Figure 3, case a) are predicted by a 1-D linear interpolation in the x direction from values at odd points $i_x = [1, 3, 5, 7, 9]$. Prediction is subtracted from true values at even points; odd point values are retained as is. This transformation can be denoted as a linear operation:

$$\mathbf{d}_{x1} = \widehat{\mathbf{W}}_{x1} \mathbf{d} \tag{2.24}$$

where $\widehat{\mathbf{W}}_{x1}$ is the first elementary compression matrix in the x direction. Note that it is a sparse matrix. Portniaguine (1999) considers the structure of a 1-D elementary compression matrix in more detail.

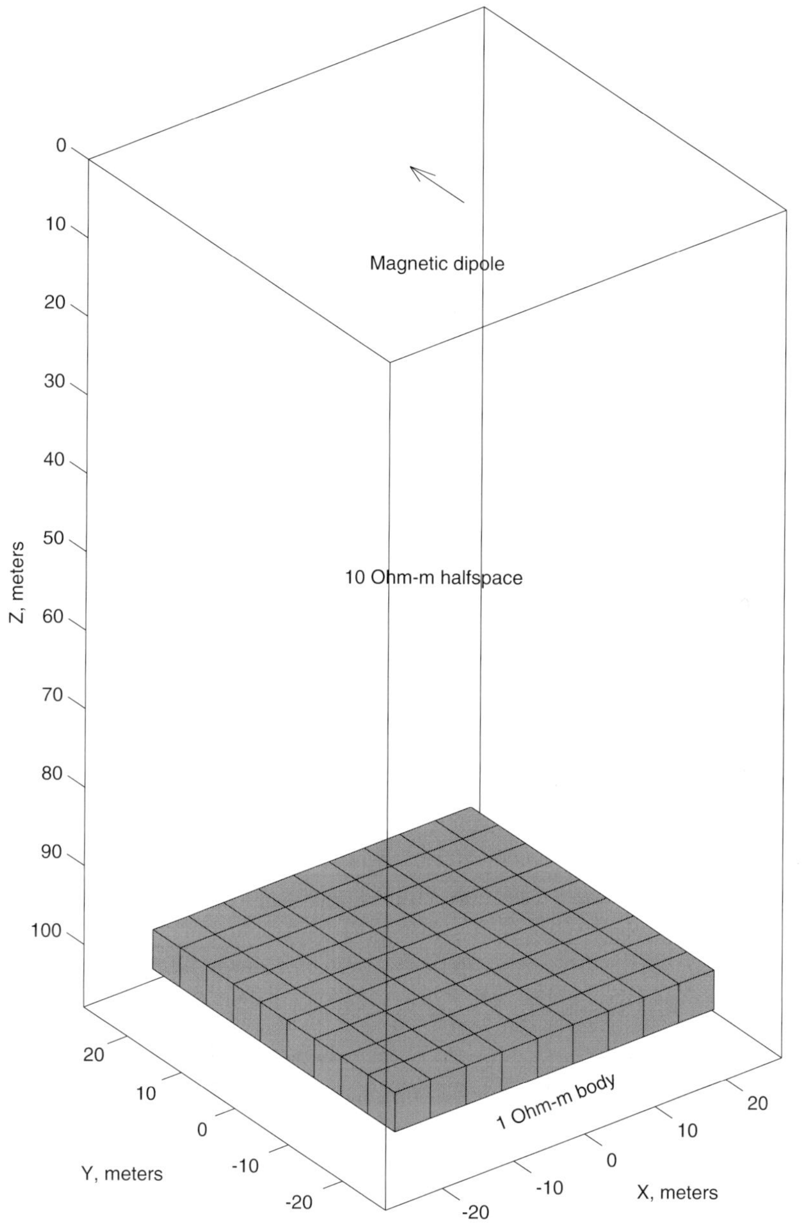

Figure 1. A model used to illustrate the work of compression in multiple dimensions. A thin 1 Ohm m body in a 10 Ohm m homogeneous half-space is divided into 9 x 9 blocks in directions *x* and *y*, respectively. There is one cell in direction *z*. The body is excited by a horizontal magnetic dipole operating at 100 Hz and located at the origin of coordinates.

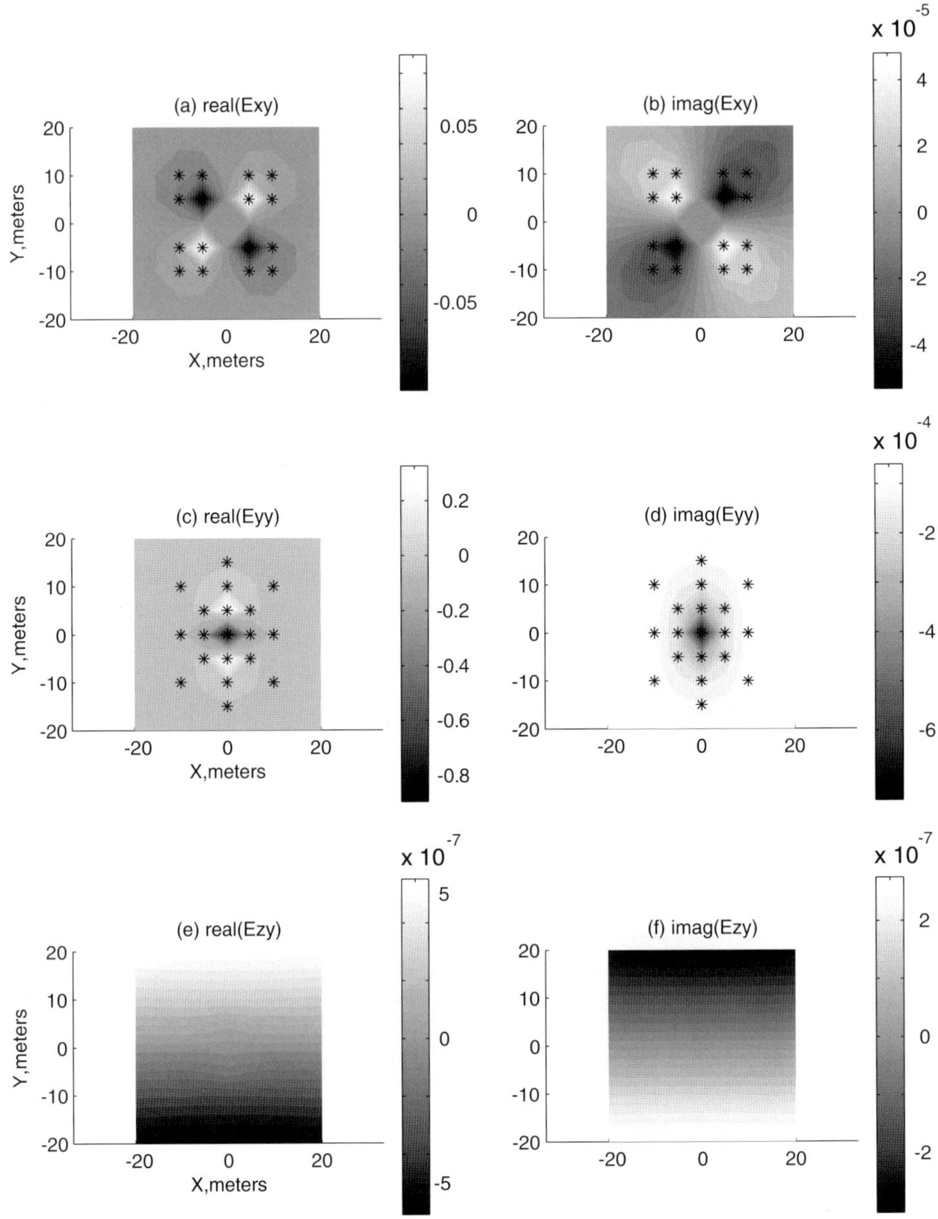

Figure 2. Components (E_{xy}, E_{yy} and E_{zy}) of electric Green's tensor for the geometry shown in Figure 1. Stars denote nodes where the values were retained after compression. The values at all other nodes were thresholded to zero; hence the compressed Green tensor matrix has become sparse.

Figure 3. Compression scheme of a function on a 9 x 9 grid. At every step, the values at starred nodes are predicted by linear interpolation from the values at circled nodes. In panels (a), (c) and (e) 1-D interpolation is done in direction x. In panels (b), (d) and (f) interpolation is done in direction y.

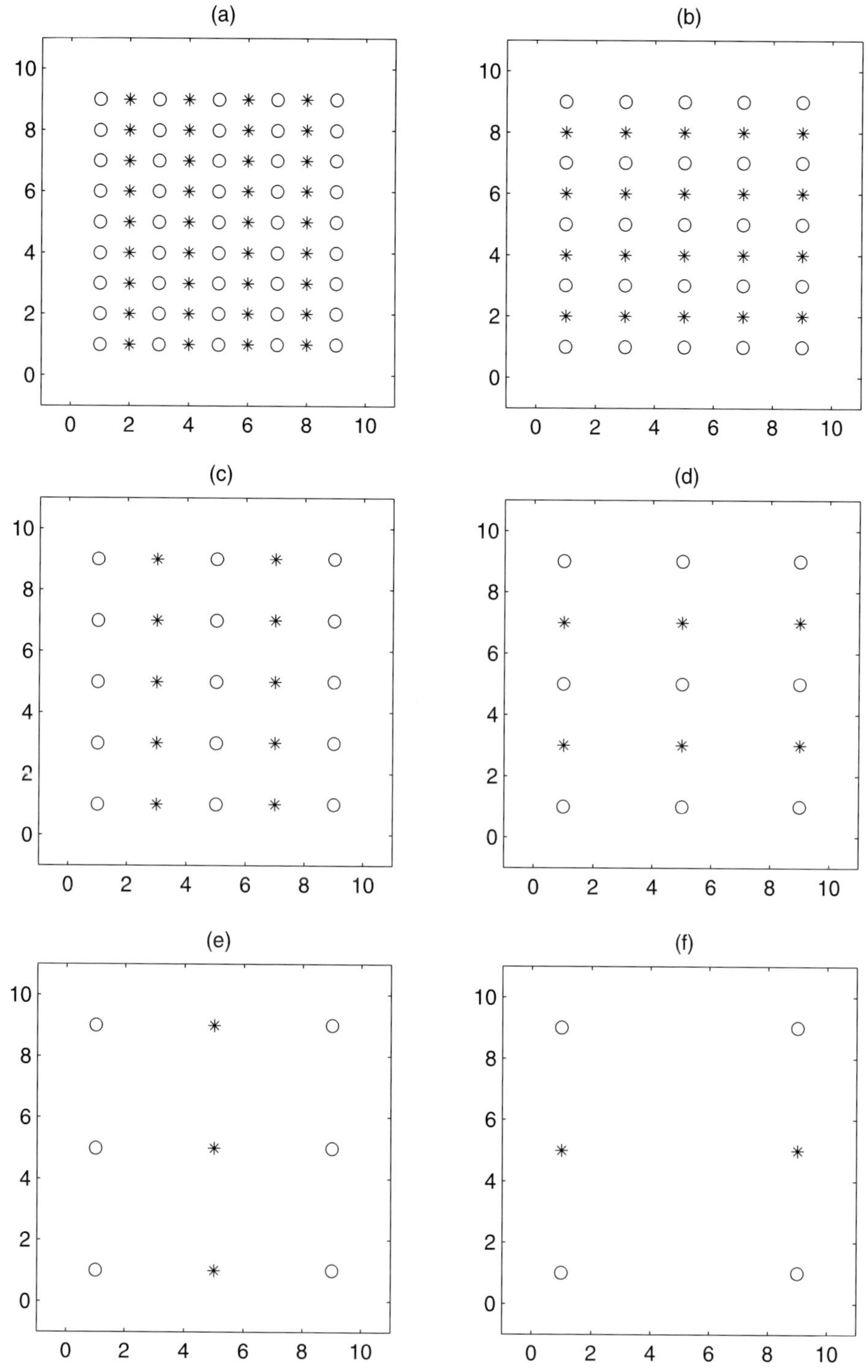
(a)
(b)
(c)
(d)
(e)
(f)
0
2
4
6
8
10

Next, the same transformation is done in direction y. Values at nodes with even y indices $i_y = 2,4,6,8$ (denoted by stars in Figure 3, case b) are predicted by a 1-D linear interpolation in the direction y from the values at odd points $i_y = [1,3,5,7,9]$. Prediction is subtracted from true values at even points; odd point values are retained as is. This transformation, again, can be described as a linear operation:

$$\mathbf{d}_{y1} = \widehat{\mathbf{W}}_{y1}\mathbf{d}_{x1} \tag{2.25}$$

In general cases, the same should be repeated in the direction z after step (2.25):

$$\mathbf{d}_{z1} = \widehat{\mathbf{W}}_{z1}\mathbf{d}_{y1} \tag{2.26}$$

In our specific example the direction z is a singleton, that is $N_z = 1$. For generality, we will retain this direction in our formulas. For the singleton dimension, the compression matrix is simply a unit matrix.

Note that in the case of multiple dimensions, compression on one pyramid level consists of a series of successive 1-D interpolations in all dimensions.

On the next pyramid level we return to the x direction and repeat the same transformation starting on a coarser grid (Figure 3, case c):

$$\mathbf{d}_{x2} = \widehat{\mathbf{W}}_{x2}\mathbf{d}_{z1}. \tag{2.27}$$

The same along the direction y (Figure 3, case d):

$$\mathbf{d}_{y2} = \widehat{\mathbf{W}}_{y2}\mathbf{d}_{x2}, \tag{2.28}$$

and the same along the direction z:

$$\mathbf{d}_{z2} = \widehat{\mathbf{W}}_{z2}\mathbf{d}_{y2}. \tag{2.29}$$

This process should be repeated for as many pyramid levels as necessary. The number of levels N_l in interpolation pyramid is:

$$N_l = 1 + \mathrm{int}(\log_2(\max([N_x, N_y, N_z]))) \tag{2.30}$$

In our example, the third level is the last one. Again, we start from the direction x (mesh is shown in Figure 3, case e):

$$\mathbf{d}_{x3} = \widehat{\mathbf{W}}_{x3}\mathbf{d}_{z2}. \tag{2.31}$$

The same in the y direction (mesh is shown in Figure 3, case f):

$$\mathbf{d}_{y3} = \widehat{\mathbf{W}}_{y3} d_{x3}, \tag{2.32}$$

and in the z direction:

$$\mathbf{d}_{z3} = \widehat{\mathbf{W}}_{z3}\mathbf{d}_{y3}. \tag{2.33}$$

Let us put the whole process together:

$$\mathbf{d}_{z3} = \widehat{\mathbf{W}}_{z3}\widehat{\mathbf{W}}_{y3}\widehat{\mathbf{W}}_{x3}\widehat{\mathbf{W}}_{z2}\widehat{\mathbf{W}}_{y2}\widehat{\mathbf{W}}_{x2}\widehat{\mathbf{W}}_{z1}\widehat{\mathbf{W}}_{y1}\widehat{\mathbf{W}}_{x1}\mathbf{d}. \tag{2.34}$$

In general cases, where the number of compression levels equals N_l, we have:

$$\mathbf{d}_{zN_l} = \prod_{k=1}^{N_l}(\widehat{\mathbf{W}}_{zk}\widehat{\mathbf{W}}_{yk}\widehat{\mathbf{W}}_{xk})\mathbf{d}, \tag{2.35}$$

or,

$$\mathbf{d}_{zN_1} = \widehat{\mathbf{W}}_c \mathbf{d}, \tag{2.36}$$

where $\widehat{\mathbf{W}}_c$ is a sparse compression matrix:

$$\widehat{\mathbf{W}}_c = \prod_{k=1}^{N_1} (\widehat{\mathbf{W}}_{zk} \widehat{\mathbf{W}}_{yk} \widehat{\mathbf{W}}_{xk}). \tag{2.37}$$

If information in the original vector $\mathbf{d}$ is redundant, then interpolation predicts intermediate values accurately. In this case the result of transformation (2.37), vector $\mathbf{d}_{zN_1}$, contains many small values and therefore can be approximated by a sparse vector $\mathbf{d}_c$:

$$\mathbf{d}_c = \mathit{threshold}(\widehat{\mathbf{W}}_c \mathbf{d}, \varepsilon) \tag{2.38}$$

where $\mathit{threshold}(, \varepsilon)$ denotes threshold transformation with the threshold level ε.

5. COMPRESSION AS PRECONDITIONER TO THE INTEGRAL EQUATION

The notion of a compression matrix gives crucial advantage for analytical work and simplifies coding. The advantage of this approach is significant in its application to forward problem solution.

Consider again a discrete 3-D electromagnetic integral Equation (2.11). Full matrix $\widehat{\mathbf{G}}$ in integral Equation (2.11) consists of nine blocks. Each of the blocks is a smaller matrix containing corresponding components of Green's tensor:

$$\widehat{\mathbf{G}} = \begin{bmatrix} \widehat{\mathbf{G}}_{xx} & \widehat{\mathbf{G}}_{xy} & \widehat{\mathbf{G}}_{xz} \\ \widehat{\mathbf{G}}_{yx} & \widehat{\mathbf{G}}_{yy} & \widehat{\mathbf{G}}_{yz} \\ \widehat{\mathbf{G}}_{zx} & \widehat{\mathbf{G}}_{zy} & \widehat{\mathbf{G}}_{zz} \end{bmatrix}. \tag{2.39}$$

Our compression matrix $\widehat{\mathbf{W}}_c$ (2.37) applies to one component of $\widehat{\mathbf{G}}$. Using $\widehat{\mathbf{W}}_c$, we may establish the sparse global compression matrix $\widehat{\mathbf{W}}_g$ to be used as preconditioner to Equation (2.11):

$$\widehat{\mathbf{W}}_g = \begin{bmatrix} \widehat{\mathbf{W}}_c & 0 & 0 \\ 0 & \widehat{\mathbf{W}}_c & 0 \\ 0 & 0 & \widehat{\mathbf{W}}_c \end{bmatrix}. \tag{2.40}$$

Using $\widehat{\mathbf{W}}_g$ as a preconditioner to Equation (2.11), we have

$$\widehat{\mathbf{W}}_g \mathbf{e} = \widehat{\mathbf{W}}_g \widehat{\mathbf{G}} \widehat{\mathbf{S}} \mathbf{e} + \widehat{\mathbf{W}}_g \mathbf{b}. \tag{2.41}$$

Rearranging (2.41), we arrive at

$$(\widehat{\mathbf{W}}_g - \widehat{\mathbf{W}}_g \widehat{\mathbf{G}} \widehat{\mathbf{S}}) \mathbf{e} = \widehat{\mathbf{W}}_g \mathbf{b}. \tag{2.42}$$

Now, we can approximate (2.42) by applying a threshold to $\widehat{\mathbf{W}}_g \widehat{\mathbf{G}}$ and $\widehat{\mathbf{W}}_g \mathbf{b}$:

$$(\widehat{\mathbf{W}}_g - \mathit{threshold}(\widehat{\mathbf{W}}_g \widehat{\mathbf{G}}, \varepsilon) \widehat{\mathbf{S}}) \hat{\mathbf{e}} \approx \mathit{threshold}(\widehat{\mathbf{W}}_g \mathbf{b}, \varepsilon). \tag{2.43}$$

Replacing

$$\mathit{threshold}(\widehat{\mathbf{W}}_g \widehat{\mathbf{G}}, \varepsilon) = \widehat{\mathbf{G}}_c \tag{2.44}$$

and

$$threshold(\widehat{\mathbf{W}}_g \mathbf{b}, \varepsilon) = \mathbf{b}_c,$$

we obtain

$$(\widehat{\mathbf{W}}_g - \widehat{\mathbf{G}}_c \widehat{\mathbf{S}})\mathbf{e} \approx \mathbf{b}_c. \tag{2.45}$$

In the compressed Equation (2.45) all matrices are sparse. Thus, we have managed to approximate the originally full system of equations (2.11) by a sparse system (2.45). This reduces memory requirements and speeds up computations. The downside is the possible loss of accuracy, which is connected to the threshold level.

6. THE ILU PRECONDITIONED CONJUGATE GRADIENT METHOD

Assume we are solving a linear problem:

$$\widehat{\mathbf{A}}\mathbf{m} = \mathbf{d}. \tag{2.46}$$

The objective is to find the parameters $\mathbf{m}$ given the matrix $\widehat{\mathbf{A}}$ and the right-hand side vector $\mathbf{d}$. In our notations, $\widehat{\mathbf{A}}$ stands for $\widehat{\mathbf{W}}_g - \widehat{\mathbf{G}}_c \widehat{\mathbf{S}}$ from Equation (2.45). The vector of parameters $\mathbf{m}$ stands for the electric field $\mathbf{e}$, and the vector of data $\mathbf{d}$ stands for the right-hand side vector $\mathbf{b}_c$ from Equation (2.45).

For 3-D cases, matrix $\widehat{\mathbf{A}}$ is large, so iterative methods have to be employed to solve Equation (2.46).

One of the methods is the conjugate gradient (CG) method (Fletcher, 1981; Gill et al., 1981; Press et al., 1996).

In the framework of the CG method, the solution of (2.46) is found iteratively, according to the following formulas:

$$
\begin{aligned}
&\mathbf{l}_i = \widehat{\mathbf{A}}^T \mathbf{r}_{i-1} && \text{(a)} \\
&s_i = \mathbf{l}_i^T \mathbf{l}_i && \text{(b)} \\
&\mathbf{h}_i = \mathbf{l}_i + \mathbf{h}_{i-1} \frac{s_i}{s_{i-1}} && \text{(c)} \\
&\mathbf{f}_i = \widehat{\mathbf{A}}\mathbf{h}_i && \text{(d)} \\
&k_i = \frac{\mathbf{f}_i^T \mathbf{r}_i}{\mathbf{f}_i^T \mathbf{f}_i} && \text{(e)} \\
&\mathbf{m}_i = \mathbf{m}_{i-1} - k_i \mathbf{h}_i && \text{(f)} \\
&\mathbf{r}_i = \mathbf{r}_{i-1} - k_i \mathbf{f}_i && \text{(g)},
\end{aligned}
\tag{2.47}
$$

where i is the iteration number, $\mathbf{r}$ is the residual vector, $\mathbf{l}$ is the gradient vector, s is its length, $\mathbf{h}$ is the conjugate direction vector in the solution space, $\mathbf{f}$ is its projection to the right-hand side, and k is the step length, a scalar. The starting values (for $i = 0$) are

$$
\begin{aligned}
&\mathbf{x}_0 = 0 && \text{(a)} \\
&\mathbf{r}_0 = \widehat{\mathbf{A}}\mathbf{m}_0 - \mathbf{d} = -\mathbf{d} && \text{(b)} \\
&\mathbf{s}_0 = 1 && \text{(c)}.
\end{aligned}
\tag{2.48}
$$

Concerning the performance of (2.47), two principal cases should be noted. The first case is when matrix $\widehat{\mathbf{A}}$ arises from the inverse problem, where $\widehat{\mathbf{A}}$ can be rectangular, and most often, even singular. In this case the CG method converges to the least-squares or minimum norm solution. It requires as many iterations as there are non-zero singular values of $\widehat{\mathbf{A}}$. The case with matrix $\widehat{\mathbf{A}}$ having all rows equal (and therefore only one non-zero singular value) illustrates this property, where CG converges to a solution in just one iteration.

The forward problem represents exactly the opposite case, where the matrix $\widehat{\mathbf{A}}$ is of full rank. This case corresponds to the finite-difference or finite-element solution of a PDE, or, in our case, the compressed IE. That means the problem is well-posed, with a unique solution, and matrix $\widehat{\mathbf{A}}$ is $N \times N$ square matrix.

For this case the standard method (2.47) works very slowly. To fully converge, it requires N iterations, which in 3-D cases is a large number. An illustration to this property is a case when matrix $\widehat{\mathbf{A}}$ is square and strictly diagonal, with all non-zero values on the main diagonal (and therefore non-singular). For this simple case CG requires N iterations to converge.

For the case, when matrix $\widehat{\mathbf{A}}$ arises from the forward problem, we apply the CG method, preconditioned with incomplete LU decomposition of $\widehat{\mathbf{A}}$ (ILU) (Freund, 1992; Van der Vorst, 1992; Chan et al., 1994; Bank and Xu, 1994).

The basic idea of this method can be described as follows. Assume we are given an approximate inverse matrix to $\widehat{\mathbf{A}}$ in the form of LU decomposition. This approximate inverse matrix is called incomplete LU, or ILU decomposition. Later we will consider various forms of ILU decomposition. Now, we concentrate on the advantages of ILU decomposition in the iterative algorithm.

Consider the non-singular matrices $\widehat{\mathbf{L}}$ and $\widehat{\mathbf{U}}$ with the property:

$$\widehat{\mathbf{A}} \approx \widehat{\mathbf{L}}\widehat{\mathbf{U}}, \tag{2.49}$$

It follows from (2.49) that

$$\mathbf{L}^{-1}\widehat{\mathbf{A}}\widehat{\mathbf{U}}^{-1} \approx \widehat{\mathbf{I}}. \tag{2.50}$$

Since the problem (2.46) is well-posed, preconditioning and post-conditioning with any non-singular matrix does not change the solution.

Thus, we replace (2.46) with

$$\widehat{\mathbf{L}}^{-1}\widehat{\mathbf{A}}\widehat{\mathbf{U}}^{-1}\widehat{\mathbf{U}}\mathbf{m} = \widehat{\mathbf{L}}^{-1}\mathbf{d}, \tag{2.51}$$

and introduce new variables

$$\mathbf{x} = \widehat{\mathbf{U}}\mathbf{m}, \tag{2.52}$$

$$\mathbf{b} = \widehat{\mathbf{L}}^{-1}\mathbf{d}, \tag{2.53}$$

arriving at

$$\widehat{\mathbf{L}}^{-1}\widehat{\mathbf{A}}\widehat{\mathbf{U}}^{-1}\mathbf{x} = \mathbf{b}. \tag{2.54}$$

In Equation (2.54) the matrix of the problem is stored as a factorization of three matrices. The standard CG algorithm is applied to (2.54). In this case, the matrix of the problem is close to the unit matrix $\widehat{\mathbf{I}}$, according to (2.50). Since it is equal to $\widehat{\mathbf{I}}$ only approximately, the CG still requires N iterations to fully converge. But the rate of convergence is much faster than for the original algorithm without preconditioning.

Note also, that the application of the inverse triangular matrix to a vector can be readily implemented via Gaussian backsubstitution. Let us denote Gaussian backsubstitution as backslash operation '\':

$$\widehat{\mathbf{L}}^{-1}\mathbf{d} = \widehat{\mathbf{L}}\backslash\mathbf{d} \tag{2.55}$$

We now cast the ILU preconditioned CG algorithm as follows:

$$\begin{aligned}
\mathbf{l}_i &= \widehat{\mathbf{U}}^{\mathrm{T}}\backslash(\widehat{\mathbf{A}}^{\mathrm{T}}(\mathbf{L}^{\mathrm{T}}\backslash\mathbf{r}_{i-1})) && \text{(a)}\\
s_i &= \mathbf{l}_i^{\mathrm{T}}\mathbf{l}_i && \text{(b)}\\
\mathbf{h}_i &= \mathbf{l}_i + \mathbf{h}_{i-1}\frac{s_i}{s_{i-1}} && \text{(c)}\\
\mathbf{f}_i &= \mathbf{L}\backslash(\widehat{\mathbf{A}}(\widehat{\mathbf{U}}\backslash\mathbf{h}_i)) && \text{(d)}\\
k_i &= \frac{\mathbf{f}_i^{\mathrm{T}}\mathbf{r}_i}{\mathbf{f}_i^{\mathrm{T}}\mathbf{f}_i} && \text{(e)}\\
\mathbf{x}_i &= \mathbf{x}_{i-1} - k_i\mathbf{h}_i && \text{(f)}\\
\mathbf{r}_i &= \mathbf{r}_{i-1} - k_i\mathbf{f}_i && \text{(g),}
\end{aligned} \tag{2.56}$$

where initial values are:

$$\begin{aligned}
\mathbf{x}_0 &= 0 && \text{(a)}\\
\mathbf{r}_0 &= -\widehat{\mathbf{L}}\backslash\mathbf{d} && \text{(b)}\\
\mathbf{s}_0 &= 1 && \text{(c).}
\end{aligned} \tag{2.57}$$

After the method converges to the solution $\mathbf{x}_n$, we have to return to original space of $\mathbf{m}$ using

$$\mathbf{m} = \widehat{\mathbf{U}}\backslash\mathbf{x}_n \tag{2.58}$$

If we use a complete LU decomposition in the algorithm, it will converge in one step. The downside of this technique, however, is the need to compute full LU, which may be very costly for large matrices. In addition, for a sparse matrix $\widehat{\mathbf{A}}$, the complete decomposition produces full matrices $\widehat{\mathbf{L}}$ and $\widehat{\mathbf{U}}$.

It is possible, to some extent, to avoid these difficulties using incomplete LU decomposition with the partial fill-up. The complete decomposition of $\widehat{\mathbf{A}}$ often produces many small terms in $\widehat{\mathbf{L}}$ and $\widehat{\mathbf{U}}$. The ILU decomposition with the partial fill-up is based on thresholding these matrices, producing sparse $\widehat{\mathbf{L}}$ and $\widehat{\mathbf{U}}$.

This method performs faster than the complete LU decomposition, because the construction of incomplete LU is faster. Still, precomputing and storing matrices $\widehat{\mathbf{L}}$ and $\widehat{\mathbf{U}}$ along with the matrix $\widehat{\mathbf{A}}$ introduces significant overhead into the algorithm.

Another form of ILU decomposition, which we call 'compositional ILU', does not have these disadvantages. Below, we discuss it in detail.

We consider separate parts of matrix $\widehat{\mathbf{A}}$, namely the lower part $\widehat{\mathbf{L}}_{\mathbf{a}}$ and the upper part $\widehat{\mathbf{U}}_{\mathbf{a}}$ (without the diagonal), and the diagonal $\widehat{\mathbf{D}}$:

$$\widehat{\mathbf{A}} = \widehat{\mathbf{L}}_{\mathbf{a}} + \widehat{\mathbf{D}} + \widehat{\mathbf{U}}_{\mathbf{a}}. \tag{2.59}$$

If matrix $\widehat{\mathbf{A}}$ is sparse and diagonally dominant, then the ILU property holds:

$$\widehat{\mathbf{A}} \approx \widehat{\mathbf{L}}\widehat{\mathbf{U}}, \tag{2.60}$$

where, if we define

$$\widehat{\mathbf{L}} = \widehat{\mathbf{L}}_{\mathbf{a}} + \widehat{\mathbf{D}}, \tag{2.61}$$

then

$$\widehat{\mathbf{U}} = \widehat{\mathbf{D}}^{-1}\widehat{\mathbf{U}}_{\mathbf{a}} + \widehat{\mathbf{I}}. \tag{2.62}$$

If $\widehat{\mathbf{A}}$ is diagonally dominant, from (2.61) and (2.62) it is clear that matrices $\widehat{\mathbf{L}}$ and $\widehat{\mathbf{U}}$ are diagonally dominant as well. Since they are triangular, they are non-singular as well.

Equation (2.60) is an incomplete factorization of the matrix $\widehat{\mathbf{A}}$. An important advantage of this equation over other types of ILU decompositions is that it is computationally cheap, since it requires only separation of upper and lower parts of $\widehat{\mathbf{A}}$ and inversion of the diagonal matrix $\widehat{\mathbf{D}}$. Moreover, compositional ILU does not require extra storage space for saving $\widehat{\mathbf{A}}$ and separately $\widehat{\mathbf{L}}$ and $\widehat{\mathbf{U}}$, since we can represent matrix $\widehat{\mathbf{A}}$ via matrices $\widehat{\mathbf{L}}$ and $\widehat{\mathbf{U}}$ as

$$\widehat{\mathbf{A}} = \widehat{\mathbf{D}}\widehat{\mathbf{U}} + \widehat{\mathbf{L}} - \widehat{\mathbf{D}}. \tag{2.63}$$

The CG algorithm using (2.63) becomes

$$\begin{aligned}
&\mathbf{t}_i = \mathbf{r}^{\mathrm{T}}\widehat{\mathbf{L}}^{-1}\widehat{\mathbf{D}} && \text{(a)}\\
&\mathbf{l}_i = (\mathbf{t}_i + (\mathbf{r}^{\mathrm{T}} - \mathbf{t}_i)\widehat{\mathbf{U}}^{-1})^{\mathrm{T}} && \text{(b)}\\
&s_i = \mathbf{l}_i^T\mathbf{l}_i && \text{(c)}\\
&\mathbf{h}_i = \mathbf{l}_i + \mathbf{h}_{i-1}\frac{s_i}{s_{i-1}} && \text{(d)}\\
&\mathbf{p}_i = \widehat{\mathbf{U}}^{-1}\mathbf{h}_i && \text{(e)}\\
&\mathbf{f}_i = \mathbf{p}_i + \widehat{\mathbf{L}}^{-1}\widehat{\mathbf{D}}(\mathbf{h} - \mathbf{p}_i) && \text{(f)}\\
&k_i = \frac{\mathbf{f}_i^{\mathrm{T}}\mathbf{r}_i}{\mathbf{f}_i^{\mathrm{T}}\mathbf{f}_i} && \text{(g)}\\
&\mathbf{x}_i = \mathbf{x}_{i-1} - k_i\mathbf{h}_i && \text{(h)}\\
&\mathbf{r}_i = \mathbf{r}_{i-1} - k_i\mathbf{f}_i && \text{(i)},
\end{aligned} \tag{2.64}$$

Note that algorithm (2.64) requires two backsubstitutions for matrix $\widehat{\mathbf{U}}^{-1}$ and two for $\widehat{\mathbf{L}}^{-1}$ (items (a, b, e, f) in algorithm (2.64)). This is equivalent of two matrix multiplications with $\widehat{\mathbf{A}}$.

The performance of the CG method, preconditioned with the compositional ILU, is better, if the degree of sparsity is higher, or diagonal dominance of $\widehat{\mathbf{A}}$ is stronger, since these properties improve the accuracy of incomplete factorization (2.60). This is exactly the case with finite-difference and finite-element problems, and also with the compressed integral equation method.

It should be noted that if $\widehat{\mathbf{A}}$ is not diagonally dominant, the method still works because Equation (2.54) is exact. It converges slower, however.

7. MODELING EXAMPLES

We have performed a comparison of compressed and non-compressed problem solution for a model shown in Figure 4. This is Model 3D-1 of COMMEMI project, excited by a rectangular loop operating at 10 Hz.

The experiments were performed with the body divided to 256, 512 and 1024 cells. For each case, three experiments with different degrees of compression were performed. The non-compressed case corresponds to 0 threshold level. Compressed cases include 0.0001 and 0.001 threshold levels.

The experiments were performed on ULTRA SPARC workstation with 160 MHz processor frequency and 128 Mbytes of memory. Table 1 summarizes the result of comparison, in terms of memory usage, CPU time and accuracy of the solution. The accuracy was measured as the L_2 norm of the difference between the compressed and uncompressed solutions (anomalous field inside the body) for the same number of cells.

We can see that compression speeds up the solution up to two orders of magnitude and decreases memory requirements one order of magnitude. The drawback, however, is loss of accuracy. Higher threshold level leads to less accurate solution. The worst case

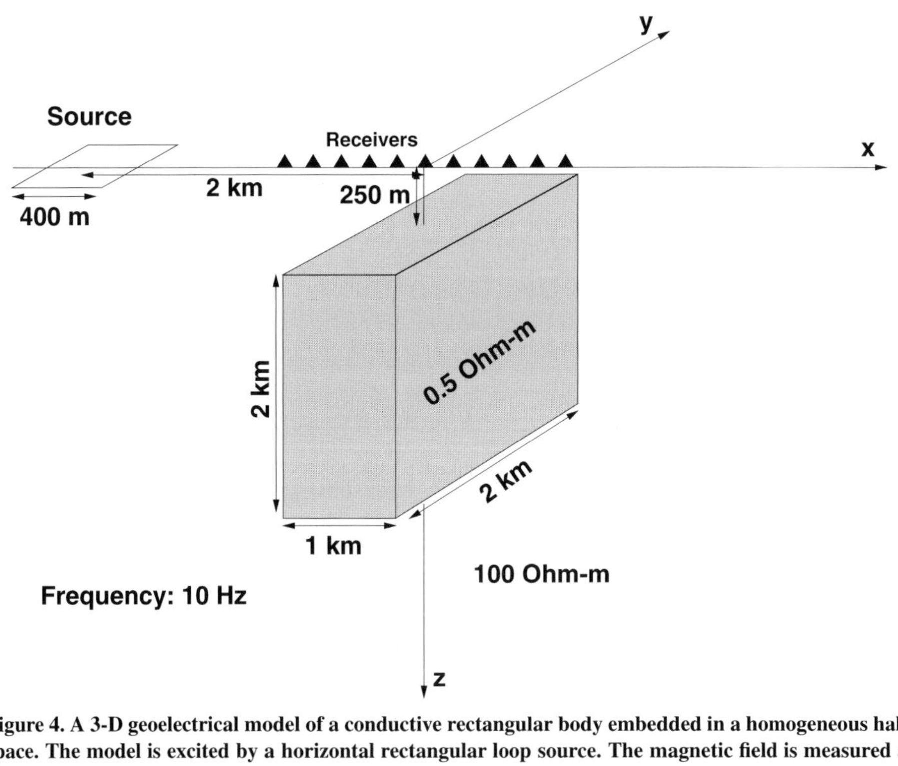

Figure 4. A 3-D geoelectrical model of a conductive rectangular body embedded in a homogeneous half-space. The model is excited by a horizontal rectangular loop source. The magnetic field is measured at the receivers marked by triangles along the *x* axis.

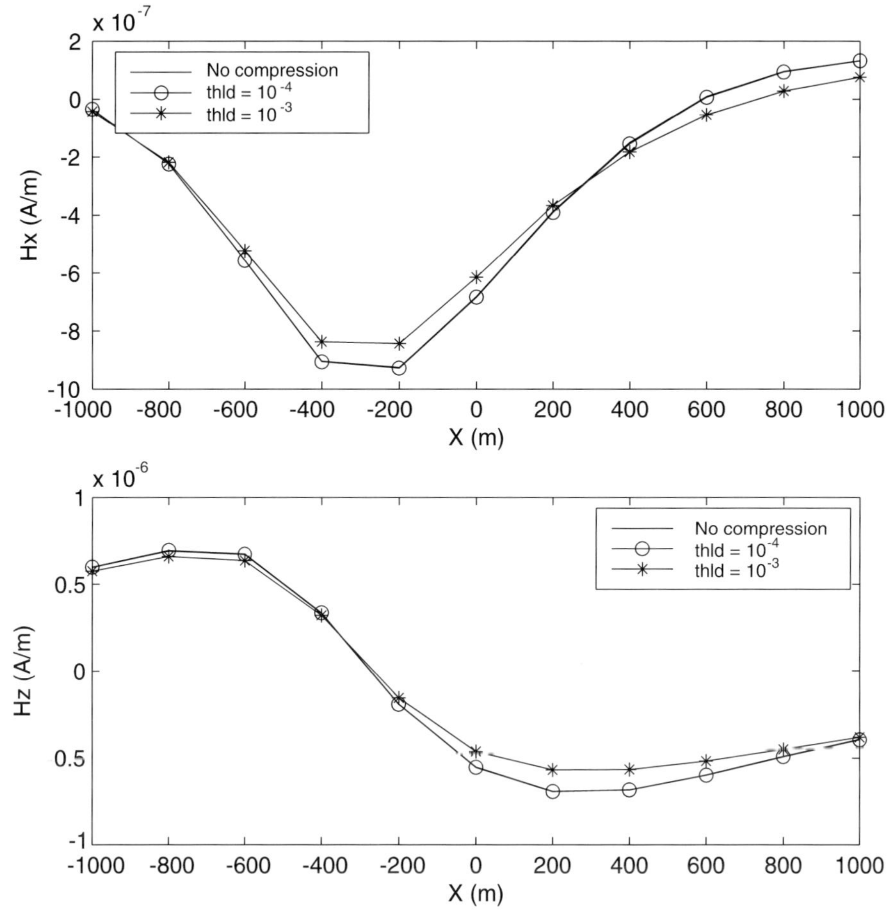

Figure 5. Behavior of the anomalous magnetic field obtained by IE solution without and with compression using different threshold levels.

here is 3% accuracy loss. The difference between the non-compressed solution and the solution compressed with 0.001 threshold level is shown in Figure 5.

Another model, consisting of two conductive bodies, is shown in Figure 6. The observation array is a cross-borehole frequency-domain system. It consists of twenty receivers located in one well. Twenty transmitters located in another well operate at 33 kHz frequency. Transmitters are shown as circles, and receivers are shown as stars.

To build such a model, we first establish a large grid that covers the entire domain between boreholes (Figure 7). The discretized integral equation for this domain is given by Formula (2.11). After compression, the size of the problem matrix was reduced in about five times. The solution of this forward EM problem (an axial magnetic field component) for a cross-borehole method is shown in Figure 8. Note that existing versions of the IE codes without compression cannot compute this model.

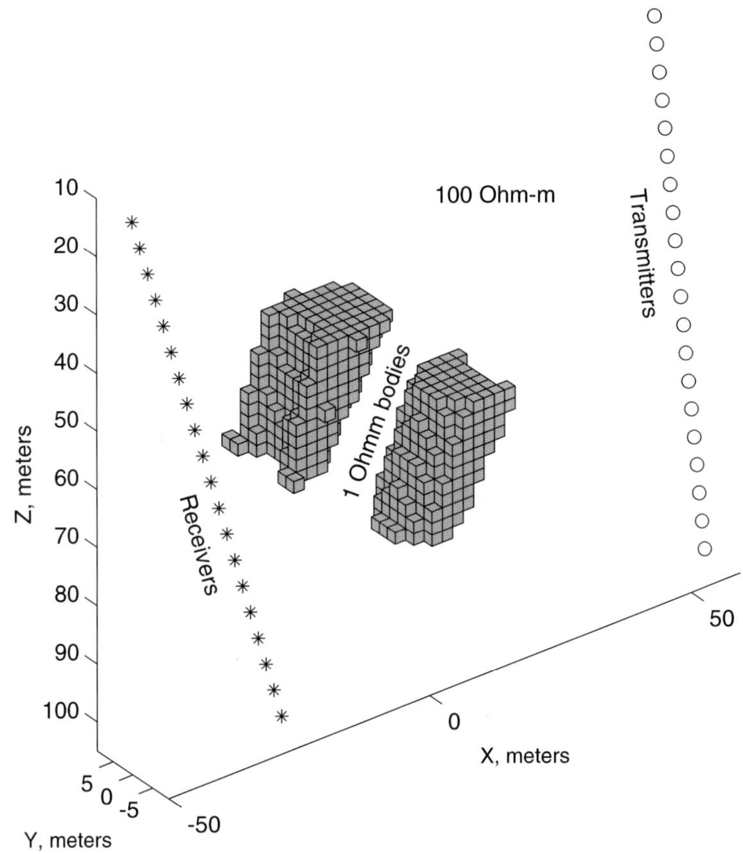

Figure 6. A model with two local conductive bodies and a cross-borehole observation system. Stars show receivers, circles show transmitters.

Table 1. Results of comparison

Threshold level	256 cells	512 cells	1024 cells
Memory usage, megabytes			
0	9.44	37.75	151
0.0001	5.47	16.61	46.81
0.001	1.52	4.81	11.22
CPU time, seconds			
0	38	538	16047
0.0001	21	188	1952
0.001	6.58	53	203
Relative error			
0	0	0	0
0.0001	0.0009	0.0033	0.0038
0.001	0.013	0.028	0.032

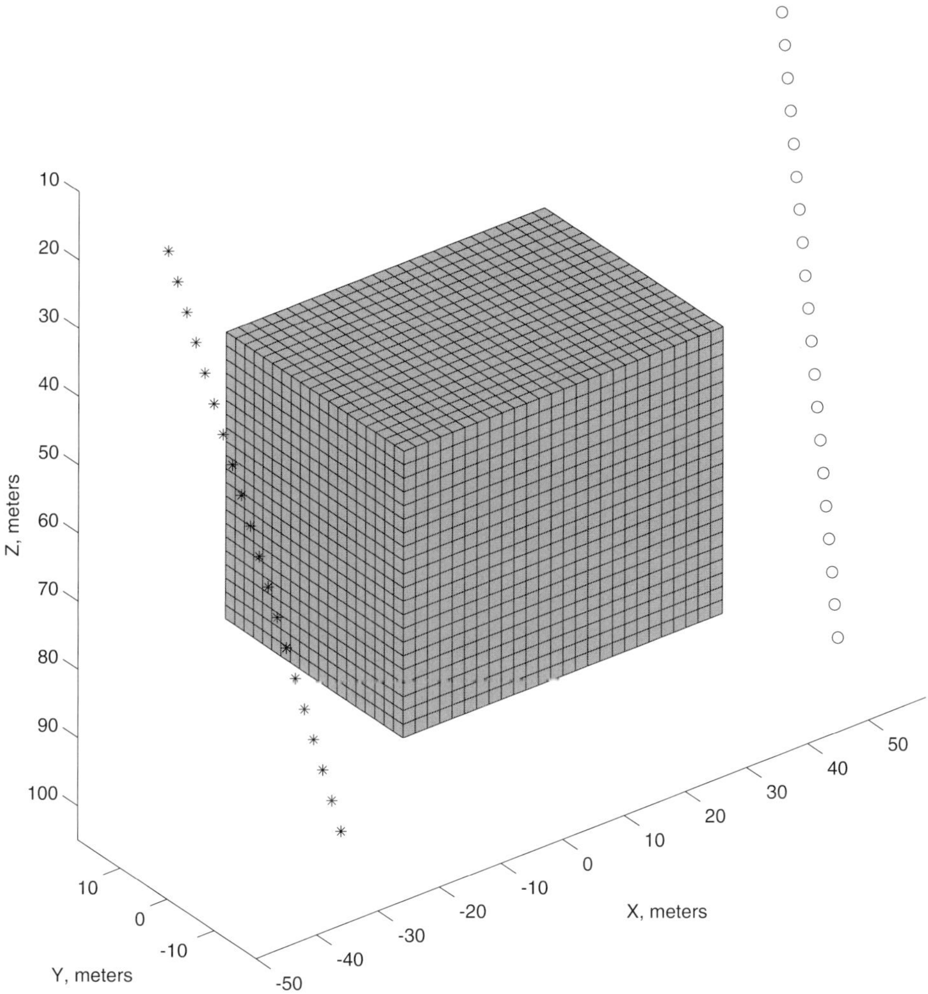

Figure 7. A domain grid for two local bodies and a cross-borehole observation system for the model with two conductive bodies shown in Figure 6. Most of the anomalous conductivity values for the cells in this grid are equal to zero. The whole grid, which is shown in this figure, is used to set up the model. This allows greater flexibility of the model's geometry. However, the anomalous field values are found only inside the bodies shown in Figure 6.

8. CONCLUSIONS

We have demonstrated the application of compression to the solution of a 3-D EM forward problem with IE. Numerical study indicates that the method can be applied to speed up computations and enable solution of larger problems.

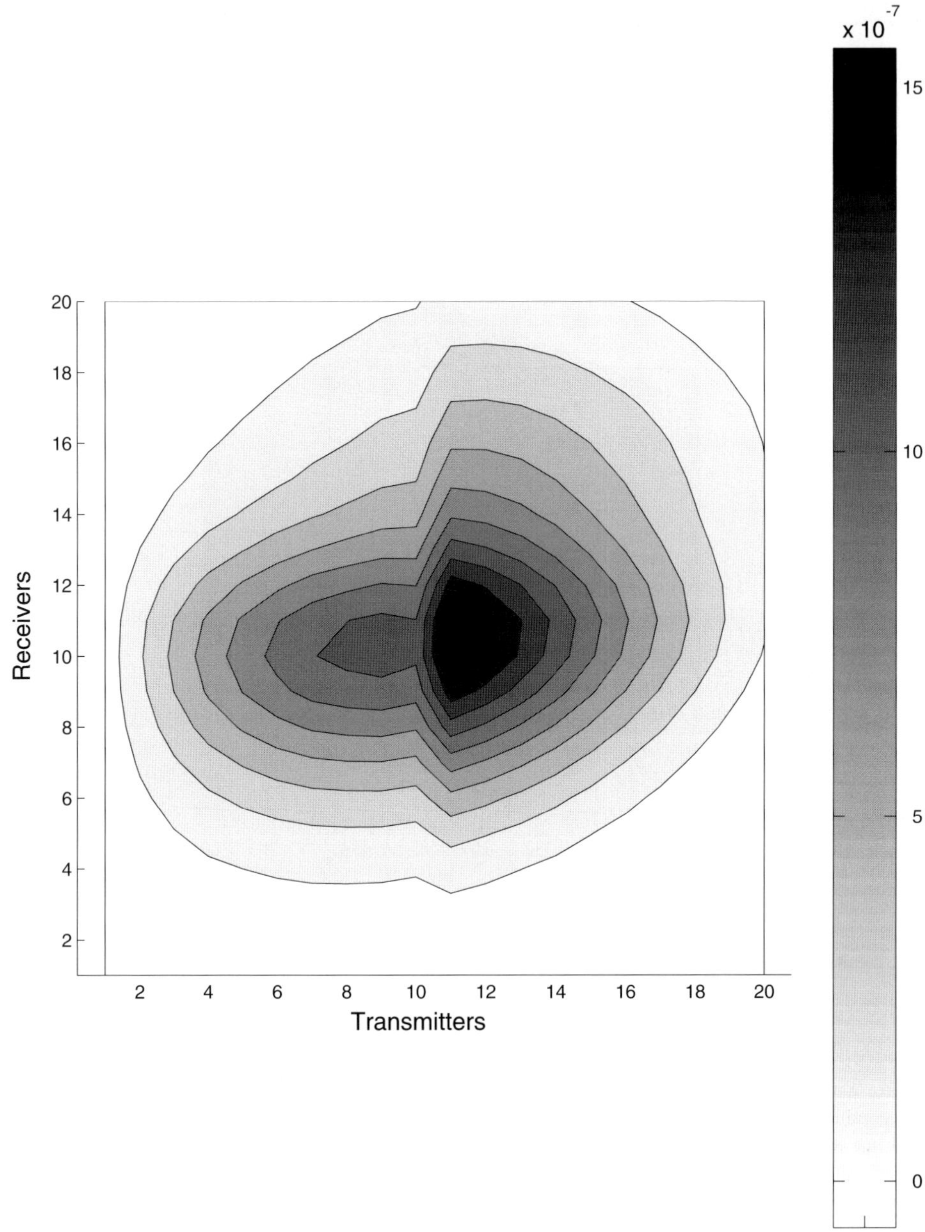

Figure 8. Axial magnetic field, the result of cross-borehole simulation for a model with two anomalous bodies shown in Figure 6. Gray scale shows magnetic field magnitude in A/m.

As a result, we have developed a new generation of IE methods that possess flexibility in forward simulations compatible with the flexibility of FD modeling, but preserves the advantages of IE techniques. There are several promising directions for further research.

It seems possible to develop fast exact solutions based on iterative series consisting of models compressed with different levels of accuracy. It is also possible to apply compression to build fast 3-D inverse solutions.

ACKNOWLEDGEMENTS

The financial support for this work was provided by the National Science Foundation under the grant No. ECS-9987779. The authors also acknowledge the support of the University of Utah Consortium for Electromagnetic Modeling and Inversion (CEMI), which includes Advanced Power Technologies Inc., Baker Atlas Logging Services, BHP Minerals, ExxonMobil Upstream Research Company, INCO Exploration, Japan National Oil Corporation, MINDECO, Naval Research Laboratory, Newmont Gold Company, Rio Tinto, Shell International Exploration and Production, Schlumberger–Doll Research, Unocal Geothermal Corporation, and Zonge Engineering.

REFERENCES

Ainsworth, M., Levesley, J., Ligth, W.A. and Marletta, M., 1997. Wavelets, Multilevel Methods and Elliptic PDEs. Oxford Science Publications, 302 pp.

Bank, R.E. and Xu, J., 1994. The hierarchical basis multigrid method and incomplete LU decomposition. In: D. Keyes and J. Xu (Eds.), Seventh International Symposium on Domain Decomposition Methods for Partial Differential Equations. AMS, Providence, RI, pp. 163–173.

Bhaskaran, V. and Konstantinides, K., 1997. Image and Video Compression Standards: Algorithms and Architectures. Kluwer Academic Publishers, Hingham, MA, 454 pp.

Chan, T.F., Gallopoulos, E. and Simoncini, V., 1994. A quasi-minimum residual variant of the Bi-CGSTAB algorithm for non-symmetric systems. SIAM J. Sci. Statist. Comput., 15, 338–347.

Fletcher, R., 1981. Practical Methods of Optimization. Wiley, New York, NY, 680 pp.

Freund, R., 1992. Conjugate gradient type methods for linear systems with complex symmetric coefficient matrices. SIAM J. Sci. Statist. Comput., 13, 425–448.

Gill, P., Murray, W. and Wright, M., 1981. Practical Optimization. Academic Press, London, 561 pp.

Henson, V., Limber, M., McCormick, S. and Robinson, B., 1996. Multilevel image reconstruction with natural pixels. SIAM J. Sci. Statist. Comput., 17, 193–216.

Habashi, T.M., Groom, R.W. and Spies, B.R., 1993. Beyond the Born and Rytov approximations: a non-linear approach to electromagnetic scattering. J. Geophys. Res., 98, 1759–1775.

Hohmann, G.W., 1975. Three-dimensional induced polarization and EM modeling. Geophysics, 40, 309–324.

Klarke, R.J., 1995. Digital Compression of Still Images and Video. Academic Press, London, 453 pp.

Losano, V. and Laget, B., 1996. Fractional pyramids for color image segmentation. Proceedings of the IEEE Southwest Symposium on Image Analysis and Interpretation, April 8–9, San Antonio, TX, pp. 13–17.

McCormick, S., 1994. Multilevel adaptive methods for elliptic eigenproblems: two-level convergence theory. SIAM J. Numer. Anal., 31, 1731–1745.

Manteuffel, T., McCormick, S., Morel, J., Oliviera, S. and Yang, G., 1994. A fast multigrid algorithm for isotropic transport problems, part I. Pure scattering. SIAM J. Sci. Statist. Comput., 15, 474–493.

Natravali, A.N., 1995. Digital Pictures: Representation, Compression and Standards. AT&T Bell Laboratories, 686 pp.

Portniaguine, O., 1999. Image Focusing and Data Compression in the Solution of Geophysical Inverse Problems. PhD dissertation, University of Utah, 116 pp.

Press, W.H., Teukolsky, S.A., Vetterling, W.T. and Flannery, B.P., 1996. Numerical Recipes in C, The Art of Scientific Computing. Cambridge University Press, 785 pp.
Ramstad, T.A., Aase, S.O. and Husoy, J.H., 1995. Subband Compression of Images: Principles and Examples. Elsevier, Amsterdam, 379 pp.
Van der Vorst, H.A., 1992. Bi-CGSTAB: a fast and smoothly convergent variant of Bi-CG for the solution of non-symmetric linear systems. SIAM J. Sci. Statist. Comput., 13, 631–644.
Wannamaker, P.E., 1991, Advances in three-dimensional magnetotelluric modeling using integral equations. Geophysics, 56, 1716–1728.
Weidelt, P., 1975. EM induction in three-dimensional structures. Geophysics, 41, 85–109.
Xiong, Z., 1992. EM modeling of three-dimensional structures by the method of system iteration using integral equations. Geophysics, 57, 1556–1561.
Zhdanov, M.S. and Fang, S., 1996a. Quasi-linear approximation in 3-D EM modeling. Geophysics, 61, 646–665.
Zhdanov, M.S. and Fang, S., 1996b. 3-D quasi-linear electromagnetic inversion: Radio Sci., 31, 741–754.
Zhdanov, M.S. and Fang, S., 1997. Quasi-linear series in 3-D EM modeling. Radio Sci., 32, 2167–2188.
Zhdanov, M.S., Dmitriev, V.I., Fang, S. and Hursan, G., 2000. Quasi-analytical approximations and series in electromagnetic modeling. Geophysics, 65, 1746–1757.

Chapter 3

MODELLING ELECTROMAGNETIC FIELDS IN A 3D SPHERICAL EARTH USING A FAST INTEGRAL EQUATION APPROACH

A.V. Kuvshinov [a], D.B. Avdeev [a], O.V. Pankratov [a], S.A. Golyshev [b] and N. Olsen [c]

[a] *Geoelectromagnetic Research Institute and* [b] *Institute of Terrestrial Magnetism, Ionosphere and Radiowave Propagation, Russian Academy of Sciences, 142092 Troitsk, Moscow Region, Russia*
[c] *Danish Space Research Institute, DK-2100 Copenhagen, Denmark*

Abstract: We present a numerical solution for the global induction problem. The solution calculates the electromagnetic fields in spherical three-dimensional (3D) earth models that are excited by external or internal currents. The models include a number of 3D isotropic (or anisotropic) inhomogeneities that reside in a radially symmetric section. The solution exploits the modified iterative–dissipative method. This fast integral equation approach allows the recovery of an accurate solution, even for large conductivity contrasts. In order to verify our solution we compare it with a staggered-grid finite-difference solution for a model with two (surface and mantle) laterally inhomogeneous conducting thin layers, and also with a Cartesian integral equation solution for a model with deep-seated local 3D anomaly. Both comparisons demonstrate very good agreement.

1. INTRODUCTION

Important information about the Earth's upper mantle comes from the induced part of long-period geomagnetic variations, mostly originating in the ionosphere and magnetosphere. These variations can be utilized for resolving mantle electrical conductivity in the depth range from hundreds to thousands kilometres. Most of the works on the subject assume radially symmetric earth models. However, recent investigations (cf. Schultz, 1990; Egbert and Booker, 1992; Bahr et al., 1993; Schultz et al., 1993; Fujii et al., 1995; Lizarralde et al., 1995; Olsen, 1999) indicate a substantial level of lateral electrical heterogeneity in the upper mantle.

Clearly, the solution of the inverse problem of inferring the mantle heterogeneities from the observed geomagnetic field variations has to be based on solving the three-dimensional (3D) forward problem. A number of numerical solutions for EM induction in spherical 3D earth models have been developed recently, including finite-element (Everett and Schultz, 1996; Weiss and Everett, 1998), spectral finite-element (Martinec, 1999), integral equation (Koyama and Utada, 1998), spectral decomposition (Tarits, 1994), and staggered-grid finite-difference (Uyeshima and Schultz, 2000) solutions.

Here we present a new numerical solution for the global induction problem in 3D spherical earth models. It exploits the modified iterative–dissipative method which was originally presented by Singer (1995) for a quasi-static field and for isotropic formations, and later extended to formations with displacement currents and anisotropy in (Pankratov et al., 1995, 1997; Singer and Fainberg, 1995, 1997). This volume integral equation (IE) approach is based on an inequality for the Green's function of Maxwell's equations that follows from the energy conservation law. The inequality allows the reduction of Maxwell's equations to a scattering equation with a contracting kernel, and then the solution of this equation by summation of its Neumann series.

In this paper we first present, following Pankratov et al. (1997), the governing integral equations as applied to spherical anisotropic 3D earth models. Then, we compare our solution with a staggered-grid finite-difference (FD) solution by Uyeshima and Schultz (2000), and with a 'Cartesian' integral equation solution by Avdeev et al. (1997).

2. GOVERNING EQUATIONS

Assuming a time harmonic dependence of $e^{-i\omega t}$, electromagnetic (EM) fields in spherical earth models are given by Maxwell's equations

$$\nabla \times \mathbf{H} = \sigma(\theta,\varphi,r)\mathbf{E} + \mathbf{j}^{\mathrm{p}}, \tag{3.1}$$

$$\nabla \times \mathbf{E} = i\omega\mu_0\mathbf{H}. \tag{3.2}$$

Here $\mathbf{j}^{\mathrm{p}}$ is the impressed current, θ is the colatitude, φ is the longitude, r is the distance from the earth's centre, $i = \sqrt{-1}$, $\mu_0 = 4\pi \times 10^{-7}$ H/m, σ is the 3×3 matrix of conductivity. Following the modified iterative–dissipative method, the solution of Equations (3.1) and (3.2) is found as follows.

Introducing a 'reference' radially symmetric section of isotropic conductivity $\sigma_0(r)$, we write Maxwell's Equations (3.1) and (3.2) for the scattered fields $\mathbf{E}^{\mathrm{s}} = \mathbf{E} - \mathbf{E}^0$, $\mathbf{H}^{\mathrm{s}} = \mathbf{H} - \mathbf{H}^0$ in the form

$$\nabla \times \mathbf{H}^{\mathrm{s}} = \sigma_0\mathbf{E}^{\mathrm{s}} + \mathbf{j}^{\mathrm{q}}, \tag{3.3}$$

$$\nabla \times \mathbf{E}^{\mathrm{s}} = i\omega\mu_0\mathbf{H}^{\mathrm{s}}, \tag{3.4}$$

where

$$\mathbf{j}^{\mathrm{q}} = \mathbf{j}^{\mathrm{s}} + (\sigma - \sigma_0\boldsymbol{I})\mathbf{E}^{\mathrm{s}}, \tag{3.5}$$

and

$$\mathbf{j}^{\mathrm{s}} = (\sigma - \sigma_0\boldsymbol{I})\mathbf{E}^0. \tag{3.6}$$

Here $\mathbf{E}^0$, $\mathbf{H}^0$ are the electric and magnetic fields excited by the impressed current $\mathbf{j}^{\mathrm{p}}$ in the reference section of conductivity $\sigma_0(r)$, and $\boldsymbol{I}$ is the identity matrix. Equations (3.3) and (3.4) lead to the conventional scattering equation with respect to the unknown field $\mathbf{E}^{\mathrm{s}}(\mathbf{r})$

$$\mathbf{E}^{\mathrm{s}}(\mathbf{r}) = \mathbf{E}_0(\mathbf{r}) + \int_V \mathcal{G}^{\mathrm{ee}}(\mathbf{r},\mathbf{r}')\left(\sigma(\mathbf{r}') - \sigma_0(r')\boldsymbol{I}\right)\mathbf{E}^{\mathrm{s}}(\mathbf{r}')\,\mathrm{d}v', \tag{3.7}$$

where the free term $\mathbf{E}_0(\mathbf{r}) = \int_V \mathcal{G}_0^{ee}(\mathbf{r},\mathbf{r}')\mathbf{j}^s(\mathbf{r}')\,dv'$, $\mathcal{G}^{ee}(\mathbf{r},\mathbf{r}')$ is the 3×3 dyadic 'electric-to-electric' Green's function of the reference section (the explicit form of $\mathcal{G}^{ee}$ is presented in the Appendix), $\mathbf{r} = (\theta,\varphi,r)$, $\mathbf{r}' = (\theta',\varphi',r')$, and V is volume where $(\sigma - \sigma_0\boldsymbol{I})$ differs from zero. The formal solution of Equation (3.7) can be written as an infinite Neumann series, starting from $\mathbf{E}_0(\mathbf{r})$ as zero'th order approximation to $\mathbf{E}^s(\mathbf{r})$

$$\mathbf{E}^s(\mathbf{r}) = \mathbf{E}_0(\mathbf{r}) + \int_V \mathcal{G}^{ee}(\mathbf{r},\mathbf{r}')(\sigma(\mathbf{r}') - \sigma_0(r')\boldsymbol{I})\mathbf{E}_0(\mathbf{r}')\,dv'$$
$$+ \int_V \mathcal{G}^{ee}(\mathbf{r},\mathbf{r}')(\sigma(\mathbf{r}') - \sigma_0(r')\boldsymbol{I})\left(\int_V \mathcal{G}^{ee}(\mathbf{r}',\mathbf{r}'')(\sigma(\mathbf{r}'') - \sigma_0(r'')\boldsymbol{I})\mathbf{E}_0(\mathbf{r}'')dv''\right)dv' + \cdots \tag{3.8}$$

It is known that convergence of such a series is not guaranteed, and for a scattered field very different from $\mathbf{E}_0(\mathbf{r})$ this series does not converge at all. But if we add term $(\sigma - \sigma_0\boldsymbol{I})\mathbf{E}^s/2\sigma_0$ to both sides of Equation (3.7) and then change the variable as

$$\chi(\mathbf{r}) = ((\sigma + \sigma_0\boldsymbol{I})\mathbf{E}^s + \mathbf{j}^s)/2\sqrt{\sigma_0}, \tag{3.9}$$

we derive the scattering equation (cf. Pankratov et al., 1997, equation (33)) with contracting integral kernel

$$\chi(\mathbf{r}) = \chi_0(\mathbf{r}) + \int_V \mathcal{K}(\mathbf{r},\mathbf{r}')\mathcal{R}(\mathbf{r}')\chi(\mathbf{r}')\,dv', \tag{3.10}$$

which allows writing its solution in the form of always-convergent Neumann series (NS)

$$\chi(\mathbf{r}) = \chi_0(\mathbf{r}) + \int_V \mathcal{K}(\mathbf{r},\mathbf{r}')\mathcal{R}(\mathbf{r}')\chi_0(\mathbf{r}')\,dv'$$
$$+ \int_V \mathcal{K}(\mathbf{r},\mathbf{r}')\mathcal{R}(\mathbf{r}')\left(\int_V \mathcal{K}(\mathbf{r}',\mathbf{r}'')\mathcal{R}(\mathbf{r}'')\chi_0(\mathbf{r}'')\,dv''\right)dv' + \cdots. \tag{3.11}$$

Here

$$\chi_0(\mathbf{r}) = \int_V \mathcal{K}(\mathbf{r},\mathbf{r}')\mathcal{R}(\mathbf{r}')\sqrt{\sigma_0(r')}\,\mathbf{E}^0(\mathbf{r}')\,dv', \tag{3.12}$$

$$\mathcal{K}(\mathbf{r},\mathbf{r}') = \delta(\mathbf{r}-\mathbf{r}')\boldsymbol{I} + 2\sqrt{\sigma_0(r)}\,\mathcal{G}^{ee}(\mathbf{r},\mathbf{r}')\sqrt{\sigma_0(r')}, \tag{3.13}$$

$$\mathcal{R}(\mathbf{r}) = (\sigma - \sigma_0\boldsymbol{I})(\sigma + \sigma_0\boldsymbol{I})^{-1}, \tag{3.14}$$

where $\delta(\mathbf{r})$ is Dirac's delta function. With χ, $\mathbf{E}^s$ and $\mathbf{j}^q$ known at V from Equations (3.11), (3.9) and (3.5) respectively, one determines the scattered fields $\mathbf{E}^s$ and $\mathbf{H}^s$ for any position $\mathbf{r}$ by

$$\mathbf{E}^s(\mathbf{r}) = \int_V \mathcal{G}^{ee}(\mathbf{r},\mathbf{r}')\mathbf{j}^q(\mathbf{r}')\,dv', \tag{3.15}$$

$$\mathbf{H}^s(\mathbf{r}) = \int_V \mathcal{G}^{me}(\mathbf{r},\mathbf{r}')\mathbf{j}^q(\mathbf{r}')\,dv', \tag{3.16}$$

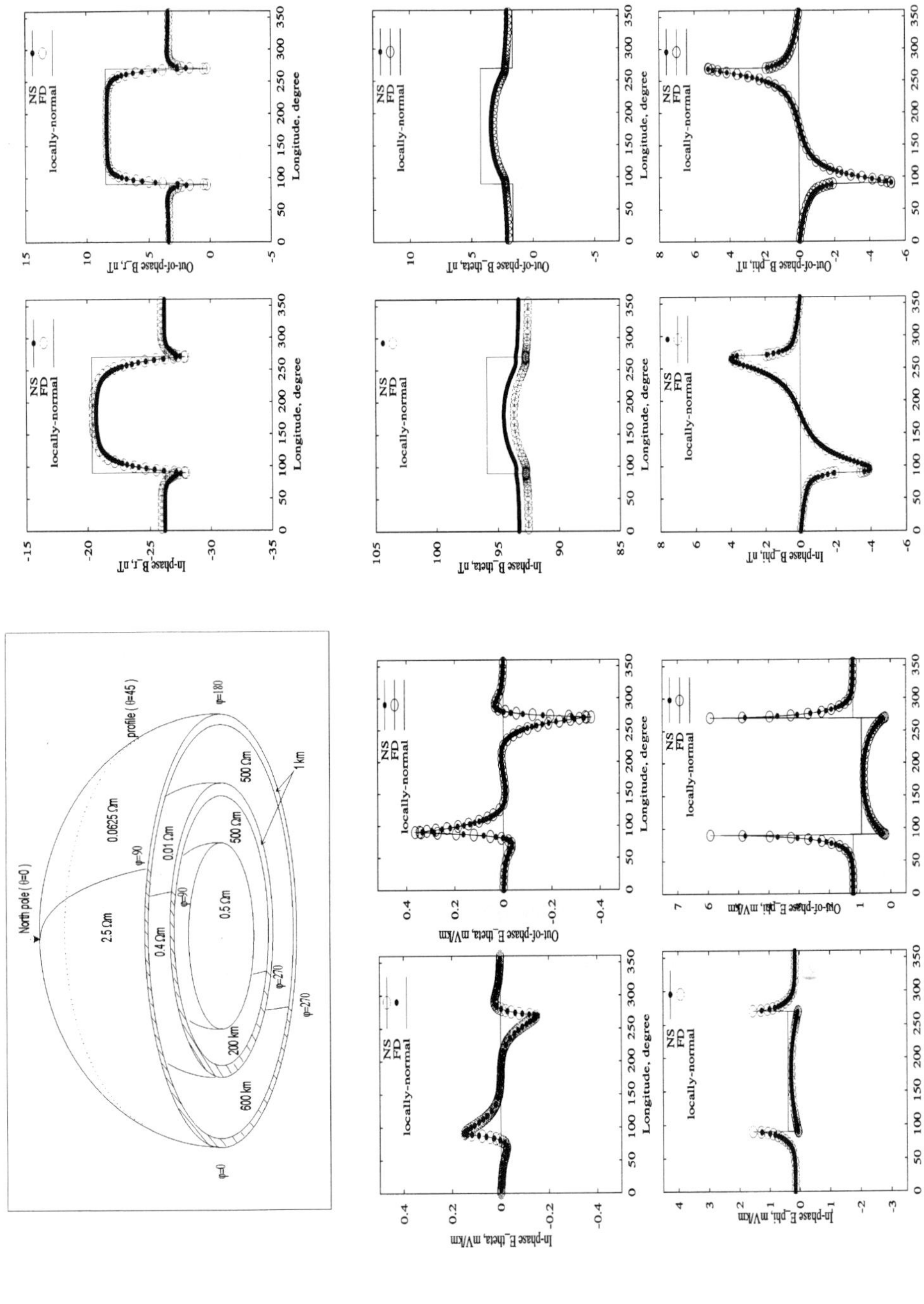
North pole (θ=0)
profile (θ=45)
2.5 Ωm
0.0625 Ωm
0.4 Ωm
0.01 Ωm
0.5 Ωm
500 Ωm
500 Ωm
600 km
200 km
1 km
φ=0
φ=90
φ=180
φ=270
NS
FD
locally-normal
Longitude, degree
In-phase B_r, nT
Out-of-phase B_r, nT
In-phase B_theta, nT
Out-of-phase B_theta, nT
In-phase B_phi, nT
Out-of-phase B_phi, nT
In-phase E_theta, mV/km
Out-of-phase E_theta, mV/km
In-phase E_phi, mV/km
Out-of-phase E_phi, mV/km

where $\mathcal{G}^{me}(\mathbf{r},\mathbf{r}')$ is the 3×3 dyadic 'electric-to-magnetic' Green's function of the reference section (the explicit form of $\mathcal{G}^{me}$ is presented in the Appendix). The total fields **E** and **H** are obtained by adding the appropriate reference and scattered fields.

3. SOLUTION VERIFICATION

In order to verify the accuracy of our spherical solution (hereafter referred to as a spherical NS solution) we compared our results with those from a spherical FD solution by Uyeshima and Schultz (2000), and from a Cartesian NS solution by Avdeev et al. (1997). The comparisons are restricted by a Dst-type impressed current

$$\mathbf{j}^{\mathrm{p}} = -(3/2)\varepsilon_1^0 \sin\theta\, \mathbf{e}_\varphi\, \delta(r - r_0 + 0)/\mu_0, \tag{3.17}$$

where r_0 is the earth's radius, $\mathbf{e}_\varphi$ is the unit vector, and $\varepsilon_1^0 = 100$ nT. The expression for $\mathbf{j}^{\mathrm{p}}$ in Equation (3.17) is consistent with the expression $V^{\mathrm{ext}}(\theta, r_0) = \varepsilon_1^0 r_0 \cos\theta$ for a potential function V^{ext} of the external magnetic field, $\mathbf{B}^{\mathrm{ext}} = -$ grad V^{ext}, at the earth's surface.

3.1. Model with surface and deep thin layers

The first model is shown in Figure 1 (upper left panel). This model comprises, at the surface and at depth 600 km, two spherical and laterally inhomogeneous thin conducting layers which are embedded in a radially symmetric background section. Each conducting layer is 1 km in thickness. A colatitudinal plane divides each layer onto two hemispheres. The 'western' ($90° < \varphi < 270°$) hemispheres, both surface and deep, are good conductors of 0.0625 Ω m and 0.01 Ω m resistivity respectively. The 'eastern' hemispheres are less conductive, being of 2.5 Ω m for the surface layer and of 0.4 Ω m for the deep layer. The surface layer represents hemispherical ocean and continent with respective conductances 16,000 S and 400 S. The deep layer approximates hypothetical inhomogeneities in the mid-mantle with a conductance of 100,000 S beneath the ocean and of 2500 S beneath the continent. The depth of the mantle layer is chosen to be consistent with the results of geomagnetic deep sounding (e.g. Schultz, 1990). The background section has a resistivity of 500 Ω m down to 800 km depth, and 0.5 Ω m beneath this depth. Section of this type is known to reproduce the essentials of the internal part of Dst variations.

Each spherical layer was discretized by 180×90 cells with angular dimensions of $2° \times 2°$. The spherical NS code requires 11 terms of the Neumann series expansion (3.11) to compute the solution, and takes 550 Mbyte of storage and 1.2 h on a

---

Figure 1. Comparison of the EM fields obtained from the FD solution of Uyeshima and Schultz (2000) (open circles) with those produced from the present spherical NS solution (black circles). The upper left panel illustrates the model, which has two spherical laterally inhomogeneous conducting thin layers; due to the symmetry of the model only the northern hemisphere, $0° \leq \theta \leq 90°$, is shown. Other panels illustrate the results for five components of the EM field for a period of EM field variation of 5 days.

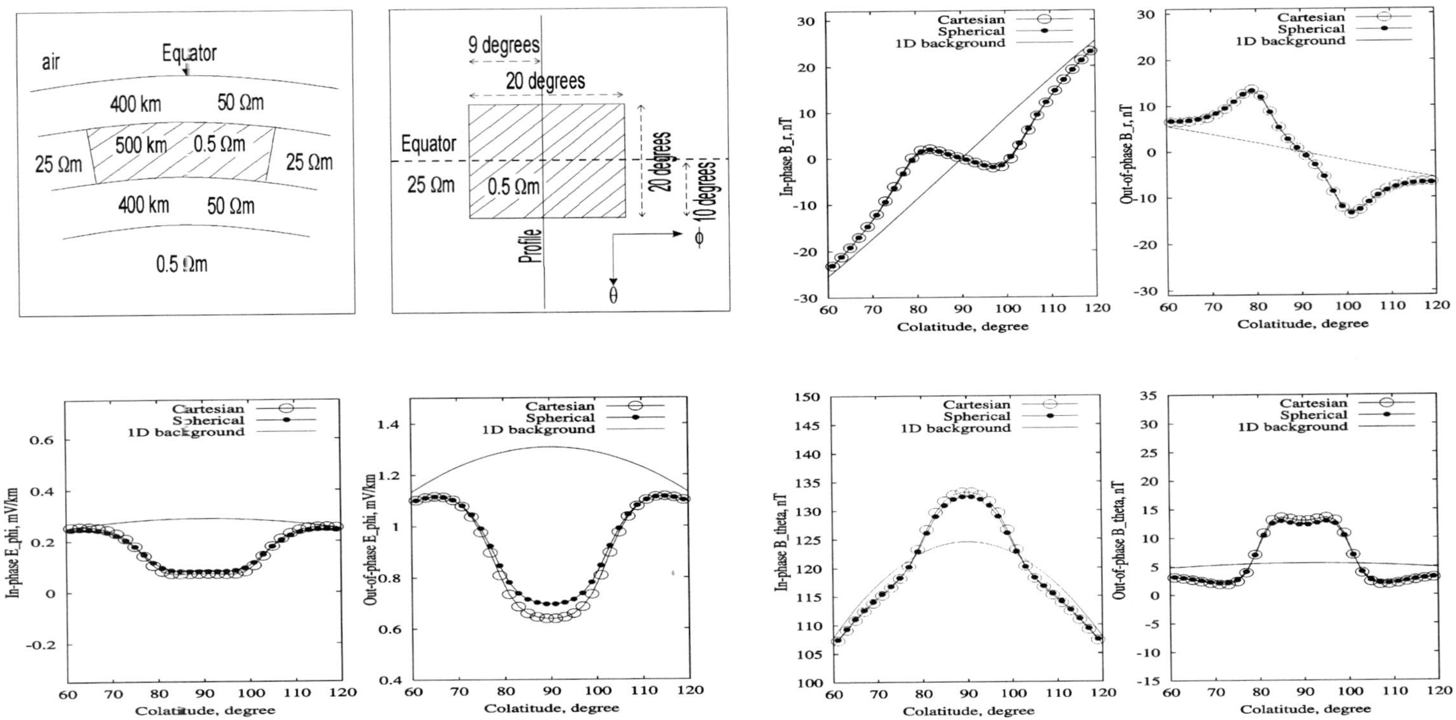

Figure 2. Comparison of the EM fields obtained from the Cartesian NS solution of Avdeev et al. (1997) (open circles) with those produced from the present spherical NS solution (black circles). The upper left panel illustrates the model, which has a deep-seated local 3D anomaly. Other panels illustrate the results for the fields B_r, B_θ, and E_φ for a period of EM field variation of 9 days.

Pentium-II/350 MHz computer. The other panels in Figure 1 compare the surface EM fields from Uyeshima and Schultz's (2000) FD solver ('FD'), with the results from our solver ('NS') for a time-variation period of 5 days. Hereafter we give magnetic fields in nT and electric fields in mV/km. For comparison we show by a solid line 'locally normal' field values which we would get for uniform layer models (with the same background section), when the local values of conductivity are taken as their uniform conductivities.

The results are presented along colatitude $\theta = 45°$. Note that hereafter out-of-phase curves are plotted with a negative sign. It is seen from Figure 1 that there is very good to excellent agreement between the NS and FD solutions. In spite of some visible discrepancy in the B_r and B_θ components, the overall disagreement does not exceed 1.5%.

It is seen in Figure 1 that anomalous effects are strongest where the hemispheres meet, as is expected; this is obviously explained by a concentration of currents parallel to the contact on the oceanic side. It is noteworthy that the oceanic portion of the model proves to be too 'small' in width for B_θ to reach its locally normal value.

3.2. Model with local 3D anomaly

Next we consider a model with a deep-seated 3D local anomaly (Figure 2; upper left panel). The model consists of a 0.5 Ω m block with dimensions of $20° \times 20° \times 500$ km which approximates hypothetical 3D inhomogeneity in the mid-mantle. The block resides in the second layer of a four-layered earth of 50, 25, 50 and 0.5 Ω m resistivities, and with respective thicknesses of 400, 500 and 400 km.

The layer comprising the block was discretized by $180 \times 90 \times 4$ cells with dimensions of $2° \times 2° \times 125$ km. The spherical NS code then requires 15 terms of Neumann series expansion (3.11), and takes 2.2 Gbyte of storage and 6 h on a Pentium-II/350 MHz computer. The other panels in Figure 2 compare the surface EM fields from the spherical NS code with the Cartesian NS code by Avdeev et al. (1997), for a time-variation period of 9 days. By a solid line, we also show 'background' fields which are excited in 1D section. The results are demonstrated along the latitudinal profile above the center of anomaly, and for the colatitudes where the total fields visibly differ from the background ones. The B_r, B_θ and E_φ components are presented only, since B_φ and E_θ components are negligibly small along the profile.

For the Cartesian code we chose the model and the source as similar to the spherical case as possible. In the Cartesian model the block was discretized by $10 \times 10 \times 4$ cells with dimensions of $222 \times 222 \times 125$ km^3. One can see from the plots in Figure 2 good to very good agreement between both solutions. The 10% discrepancy in the out-of-phase part of the E_φ component we explain as difference between the spherical and the Cartesian models. But, as a whole, the agreement is encouraging.

It is seen in Figure 2 that the anomalous effect in surface fields from deep-seated heterogeneity is clearly detectable. Say, for a vertical magnetic component, it reaches 40% relative to its background field.

4. CONCLUDING REMARKS

The NS solution based on fast integral equation approach may serve as an alternative tool to calculate EM fields in full 3D spherical earth models. The solution accounts for anisotropy of conductivity, and permits the examination of the induction pattern caused by external (magnetospheric, ionospheric) or/and internal (core or water motion) electric currents. Comparison of the spherical NS solution, presented in this paper with spherical FD and Cartesian NS solutions demonstrates generally good agreement which verifies all the solutions.

ACKNOWLEDGEMENTS

The authors would like to acknowledge Dr. Makoto Uyeshima of the Earthquake Research Institute, University of Tokyo, for providing the FD responses. They are also grateful to Mark Everett, Phil Wannamaker and an anonymous reviewer for their helpful comments on the manuscript. The research described in this publication was made possible in part by grant no. 97-0157 from the INTAS.

Appendix A. DYADIC GREEN'S FUNCTIONS OF RADIALLY SYMMETRIC SECTION

In accordance with an approach described in (Avdeev et al., 1997, appendices A and B) we present here the explicit forms of 3 × 3 dyadic Green's functions. These forms express electric and magnetic fields induced in a radially symmetric section of conductivity $\sigma_0(r)$ by an impressed current $\mathbf{j}$

$$\mathbf{E}(r,\theta,\varphi) = \int_\Omega\int_0^\infty \mathcal{G}^{ee}(r,r',\theta,\theta',\varphi-\varphi')\,\mathbf{j}(r',\theta',\varphi')\,\mathrm{d}\Omega'\,\mathrm{d}r', \tag{3.A1}$$

$$\mathbf{H}(r,\theta,\varphi) = \int_\Omega\int_0^\infty \mathcal{G}^{me}(r,r',\theta,\theta',\varphi-\varphi')\,\mathbf{j}(r',\theta',\varphi')\,\mathrm{d}\Omega'\,\mathrm{d}r'. \tag{3.A2}$$

Here Ω is a full solid angle, $\mathrm{d}\Omega = \sin\theta\mathrm{d}\theta\mathrm{d}\varphi$. Components of the 3 × 3 dyadic 'electric-to-electric' Green's function $\mathcal{G}^{ee}$ can be written in the following form

$$\mathcal{G}^{ee}_{\theta\theta} = \frac{1}{\sin\theta}\frac{1}{\sin\theta'}\partial_\varphi\partial_{\varphi'}\mathcal{P}\left[\frac{r'}{r}\frac{G^{t}}{n(n+1)}\right] + \partial_\theta\partial_{\theta'}\mathcal{P}\left[\frac{r'}{r}\frac{G^{p}}{n(n+1)}\right],$$

$$\mathcal{G}^{ee}_{\theta\varphi} = -\frac{1}{\sin\theta}\partial_\varphi\partial_{\theta'}\mathcal{P}\left[\frac{r'}{r}\frac{G^{t}}{n(n+1)}\right] + \frac{1}{\sin\theta'}\partial_\theta\partial_{\varphi'}\mathcal{P}\left[\frac{r'}{r}\frac{G^{p}}{n(n+1)}\right],$$

$$\mathcal{G}^{ee}_{\theta r} = -\partial_\theta\mathcal{P}\left[\frac{\beta^{p}G^{p}}{r\sigma_0(r')}\right],$$

$$
\begin{aligned}
\mathcal{G}^{\mathrm{ee}}_{\varphi\theta} &= -\frac{1}{\sin\theta'}\partial_\theta\partial_{\varphi'}\mathcal{P}\left[\frac{r'}{r}\frac{G^{\mathrm{t}}}{n(n+1)}\right] + \frac{1}{\sin\theta}\partial_\varphi\partial_{\theta'}\mathcal{P}\left[\frac{r'}{r}\frac{G^{\mathrm{p}}}{n(n+1)}\right],\\
\mathcal{G}^{\mathrm{ee}}_{\varphi\varphi} &= \partial_\theta\partial_{\theta'}\mathcal{P}\left[\frac{r'}{r}\frac{G^{\mathrm{t}}}{n(n+1)}\right] + \frac{1}{\sin\theta}\frac{1}{\sin\theta'}\partial_\varphi\partial_{\varphi'}\mathcal{P}\left[\frac{r'}{r}\frac{G^{\mathrm{p}}}{n(n+1)}\right],\\
\mathcal{G}^{\mathrm{ee}}_{\varphi r} &= -\frac{1}{\sin\theta}\partial_\varphi\mathcal{P}\left[\frac{\beta^{\mathrm{p}}G^{\mathrm{p}}}{r\sigma_0(r')}\right],\\
\mathcal{G}^{\mathrm{ee}}_{r\theta} &= \partial_{\theta'}\mathcal{P}\left[\frac{r'}{r^2}\frac{\alpha^{\mathrm{p}}G^{\mathrm{p}}}{\sigma_0(r)}\right],\\
\mathcal{G}^{\mathrm{ee}}_{r\varphi} &= \frac{1}{\sin\theta'}\partial_{\varphi'}\mathcal{P}\left[\frac{r'}{r^2}\frac{\alpha^{\mathrm{p}}G^{\mathrm{p}}}{\sigma_0(r)}\right],\\
\mathcal{G}^{\mathrm{ee}}_{rr} &= -\frac{1}{\sigma_0(r)}\delta(r-r')\delta(\theta-\theta')\delta(\varphi-\varphi') + \mathcal{P}\left[\frac{1}{r^2}\frac{n(n+1)\alpha^{\mathrm{p}}\beta^{\mathrm{p}}G^{\mathrm{p}}}{\sigma_0(r)\sigma_0(r')}\right],
\end{aligned}
\tag{3.A3}
$$

whereas the components of the 3 × 3 dyadic 'electric-to-magnetic' Green's function $\mathcal{G}^{\mathrm{me}}$ are

$$
\begin{aligned}
\mathcal{G}^{\mathrm{me}}_{\theta\theta} &= -\frac{1}{\sin\theta'}\partial_\theta\partial_{\varphi'}\mathcal{P}\left[\frac{r'}{r}\frac{\alpha^{\mathrm{t}}G^{\mathrm{t}}}{n(n+1)}\right] - \frac{1}{\sin\theta}\partial_\varphi\partial_{\theta'}\mathcal{P}\left[\frac{r'}{r}\frac{\alpha^{\mathrm{p}}G^{\mathrm{p}}}{n(n+1)}\right],\\
\mathcal{G}^{\mathrm{me}}_{\varphi\theta} &= \partial_\theta\partial_{\theta'}\mathcal{P}\left[\frac{r'}{r}\frac{\alpha^{\mathrm{t}}G^{\mathrm{t}}}{n(n+1)}\right] - \frac{1}{\sin\theta'}\partial_\theta\partial_{\varphi'}\mathcal{P}\left[\frac{r'}{r}\frac{\alpha^{\mathrm{p}}G^{\mathrm{p}}}{n(n+1)}\right],\\
\mathcal{G}^{\mathrm{me}}_{\theta\varphi} &= -\frac{1}{\sin\theta}\frac{1}{\sin\theta'}\partial_\varphi\partial_{\varphi'}\mathcal{P}\left[\frac{r'}{r}\frac{\alpha^{\mathrm{t}}G^{\mathrm{t}}}{n(n+1)}\right] + \partial_\theta\partial_{\theta'}\mathcal{P}\left[\frac{r'}{r}\frac{\alpha^{\mathrm{p}}G^{\mathrm{p}}}{n(n+1)}\right],\\
\mathcal{G}^{\mathrm{me}}_{\varphi\varphi} &= \frac{1}{\sin\theta}\partial_\varphi\partial_{\theta'}\mathcal{P}\left[\frac{r'}{r}\frac{\alpha^{\mathrm{t}}G^{\mathrm{t}}}{n(n+1)}\right] + \frac{1}{\sin\theta'}\partial_\theta\partial_{\varphi'}\mathcal{P}\left[\frac{r'}{r}\frac{\alpha^{\mathrm{p}}G^{\mathrm{p}}}{n(n+1)}\right],\\
\mathcal{G}^{\mathrm{me}}_{\theta r} &= -\frac{1}{\sin\theta}\partial_\varphi\mathcal{P}\left[\frac{\alpha^{\mathrm{p}}\beta^{\mathrm{p}}G^{\mathrm{p}}}{r\sigma_0(r')}\right],\\
\mathcal{G}^{\mathrm{me}}_{\varphi r} &= \partial_\theta\mathcal{P}\left[\frac{\alpha^{\mathrm{p}}\beta^{\mathrm{p}}G^{\mathrm{p}}}{r\sigma_0(r')}\right],\\
\mathcal{G}^{\mathrm{me}}_{r\theta} &= \frac{1}{\sin\theta'}\partial_{\varphi'}\mathcal{P}\left[\frac{r'}{r^2}\frac{G^{\mathrm{t}}}{i\omega\mu_0}\right],\\
\mathcal{G}^{\mathrm{me}}_{r\varphi} &= -\partial_{\theta'}\mathcal{P}\left[\frac{r'}{r^2}\frac{G^{\mathrm{t}}}{i\omega\mu_0}\right],\\
\mathcal{G}^{\mathrm{me}}_{rr} &= 0.
\end{aligned}
\tag{3.A4}
$$

Note that all components of Equations (3.A3) and (3.A4) depend on $(r,r',\theta,\theta',\varphi-\varphi')$. Here the transform

$$\mathcal{P}[f](\cos\gamma,r,r') = \frac{2n+1}{4\pi}\sum_{n=1}^{\infty} f(n,r,r')P_n(\cos\gamma) \tag{3.A5}$$

converts a function of spectral number n onto a function of $\cos\gamma$, where $\cos\gamma = \cos\theta\cos\theta' + \sin\theta\sin\theta'\cos(\varphi-\varphi')$; P_n is Legendre polynomial of degree n, $n = 1,2,3,\ldots$.

The spectral functions $\alpha^{\rm t}$, $\alpha^{\rm p}$ are determined via four admittances $Y^{\rm l,t}, Y^{\rm u,t}, Y^{\rm l,p}, Y^{\rm u,p}$ as

$$\alpha(n,r,r') = \beta(n,r',r) = \begin{cases} -Y^{\rm u}(n,r), & r > r' \\ Y^{\rm l}(n,r), & r < r' \end{cases} \tag{3.A6}$$

Hereafter superscripts 't' and 'p' are omitted for clarity. The scalar Green's functions $G^{\rm t}, G^{\rm p}$ of Equations (3.A3) and (3.A4) are determined as

$$G(n,r,r') = -\frac{1}{Y^{\rm l}(n,r')+Y^{\rm u}(n,r')}\exp\left(\int_{r'}^{r} p(n,\xi)\alpha(n,\xi,r')\,d\xi\right), \tag{3.A7}$$

where $p^{\rm t} = -i\omega\mu_0$, $p^{\rm p} = -i\omega\mu_0 + \frac{n(n+1)}{r^2\sigma_0(r)}$. Four admittances $Y^{\rm l,t}(n,r)$, $Y^{\rm u,t}(n,r)$, $Y^{\rm l,p}(n,r)$, $Y^{\rm u,p}(n,r)$, and two factors $F^{\rm t}(n,r,r') = \exp\left(\int_{r'}^{r} p^{\rm t}(\xi)\alpha^{\rm t}(n,\xi,r')\,d\xi\right)$, $F^{\rm p}(n,r,r') = \exp\left(\int_{r'}^{r} p^{\rm p}(n,\xi)\alpha^{\rm p}(n,\xi,r')\,d\xi\right)$ are calculated by a unified procedure described below.

We assume that the radially symmetric reference section consists of N layers $\{r_{k+1} < r \le r_k\}_{k=1,2,\ldots,N}$. We construct the set $\{r_k\}_{k=1,2,\ldots,N}$ in such a way that it includes all levels r_j where we will calculate the admittances and the Green's scalar functions. We assume that within each layer the conductivity varies as

$$\sigma_0(r) = \sigma_k\left(\frac{r_k}{r}\right)^2, \qquad r_{k+1} < r \le r_k, \tag{3.A8}$$

where $r_1 = r_{\rm e}$ is the earth's radius, $r_{N+1} = 0$, σ_k is an appropriate constant. Since N can be taken as large as necessary, distribution (3.A8) allows the approximation of any radially symmetric conductivity section. Distribution (3.A8) (cf. Rokityansky, 1982; Fainberg et al., 1990) is chosen to make recurrences as simple as possible. We express these recurrences in the following form

$$Y_k^{\rm l,t} = \frac{1}{q_k}\frac{q_{k+1}Y_{k+1}^{\rm l,t}(b_k - 0.5\tau_k) + b_k^+ b_k^- \tau_k}{(b_k + 0.5\tau_k) + q_{k+1}\tau_k Y_{k+1}^{\rm l,t}}, \qquad k = N-1, N-2, \ldots, 1, \quad Y_N^{\rm l,t} = -\frac{b_N^+}{q_N}, \tag{3.A9}$$

$$Y_{k+1}^{\rm u,t} = \frac{1}{q_{k+1}}\frac{q_k Y_k^{\rm u,t}(b_k + 0.5\tau_k) + b_k^+ b_k^- \tau_k}{(b_k - 0.5\tau_k) + q_k\tau_k Y_k^{\rm u,t}}, \qquad k = 1,2,\ldots,N-1, \quad Y_1^{\rm u,t} = -\frac{n}{q_1}, \tag{3.A10}$$

$$Y_{k+1}^{\rm u,p} = g_k\eta_k\frac{Y_k^{\rm u,p}(b_k + 0.5\tau_k) - g_k\tau_k}{g_k(b_k - 0.5\tau_k) - b_k^+ b_k^- \tau_k Y_k^{\rm u,p}}, \qquad k = 1,2,\ldots,N-1, \quad Y_1^{\rm u,p} = 0, \tag{3.A11}$$

$$Y_k^{\mathrm{l,p}} = g_k \frac{Y_{k+1}^{l,p}(b_k - 0.5\tau_k) - g_k \eta_k \tau_k}{g_k \eta_k (b_k + 0.5\tau_k) - b_k^+ b_k^- \tau_k Y_{k+1}^{\mathrm{l,p}}}, \qquad k = N-1, N-2, \ldots, 1, \quad Y_N^{\mathrm{l,p}} = \frac{\sigma_N r_N}{b_N^-}, \tag{3.A12}$$

where Y_k stands for $Y(n, r_k)$, the layer number k is the index of the recurrence, n is spectral number, and where

$$\eta_k = \frac{r_k}{r_{k+1}}, \qquad \tau_k = \frac{1-\zeta_k}{1+\zeta_k}, \qquad \zeta_k = \eta_k^{2b_k}, \qquad g_k = \sigma_k r_k, \qquad q_k = i\omega\mu_0 r_k.$$

$$b_k^- = b_k - \tfrac{1}{2}, \qquad b_k^+ = b_k + \tfrac{1}{2}, \qquad b_k = \left\{ \left(n + \tfrac{1}{2}\right)^2 - i\omega\mu_0\sigma_k r_k^2 \right\}^{\frac{1}{2}}.$$

Finally the recurrences for scalar Green's functions $G^{\mathrm{t}}, G^{\mathrm{p}}$ are given by

$$G(r_i, r_j) = -\frac{1}{Y_j^{\mathrm{l}} + Y_j^{\mathrm{u}}} \prod_{m=j}^{i-1} F_m, \qquad r_i \le r_j,$$

$$G(r_i, r_j) = G(r_j, r_i), \qquad r_i > r_j. \tag{3.A13}$$

where factors F_m are

$$F_m^{\mathrm{t}} = \frac{1}{1+\zeta_m} \frac{2b_m \eta_m^{\,b_m^-}}{(b_m + 0.5\tau_m) + q_m \tau_m Y_{m+1}^{\mathrm{l,t}}}, \tag{3.A14}$$

for a toroidal mode, and

$$F_m^{\mathrm{p}} = \frac{1}{1+\zeta_m} \frac{2g_m b_m \eta_m^{\,b_m^-}}{g_m \eta_m (b_m + 0.5\tau_m) - b_m^+ b_m^- \tau_m Y_{m+1}^{\mathrm{l,p}}}, \tag{3.A15}$$

for a poloidal mode. Note, that in Equations (3.A9)–(3.A15) a dependence on the spectral number n is omitted but implied.

REFERENCES

Avdeev, D.B., Kuvshinov, A.V., Pankratov, O.V. and Newman, G.A., 1997. High-performance three-dimensional electromagnetic modeling using modified Neumann series. Wide-band numerical solution and examples. J. Geomagn. Geoelectr., 49, 1519–1539.

Bahr, K., Olsen, N. and Shankland, T.J., 1993. On the combination of the magnetotelluric and the geomagnetic depth sounding method for resolving an electrical conductivity increase at 400 km depth. Geophys. Res. Lett., 20, 2937–2940.

Egbert, G. and Booker, J.R., 1992. Very long period megnetotellurics at Tucson observatory: Implications for mantle conductivity, J. Geophys. Res., 97(B11), 15099–15112.

Everett, M.E. and Schultz, A., 1996. Geomagnetic induction in a heterogeneous sphere: azimutally symmetric test computations and the response of an undulating 660-km discontinuity. J. Geophys. Res., 101, 2765–2783.

Fainberg, E.B., Kuvshinov, A.V. and Singer, B.Sh. 1990. Electromagnetic induction in a spherical earth with non-uniform oceans and continents in electric contact with the underlying medium, I. Theory, method and example. Geophys. J. Int., 102, 273–281.

Fujii, I., Utada, H. and Yumoto, K., 1995. Averaged distribution of electrical conductivity of the Philippine Sea Plate. Abstracts of XXI General Assembly of IUGG, Week B, 63–64.

Koyama, T. and Utada, H., 1998. Induction modelling in the 3D global earth — Modified Neumann Series. Proc. Conductivity Anomaly Symposium, Tokyo, pp. 279–286 (in Japanese).

Lizarralde, D., Chave, A., Hirth, G. and Schultz, A., 1995. Long period magnetotelluric study using Hawaii-to-California submarine cable data: implications for mantle conductivity. J. Geophys. Res., 100(B9), 17873–17884.

Martinec, Z., 1999. Spectral-finite-element approach to three-dimensional electromagnetic induction in a spherical Earth. Geophys. J. Int., 136, 229–250.

Olsen, N., 1999. Induction studies with satellite data. Surv. Geophys., 20, 309–340.

Pankratov, O.V., Avdeev, D.B. and Kuvshinov, A.V., 1995. Electromagnetic field scattering in a heterogeneous earth: A solution to the forward problem. Phys. Solid Earth, 31, 201–209 (English ed.).

Pankratov, O.V., Kuvshinov, A.V. and Avdeev, D.B., 1997. High-performance three-dimensional electromagnetic modeling using modified Neumann series. Anisotropic case. J. Geomagn. Geoelectr., 49, 1541–1547.

Rokityansky, I.I., 1982. Geoelectromagnetic Investigation of the Earth's Crust and Mantle. Springer, Berlin.

Schultz, A., 1990. On the vertical gradient and associated and associated heterogeneity in mantle electrical conductivity. Phys. Earth Planet. Inter., 64, 68–86.

Schultz, A., Kurtz, R.D., Chave, A.D. and Jones, A.G., 1993. Conductivity discontinuities in the upper mantle beneath a stable craton. Geophys. Res. Lett., 20, 2941–2944.

Singer, B.Sh., 1995. Method for solution of Maxwell's equations in non-uniform media. Geophys. J. Int., 120, 590–598.

Singer, B.Sh. and Fainberg, E.B., 1995, Generalization of the iterative–dissipative method for modeling electromagnetic fields in nonuniform media with displacement currents. J. Appl. Geophys., 34, 41–46.

Singer, B.Sh. and Fainberg, E.B., 1997. Fast and stable method for 3-D modelling of electromagnetic field. Explor. Geophys., 28, 130–135.

Tarits, P., 1994. Electromagnetic studies of global geodynamic processes. Surv. Geophys., 15, 209–238.

Uyeshima, M. and Schultz, A., 2000. Geoelectromagnetic induction in a heterogeneous sphere: a new 3-D forward solver using a staggered-grid integral formulation. Geophys. J. Int., 140, 636–650.

Weiss, C.J. and Everett, M.E., 1998. Geomagnetic induction in a heterogeneous sphere: Three-dimensional test computation and the response of a realistic distribution of oceans and continents. Geophys. J. Int., 135, 650–662.

Chapter 4

MODELLING INDUCTION LOG RESPONSES IN 3D GEOMETRIES USING A FAST INTEGRAL EQUATION APPROACH

D.B. Avdeev [a], A.V. Kuvshinov [a], O.V. Pankratov [a], G.A. Newman [b] and B.V. Rudyak [c]

[a] *Geoelectromagnetic Research Institute, Russian Academy of Sciences, 142092 Troitsk, Moscow Region, Russia*
[b] *Sandia National Laboratories, Org. 6116, P.O. Box 5800, Albuquerque, NM 87185-0750, USA*
[c] *Research and Production Center 'TverGeophysics', 107001, Tver, Russia*

Abstract: We present modelling results for the electromagnetic logging problem. Our numerical solution calculates the log responses for three-dimensional earth models that include a deviated/tilted borehole, including invaded beds. The solution accounts for both displacement currents and electrical anisotropy of the formation. Our solution exploits the modified iterative dissipative method that is based on a convergent Neumann series expansion. To achieve solution convergence, the conventional scattering integral equation is modified using a spectral shift and a change of variable, which are based on an inequality for the Green's dyadic function. For deviated boreholes in layered formations, there are no known analytical solutions. Therefore, our solution was first validated against the mode-matching solution for a vertical borehole model in a layered formation. In this model we varied the eccentricity of the 50 MHz two-coil sonde. Comparison to the 10 kHz, 160 kHz and 5 MHz responses from a staggered-grid finite difference solution was then conducted for a 45-degree deviated borehole that crosses a formation boundary. All comparisons showed excellent agreement, and demonstrate verifiable induction log responses in the presence of displacement currents and deviated boreholes. Results for a sonde with a resistive case are also reported.

1. INTRODUCTION

Although an accurate simulation of 3D induction logs is a nontrivial calculation due to the fine meshes required for the problem, advances in both computers and numerical methods now make it possible. Encouraging examples of such simulations using the spectral Lanczos decomposition method (SLDM) have been reported by van der Horst et al. (1995) and Druskin et al. (1999).

Here we investigate the validity of an integral equation (IE) approach for solving 3D induction-logging problems. The approach that is to be employed is the modified iterative dissipative method, which was originally presented by Singer (1995) for a quasi-static field in isotropic formations, and later extended to formations with displacement currents and anisotropy (Pankratov et al., 1995, 1997; Singer and Fainberg,

1995, 1997). This method is based on an inequality for the Green's dyadic function of Maxwell's equations, which follows from an energy conservation law. This inequality allows one to reduce Maxwell's equations to a scattering integral equation with a contracting kernel, and hence to solve this equation by summation of its Neumann series.

In this paper we first review, following Pankratov et al. (1997), the IE approach, and report computational loads of our numerical solution. Then, we check the solution against the mode matching (MM) solution by Liu (1993) for vertical borehole model in a layered formation. Comparison to a staggered-grid finite difference (FD) solution (Newman and Alumbaugh, 1995; Alumbaugh et al., 1996) is then conducted for a 45-degree deviated borehole that crosses a formation boundary. We finally perform the model studies to estimate a possible effect associated with the resistive case that encloses a three-coil sonde. The responses happen to be practically insensitive to resistivity of the case regardless of sonde eccentricity.

2. THEORY

Assuming a time harmonic dependence of $e^{-i\omega t}$, the electric $\mathbf{E}(\mathbf{r},\omega)$ and magnetic $\mathbf{H}(\mathbf{r},\omega)$ fields in a 3D model are given by Maxwell's equations

$$\nabla\times\mathbf{H}=\zeta(\mathbf{r},\omega)\mathbf{E}, \tag{4.1}$$

$$\nabla\times\mathbf{E}=i\omega\mu(z)\mathbf{H}+\mathbf{m}\delta(\mathbf{r}-\mathbf{r}_{\mathrm{T}}). \tag{4.2}$$

Here $\zeta(\mathbf{r},\omega)$ is the generalized conductivity which is defined via conductivity $\sigma(\mathbf{r},\omega)$ and dielectric permittivity $\varepsilon(\mathbf{r},\omega)$ as $\zeta=\sigma-i\omega\varepsilon$, δ is the Dirac's delta-function, $\mathbf{r}=(x,y,z)$, and $i=\sqrt{-1}$. For the sake of clarity and without loss of generality, we assume that ζ is a complex-valued 3×3 diagonal matrix $\zeta=\mathrm{diag}(\zeta_{xx},\zeta_{yy},\zeta_{zz})$. The magnetic permeability $\mu(z)$ is a positive real-valued function of depth z. We assume also that the model is excited with a multi-directional magnetic transmitting dipole of moment $\mathbf{m}$, which is positioned at a point $\mathbf{r}_{\mathrm{T}}$. Hereafter dependencies on $\mathbf{r}$ and ω are omitted where they are not essential.

Following the modified iterative dissipative method, the solution of Equations (4.1) and (4.2) is found as follows. Introducing a 1D 'reference' bedding of conductivity $\zeta_0(z,\omega)=\mathrm{diag}(\zeta_{0\tau},\zeta_{0\tau},\zeta_{0z})$ we rewrite Equations (4.1) and (4.2) for the scattered fields $\mathbf{E}^s=\mathbf{E}-\mathbf{E}^0$, $\mathbf{H}^s=\mathbf{H}-\mathbf{H}^0$

$$\nabla\times\mathbf{H}^s=\zeta_0(z,\omega)\mathbf{E}^s+\mathbf{j}^s, \tag{4.3}$$

$$\nabla\times\mathbf{E}^s=i\omega\mu(z)\mathbf{H}^s, \tag{4.4}$$

where $\mathbf{j}^s=(\zeta-\zeta_0)\mathbf{E}$, and $\mathbf{E}^0$, $\mathbf{H}^0$ are the electric and magnetic 'reference' fields excited by impressed source $\mathbf{m}\delta(\mathbf{r}-\mathbf{r}_{\mathrm{T}})$ in the reference bedding. Further, Equations (4.3) and (4.4) are reduced to a scattering equation (cf. Pankratov et al., 1997, equation (33))

$$\chi(\mathbf{r})=\chi_0(\mathbf{r})+\int_{V^m}K(\mathbf{r},\mathbf{r}')R(\mathbf{r}')\chi(\mathbf{r}')\,dv'. \tag{4.5}$$

Here

$$\chi(\mathbf{r}) = (2\lambda(z))^{-1}\left((\zeta(\mathbf{r})+\zeta_0^*(z))\mathbf{E}^{\mathrm{s}}+\mathbf{j}^{\mathrm{s}}(\mathbf{r})\right), \tag{4.6}$$

$$\chi_0(\mathbf{r}) = \int\limits_{V^{\mathrm{m}}} K(\mathbf{r},\mathbf{r}')R(\mathbf{r}')\lambda(z')\mathbf{E}^0(\mathbf{r}')\mathrm{d}v', \tag{4.7}$$

$$K(\mathbf{r},\mathbf{r}') = I\delta(\mathbf{r}-\mathbf{r}')+2\lambda(z)G_0^{\mathrm{ee}}(\mathbf{r},\mathbf{r}')\lambda(z'), \tag{4.8}$$

$$R(\mathbf{r}) = \left(\zeta(\mathbf{r})-\zeta_0(z)\right)\left(\zeta(\mathbf{r})+\zeta_0^*(z)\right)^{-1}, \tag{4.9}$$

where $G_0^{\mathrm{ee}}(\mathbf{r},\mathbf{r}')$ is a 3×3 dyadic 'electric-to-electric' Green's function of 1D reference bedding (cf. Avdeev et al., 1997, appendix A), I is the identity matrix, $\lambda = \mathrm{diag}\left(\sqrt{\mathrm{Re}\,\zeta_{0\tau}}, \sqrt{\mathrm{Re}\,\zeta_{0\tau}}, \sqrt{\mathrm{Re}\,\zeta_{0z}}\right)$, ζ_0^* and $\mathrm{Re}\,\zeta_0$ are respectively complex conjugation and real part of ζ_0, and V^{m} is volume where $\zeta-\zeta_0$ differs from zero. Scattering Equation (4.5) has a contracting integral kernel, which allows writing its solution in the form of an always convergent Neumann series

$$\chi(\mathbf{r}) = \chi_0(\mathbf{r}) + \int\limits_{V^{\mathrm{m}}} K(\mathbf{r},\mathbf{r}')R(\mathbf{r}')\chi_0(\mathbf{r}')\mathrm{d}v'$$
$$+\int\limits_{V^{\mathrm{m}}} K(\mathbf{r},\mathbf{r}')R(\mathbf{r}')\left(\int\limits_{V^{\mathrm{m}}} \mathbf{K}(\mathbf{r}',\mathbf{r}'')R(\mathbf{r}'')\chi_0(\mathbf{r}'')\mathrm{d}v''\right)\mathrm{d}v' + \cdots \tag{4.10}$$

as $\chi(\mathbf{r}) = \lim_{N\to\infty}\chi^{(N)}$, where the N-th partial sum $\chi^{(N)}$ is determined iteratively by

$$\chi^{(N+1)}(\mathbf{r}) = \chi_0(\mathbf{r}) + \int\limits_{V^{\mathrm{m}}} K(\mathbf{r},\mathbf{r}')R(\mathbf{r}')\chi^{(N)}(\mathbf{r}')\mathrm{d}v',$$

$$\chi^{(1)}(\mathbf{r}) = \chi_0(\mathbf{r}), \quad N = 1,2,\ldots. \tag{4.11}$$

The optimal choice of $\zeta_0(z)$ to achieve the fastest convergence rate has been discussed in Pankratov et al. (1995) and Singer and Fainberg (1995). Stopping criterion for iteration (4.11) is based on the expression $||\chi^{(N+1)}-\chi^{(N)}||^2/||\chi^{(N)}||^2 \le 4\epsilon^2/C_{\mathrm{l}}$ for the L_2 norm. Here ϵ is the relative accuracy of the solution, i.e. $||\chi-\chi^{(N)}||/||\chi^{(N)}|| \le \epsilon$, and C_{l} is the maximum lateral contrast of $\zeta(\mathbf{r})$.

With χ, $\mathbf{E}^{\mathrm{s}}$ and $\mathbf{j}^{\mathrm{s}}$ now known at V^{m} from Equations (4.5) and (4.6), we determine $\mathbf{E}^{\mathrm{s}}$ and $\mathbf{H}^{\mathrm{s}}$ for any position $\mathbf{r}$ with

$$\mathbf{E}^{\mathrm{s}}(\mathbf{r}) = \int\limits_{V^{\mathrm{m}}} G_0^{\mathrm{ee}}(\mathbf{r},\mathbf{r}')\mathbf{j}^{\mathrm{s}}(\mathbf{r}')\mathrm{d}v', \tag{4.12}$$

$$\mathbf{H}^{\mathrm{s}}(\mathbf{r}) = \int\limits_{V^{\mathrm{m}}} G_0^{\mathrm{me}}(\mathbf{r},\mathbf{r}')\mathbf{j}^{\mathrm{s}}(\mathbf{r}')\mathrm{d}v', \tag{4.13}$$

where $G_0^{\mathrm{me}}(\mathbf{r},\mathbf{r}')$ is a 3×3 dyadic 'electric-to-magnetic' Green's function of the reference bedding. The total fields $\mathbf{E}$ and $\mathbf{H}$ are obtained by adding the appropriate reference and scattered fields.

3. COMPUTATIONAL LOAD

Our numerical implementation of the approach reviewed in Section 2, hereafter referred to as NS (Neumann series) solution, is an outgrowth of that presented in Avdeev et al. (1997, 1998). We assume that volume V^{m} is divided into $N_x \times N_y \times N_z$ rectangular prisms, and suppose that χ_0 and R in expansion (4.10) are constants within each prism. To preserve the geometry of the borehole and invasion zone we prescribe to each rectangular prism a conductivity value, that is determined from the average value of the true conductivity distribution within the prism. The number of operations, M, and memory size, S, needed to obtain the NS solution are given by

$$M \approx O(N_x N_y N_z (N N_z + (N + N_z)\log_2 N_x \log_2 N_y)), \tag{4.14}$$

$$S \approx O(N_x N_y N_z^2) \text{ bytes.} \tag{4.15}$$

Here N is the number of terms in expansion (4.10) needed to get the solution χ within a given relative accuracy ϵ. From Equation (4.14) it can be seen that M depends on the horizontal dimensions N_x and N_y as $O(N_\tau \log N_\tau)$, and depends quadratically on the vertical size N_z.

4. NUMERICAL EXAMPLES

4.1. Model with displacement currents

To verify the solution to simulate responses arising from models that include displacement currents effects, we consider the model shown in the upper panel of Figure 1. A vertical borehole crosses a three-layered formation. Resistivities and dielectric permittivities of the borehole-mud and bed shoulders are $\rho_{\mathrm{m}} = 1\ \Omega\,\mathrm{m}$ and $\varepsilon_{\mathrm{m}} = 70$, whereas those of the bed are $\rho_{\mathrm{m}} = 2\ \Omega\,\mathrm{m}$ and $\varepsilon_{\mathrm{m}} = 10$. A two-coil sonde with co-axial magnetic dipole transmitter and receiver is fixed on the borehole axis, the receiver being 25.4 cm above the transmitter. The sonde moves upward. Shown in the lower panels of Figure 1 are the 50 MHz responses from our solution (NS) and quasi-analytical mode-matching solution (MM) by Liu (1993). The responses are shown as functions of the depth, z, of the midpoint between the transmitter and the receiver (e.g. depth $z = 0$ when the sonde midpoint is opposite the centre of the bed). We also plotted the responses when the sonde is moved 12 cm off the well bore axis. The curves exhibit excellent agreement both in amplitude and phase of responses for all depths. We also have plotted the responses of the formation when the borehole is absent. The model was discretized with $18 \times 18 \times 20$ rectangular cells with dimensions of $1.4 \times 1.4 \times 1.5\ \mathrm{cm}^3$. On a PC Pentium/350 MHz the NS code takes 180 s and requires 6 terms of expansion (4.10) per sonde position to compute the series with relative accuracy $\epsilon = 0.1\%$.

4.2. Deviated borehole in a layered formation

We now present a comparison between the results calculated with the staggered-grid FD solution by Newman and Alumbaugh (1995) and our solution for the model (Figure 2,

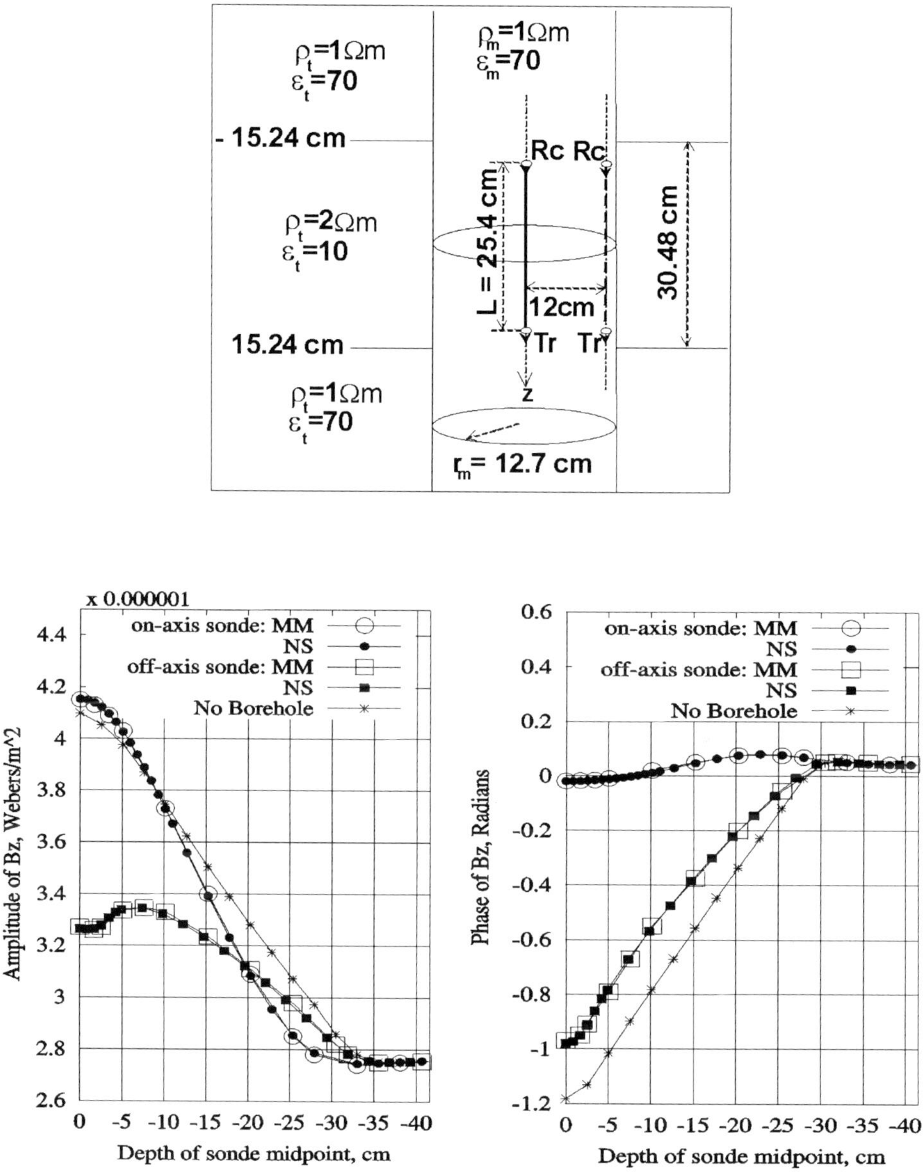

Figure 1. Comparisons of the responses obtained from NS solution (black symbols) with those produced from MM solution (open symbols; redrawn from Liu, 1993). The upper panel illustrates the model, the lower panels illustrate the 50 MHz responses for on- and off-axis positions of two-coil sonde.

upper panel) with a 45-degree deviated borehole in a two-layered formation. Figure 2 also shows the responses with the direct-coupled field removed. Also shown in the figure are the responses for the model with a non-deviated borehole. Again all curves exhibit

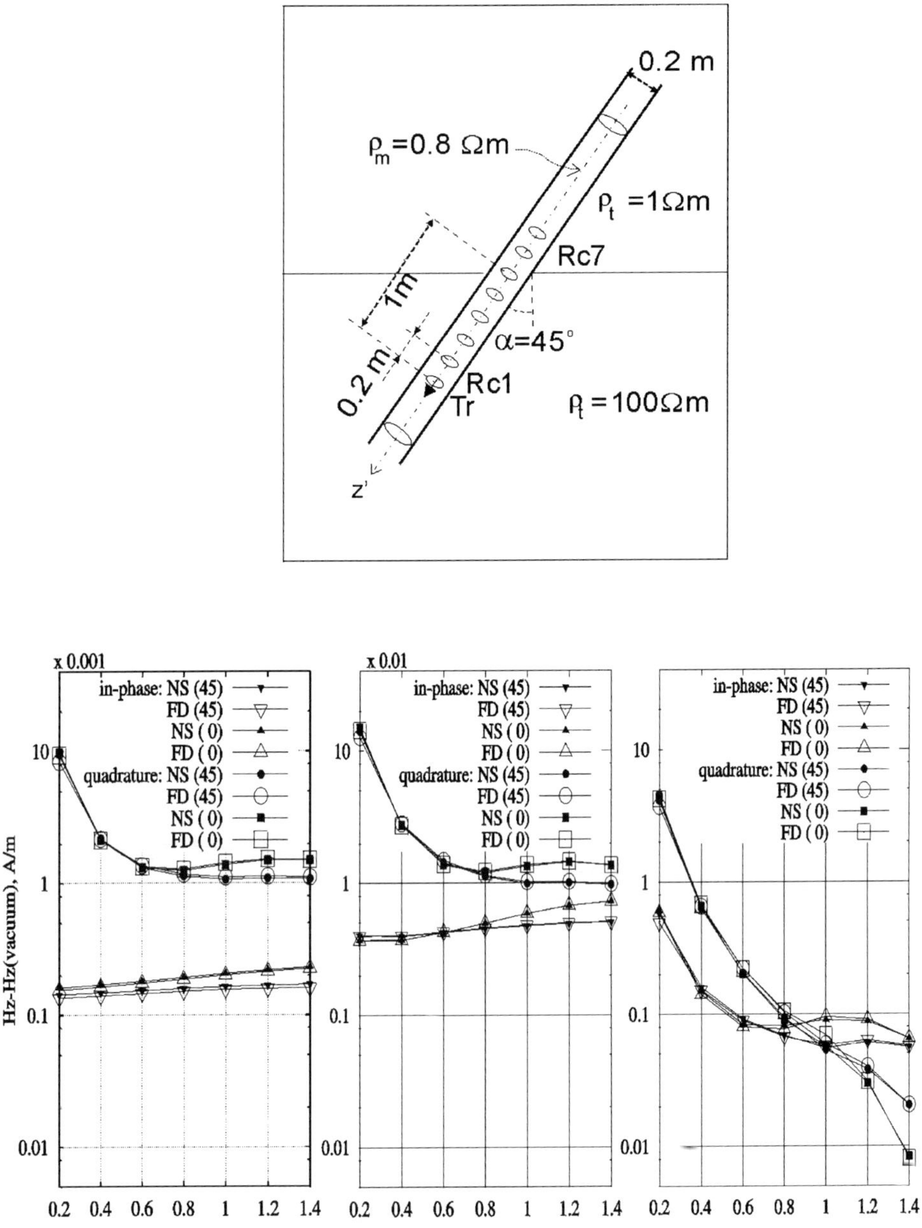

Figure 2. Comparisons of the responses obtained from the NS and FD solutions of the model of a 45 degree deviated borehole. The upper panel illustrates the model, the lower panels illustrate the 10 kHz (left), 160 kHz (centre) and 5 MHz (right) responses. Responses for a vertical borehole model are also included for comparison.

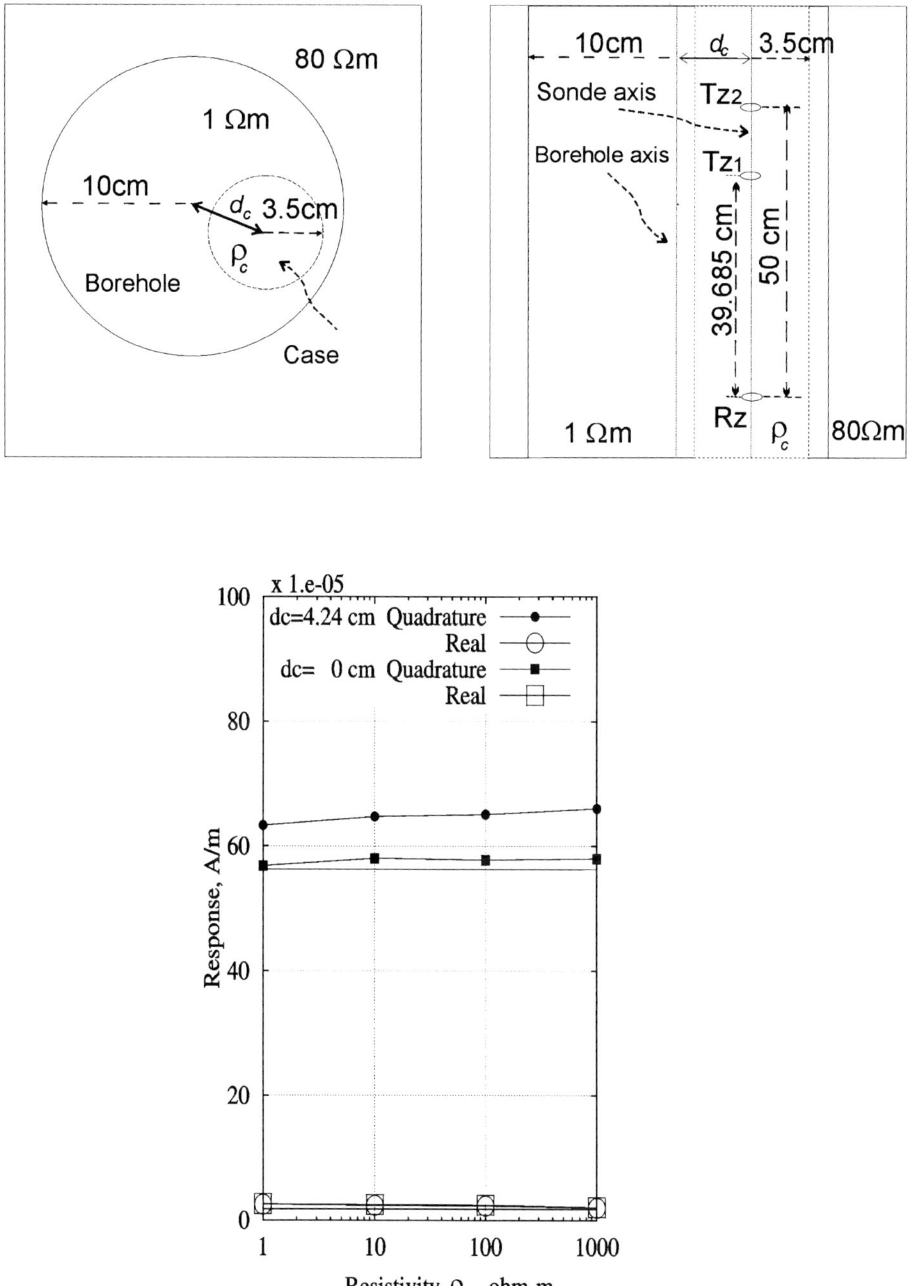

Figure 3. The three-coil sonde with a resistive case. Upper panels are plane and side views of the model, the lower panel illustrates the responses, *R*, with respect to resistivity ρ_c for the sonde eccentricity $d_c = 0$ cm and 4.24 cm.

very good agreement. The models were discretized with $31 \times 31 \times 32$ rectangular cells with dimensions of $0.95 \times 0.66 \times 4.6$ cm^3. On a PC Pentium/350 MHz, the NS code takes 7200 s and requires 40 terms of series (4.10) to compute the solution with relative accuracy $\epsilon = 0.1\%$.

4.3. Effect of the sonde case

We finally perform the model studies to estimate a possible effect associated with a resistive case that encloses three-coil sonde IKZ (Devitsin et al., 1997). This sonde of 39.685 cm and 50 cm spacing is positioned within a vertical well bore in a homogeneous formation (Figure 3, upper panels). The sonde has a cylindrical case with a diameter of 7 cm. The eccentricity, d_c, of the sonde and resistivity, ρ_c, of the case are varied. The lower panel in Figure 3 shows the 100 kHz responses, $R = H_z(z_2) - H_z(z_1)/2$, as functions of d_c and ρ_c; on the same panel, unmarked curves present the responses of the formation when the borehole is absent. It is seen that the resistive case weakly affects the responses, irrespective of eccentricity.

5. CONCLUDING REMARKS

We believe that the numerical solution presented in this paper may serve as an effective tool to calculate induction log responses in 3D geometries. Our solution is not intended for and cannot directly compete with the SLDM solution by Druskin et al. (1999), when multiple frequency responses are desired. For a single frequency, however, the solution can be quite fast and is competitive with SLDM and other FD solutions.

ACKNOWLEDGEMENTS

The research was made possible in part by grant No. 00-05-64182 from the Russian Foundation for Basic Research and Office of Basic Energy Science, administered by the United States Department of Energy under contract DE-AC04-94AL85000.

REFERENCES

Alumbaugh, D.L., Newman, G.A, Prevost, L. and Shadid, J., 1996. Three-dimensional wideband electromagnetic modeling on massively parallel computers. Radio Sci., 31, 1–23.

Avdeev, D.B., Kuvshinov, A.V., Pankratov, O.V. and Newman, G.A., 1997. High-performance three-dimensional electromagnetic modeling using modified Neumann series. Wide-band numerical solution and examples. J. Geomagn. Geoelectr., 49, 1519–1539.

Avdeev, D.B., Kuvshinov, A.V., Pankratov, O.V. and Newman, G.A., 1998. Three-dimensional frequency-domain modelling of airborne electromagnetic responses. Explor. Geophys., 29, 1–9.

Devitsin, V.A., Pantiuhin, V.A., Pasechnik, M.P., Rudyak, B.V., Snejko, O.M. and Shein, Yu.L., 1997. Multicoil induction tools (in Russian). Logger, 30, 24–32.

Druskin, V., Knizhnerman, L. and Lee P., 1999. A new spectral Lanczos decomposition method for induction modeling in arbitrary 3D geometry. Geophysics, 64, 701–706.

Liu, Q.H., 1993. Electromagnetic field generated by an off-axis source in a cylindrically layered medium with an arbitrary number of horizontal discontinuities. Geophysics, 58, 616–625.

Newman, G.A. and Alumbaugh, D.L., 1995. Frequency-domain modeling of airborne electromagnetic responses using staggered finite differences. Geophys. Prospect., 43, 1021–1042.

Pankratov, O.V., Avdeev, D.B. and Kuvshinov, A.V., 1995. Electromagnetic field scattering in a heterogeneous earth: A solution to the forward problem. Phys. Solid Earth, 31, 201–209.

Pankratov, O.V., Kuvshinov, A.V. and Avdeev, D.B., 1997. High-performance three-dimensional electromagnetic modeling using modified Neumann series. Anisotropic case. J. Geomagn. Geoelectr., 49, 1541–1547.

Singer, B.Sh., 1995. Method for solution of Maxwell's equations in non-uniform media. Geophys. J. Int., 120, 590–598.

Singer, B.Sh. and Fainberg, E.B., 1995. Generalization of the iterative–dissipative method for modeling electromagnetic fields in nonuniform media with displacement currents. J. Appl. Geophys., 34, 41–46.

Singer, B.Sh. and Fainberg, E.B., 1997. Fast and stable method for 3-D modelling of electromagnetic field. Explor. Geophys., 28, 130–135.

van der Horst, M., Druskin, V. and Knizhnerman, L., 1995. Modeling the response of induction logging tools in 3D geometries with the Spectral Lanczos Decomposition Method. In: International Symposium on Three-Dimensional Electromagnetics, Expanded Abstracts. Schlumberger–Doll Research, Ridgefield, CT, pp. 4–6.

Chapter 5

NONLINEAR APPROXIMATIONS FOR ELECTROMAGNETIC SCATTERING FROM ELECTRICAL AND MAGNETIC INHOMOGENEITIES

Arvidas Cheryauka [a], Michael S. Zhdanov [a] and Motoyuki Sato [b]

[a] *University of Utah, Department of Geology and Geophysics, Salt Lake City, UT 84112, USA*
[b] *Center for Northeast Asian Studies, Tohoku University, Sendai, Japan*

Abstract: We extend linear and nonlinear approximations for electromagnetic fields to a medium with inhomogeneous distribution of both electrical and magnetic material properties. These approximations are presented in the form of tensor integrals over a domain with anomalous parameters. The developed approximations combine the linear and nonlinear estimations depending on the ratio of complex conductivity and magnetic susceptibility perturbations. These approximations form a basis for fast EM modeling and imaging in multi-dimensional environments where joint electrical and magnetic inhomogeneity is an essential feature of the model. Numerical tests carried out for one-dimensional electromagnetic logging applications demonstrate the validity of the theory and the effectiveness of the proposed approximations.

1. INTRODUCTION

Traditionally, in modeling electromagnetic fields in geophysical explorations, one takes into account the distribution of the anomalous electrical conductivity only. However, there are many practical situations when a conductive object has significant magnetic properties as well. For example, a magnetite-containing ore body, some geological formations of sedimentary or volcanic origin, and drilling mud with heavy material ingredients are characterized by both anomalous conductivity and magnetic susceptibility, which can produce significant effects on the electromagnetic tool response.

The foundations of the integral equation method were developed in pioneering works by Hohmann (1975), Tabarovsky (1975), Weidelt (1975), etc. Recently developed localized nonlinear (Habashy et al., 1993; Torres-Verdin and Habashy, 1994), quasi-linear (Zhdanov and Fang, 1996, 1997) and quasi-analytical (Zhdanov et al., 2000) approximations in an electrically inhomogeneous medium are the basis for high-performance forward modeling and inversion methods. Some theoretical aspects of Born type approximations for magnetic properties were considered by Murray et al. (1999) and Cheryauka and Sato (1999). In this paper we extend these approaches and formulate linear and nonlinear approximations for a model with joint fluctuations of electrical and magnetic properties.

Synthetic modeling examples illustrating the comparison with analytical solutions for one-dimensional induction logging and casing-scanning problems show the areas of potential applications of these nonlinear algorithms.

2. INTEGRAL EQUATION FORMULATION

Let us consider the general 3-D EM forward problem illustrated in Figure 1. A medium with joint electrical and magnetic inhomogeneities is excited by electrical, $\mathbf{J}^{\text{inc}}$, and magnetic, $\mathbf{M}^{\text{inc}}$, harmonic currents distributed within a domain V_{inc}. Time dependence is $e^{-i\omega t}$.

A lossy unbounded medium is characterized by a complex electrical conductivity $\tilde{\sigma}(\mathbf{r}) = \sigma(\mathbf{r}) - i\omega\varepsilon(\mathbf{r})$, and a magnetic susceptibility $\chi(\mathbf{r})$, where $\sigma(\mathbf{r})$ and $\varepsilon(\mathbf{r})$ are electrical conductivity and dielectric constant respectively. To symmetrize further considerations, we introduce a complex magnetic permeability $\tilde{\mu}$:

$$\tilde{\mu}(\mathbf{r}) = i\omega\mu_0(1 + \chi(\mathbf{r})),$$

where μ_0 is the free space magnetic permeability. Note that, in general cases, $\tilde{\sigma}(\mathbf{r})$, $\tilde{\mu}(\mathbf{r})$ can be frequency-dependent and, in anisotropic media, can be represented by 3×3 dyadic functions. In this paper we will study the isotropic medium only.

We assume that the material property distributions $\tilde{\sigma}(\mathbf{r})$, $\tilde{\mu}(\mathbf{r})$ are expressed by a sum of background, $\tilde{\sigma}_b(\mathbf{r})$, $\tilde{\mu}_b(\mathbf{r})$, and anomalous, $\tilde{\sigma}_a(\mathbf{r})$, $\tilde{\mu}_a(\mathbf{r})$, distributions:

$$\begin{aligned} \tilde{\sigma}(\mathbf{r}) &= \tilde{\sigma}_b(\mathbf{r}) + \tilde{\sigma}_a(\mathbf{r}), \\ \tilde{\mu}(\mathbf{r}) &= \tilde{\mu}_b(\mathbf{r}) + \tilde{\mu}_a(\mathbf{r}). \end{aligned} \tag{5.1}$$

We assume that anomalous distributions are nonzero only within the corresponding domains V_σ and V_μ. The electromagnetic fields **E**, **H** in this model can be presented as a

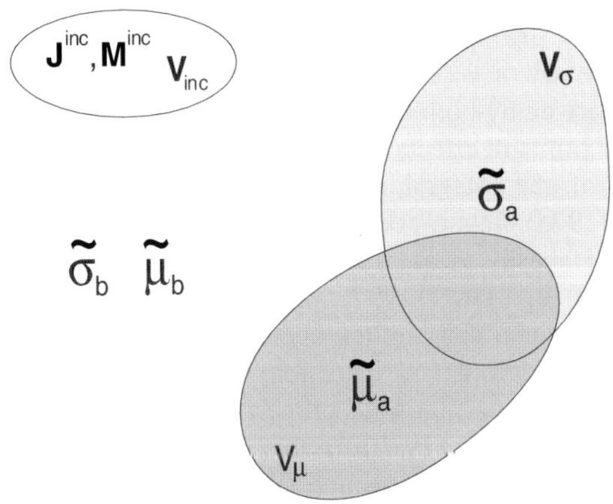

Figure 1. Model statement.

superposition of background $\mathbf{E}^{\rm b}$, $\mathbf{H}^{\rm b}$ and anomalous $\mathbf{E}^{\rm a}$, $\mathbf{H}^{\rm a}$ fields,

$$\mathbf{E} = \mathbf{E}^{\rm b} + \mathbf{E}^{\rm a},$$
$$\mathbf{H} = \mathbf{H}^{\rm b} + \mathbf{H}^{\rm a}, \tag{5.2}$$

which satisfy Maxwell equations

$$\nabla \times \mathbf{H}^{\rm b} - \tilde{\sigma}_{\rm b}(\mathbf{r})\mathbf{E}^{\rm b} = \mathbf{J}^{\rm inc},$$
$$\nabla \times \mathbf{E}^{\rm b} - \tilde{\mu}_{\rm b}(\mathbf{r})\mathbf{H}^{\rm b} = \mathbf{M}^{\rm inc}, \tag{5.3}$$

and

$$\nabla \times \mathbf{H}^{\rm a} - \tilde{\sigma}_{\rm b}(\mathbf{r})\mathbf{E}^{\rm a} = \tilde{\sigma}_{\rm a}(\mathbf{r})(\mathbf{E}^{\rm b} + \mathbf{E}^{\rm a}),$$
$$\nabla \times \mathbf{E}^{\rm a} - \tilde{\mu}_{\rm b}(\mathbf{r})\mathbf{H}^{\rm a} = \tilde{\mu}_{\rm a}(\mathbf{r})(\mathbf{H}^{\rm b} + \mathbf{H}^{\rm a}). \tag{5.4}$$

The background field can be derived using the Green's function method (Felsen and Marcuvitz, 1994):

$$\mathbf{E}^{\rm b} = \left\langle \hat{\mathbf{G}}^{JE}\mathbf{J}^{\rm inc} + \hat{\mathbf{G}}^{ME}\mathbf{M}^{\rm inc}\right\rangle_{V_{\rm inc}},$$
$$\mathbf{H}^{\rm b} = \left\langle \hat{\mathbf{G}}^{JH}\mathbf{J}^{\rm inc} + \hat{\mathbf{G}}^{MH}\mathbf{M}^{\rm inc}\right\rangle_{V_{\rm inc}}, \tag{5.5}$$

where we use the notation

$$\left\langle \hat{\mathbf{G}}\mathbf{J}\right\rangle_V = \int_V \hat{\mathbf{G}}(\mathbf{r}|\mathbf{r}')\,\mathbf{J}(\mathbf{r}')\,{\rm d}\mathbf{r}'.$$

In the last formula, $\hat{\mathbf{G}}^{JE}, \hat{\mathbf{G}}^{JH}, \hat{\mathbf{G}}^{MH}$, and $\hat{\mathbf{G}}^{ME}$ are tensor Green's functions satisfying the following second-order differential equations:

$$\nabla \times \frac{1}{\tilde{\mu}_{\rm b}(\mathbf{r})}\nabla \times \hat{\mathbf{G}}^{JE}(\mathbf{r}'|\mathbf{r}) - \tilde{\sigma}_{\rm b}(\mathbf{r})\hat{\mathbf{G}}^{JE}(\mathbf{r}'|\mathbf{r}) = \hat{\mathbf{I}}\delta(\mathbf{r} - \mathbf{r}'),$$
$$\hat{\mathbf{G}}^{JH}(\mathbf{r}'|\mathbf{r}) = \frac{1}{\tilde{\mu}_{\rm b}(\mathbf{r})}\nabla \times \hat{\mathbf{G}}^{JE}(\mathbf{r}'|\mathbf{r}),$$
$$\nabla \times \frac{1}{\tilde{\sigma}_{\rm b}(\mathbf{r})}\nabla \times \hat{\mathbf{G}}^{MH}(\mathbf{r}'|\mathbf{r}) - \tilde{\mu}_{\rm b}(\mathbf{r})\hat{\mathbf{G}}^{MH}(\mathbf{r}'|\mathbf{r}) = \hat{\mathbf{I}}\delta(\mathbf{r} - \mathbf{r}'),$$
$$\hat{\mathbf{G}}^{ME}(\mathbf{r}'|\mathbf{r}) = \frac{1}{\tilde{\sigma}_{\rm b}(\mathbf{r})}\nabla \times \hat{\mathbf{G}}^{MH}(\mathbf{r}'|\mathbf{r}),$$

where $\hat{\mathbf{I}}$ is the identity tensor, and $\delta(\mathbf{r} - \mathbf{r}')$ is the Dirac function (Tai, 1979).

The volume densities of electrical and magnetic anomalous currents $\mathbf{J}^{\rm a}$, $\mathbf{M}^{\rm a}$ are equal to

$$\mathbf{J}^{\rm a}(\mathbf{r}) = \tilde{\sigma}_{\rm a}(\mathbf{r})(\mathbf{E}^{\rm b} + \mathbf{E}^{\rm a}),$$
$$\mathbf{M}^{\rm a}(\mathbf{r}) = \tilde{\mu}_{\rm a}(\mathbf{r})(\mathbf{H}^{\rm b} + \mathbf{H}^{\rm a}). \tag{5.6}$$

Therefore, the anomalous field excited by arbitrary anomalous currents $\mathbf{J}^{\rm a}$, $\mathbf{M}^{\rm a}$ can be expressed by a formula similar to Equation (5.5):

$$\mathbf{E}^{\rm a} = \left\langle \hat{\mathbf{G}}^{JE}\mathbf{J}^{\rm a}\right\rangle_{V_\sigma} + \left\langle \hat{\mathbf{G}}^{ME}\mathbf{M}^{\rm a}\right\rangle_{V_\mu},$$
$$\mathbf{H}^{\rm a} = \left\langle \hat{\mathbf{G}}^{JH}\mathbf{J}^{\rm a}\right\rangle_{V_\sigma} + \left\langle \hat{\mathbf{G}}^{MH}\mathbf{M}^{\rm a}\right\rangle_{V_\mu}. \tag{5.7}$$

3. BORN APPROXIMATION

The conventional Born approximation, $\mathbf{E}^{\mathrm{B}}$, $\mathbf{H}^{\mathrm{B}}$, is based on an assumption that within domains V_σ and V_μ the anomalous fields $\mathbf{E}^{\mathrm{a}}$, $\mathbf{H}^{\mathrm{a}}$ are negligibly small in comparison with the background fields $\mathbf{E}^{\mathrm{b}}$, $\mathbf{H}^{\mathrm{b}}$:

$$\begin{matrix} \mathbf{E}^{\mathrm{a}}(\mathbf{r}) \approx 0, \\ \mathbf{H}^{\mathrm{a}}(\mathbf{r}) \approx 0, \end{matrix} \quad \mathbf{r} \in V_\sigma \cup V_\mu. \tag{5.8}$$

In this case, according to Equation (5.6), Formula (5.7) for anomalous fields outside domains V_σ and V_μ is simplified:

$$\begin{aligned} \mathbf{E}^{\mathrm{a}} \approx \mathbf{E}^{\mathrm{B}} &= \langle \hat{\mathbf{G}}^{JE} \mathbf{J}^{\mathrm{B}} \rangle_{V_\sigma} + \langle \hat{\mathbf{G}}^{ME} \mathbf{M}^{\mathrm{B}} \rangle_{V_\mu}, \\ \mathbf{H}^{\mathrm{a}} \approx \mathbf{H}^{\mathrm{B}} &= \langle \hat{\mathbf{G}}^{JH} \mathbf{J}^{\mathrm{B}} \rangle_{V_\sigma} + \langle \hat{\mathbf{G}}^{MH} \mathbf{M}^{\mathrm{B}} \rangle_{V_\mu}, \end{aligned} \tag{5.9}$$

where Born current densities $\mathbf{J}^{\mathrm{B}}$, $\mathbf{M}^{\mathrm{B}}$ are

$$\mathbf{J}^{\mathrm{B}} = \tilde{\sigma}_{\mathrm{a}} \mathbf{E}^{\mathrm{b}}, \qquad \mathbf{M}^{\mathrm{B}} = \tilde{\mu}_{\mathrm{a}} \mathbf{H}^{\mathrm{b}}.$$

In fact, this approximation is linear with respect to the anomalous material properties $\tilde{\sigma}_{\mathrm{a}}$, $\tilde{\mu}_{\mathrm{a}}$ and can be treated as a first-order term in a complete Born–Neumann series. These series of limited numbers of terms can be treated as nonlinear approximations; however, their convergence, in most cases, is problematical.

4. LOCALIZED APPROXIMATION

The Localized Nonlinear (LN) approximation (Habashy et al., 1993; Torres-Verdin and Habashy, 1994) is based on the assumption that the internal electrical field has a small spatial gradient, which can be neglected to zero order regardless of medium properties. As a result, the scattering tensor can be expressed in explicit form. Extending this method, we consider that variations of both electrical and magnetic fields in the vicinity of some inner point $\mathbf{r}$ of anomalous areas are sufficiently smooth:

$$\begin{aligned} \mathbf{E}(\mathbf{r}+\delta\mathbf{r}) &= \mathbf{E}(\mathbf{r}) + \delta\mathbf{r} \cdot \nabla \mathbf{E}(\mathbf{r}) + O(\delta\mathbf{r}^2), \\ \mathbf{H}(\mathbf{r}+\delta\mathbf{r}) &= \mathbf{H}(\mathbf{r}) + \delta\mathbf{r} \cdot \nabla \mathbf{H}(\mathbf{r}) + O(\delta\mathbf{r}^2), \end{aligned} \quad \mathbf{r} \in V_\sigma, V_\mu. \tag{5.10}$$

Note that, in the last formula, we consider the dyadic product of vector operator ∇ and the vectors of electric and magnetic fields, $\mathbf{E}(\mathbf{r})$ and $\mathbf{H}(\mathbf{r})$. Following Habashy et al. (1993), we obtain four scattering tensors for a model with joint electrical and magnetic inhomogeneities. Substituting expansions (5.10) into Equation (5.7) and neglecting terms of order higher than the one with respect to $\delta\mathbf{r}$, we find from Equations (5.2) and (5.7)

$$\begin{aligned} \mathbf{E} &\approx \mathbf{E}^{\mathrm{b}} + \langle \hat{\mathbf{G}}^{JE} \tilde{\sigma}_{\mathrm{a}} \rangle_{V_\sigma} \cdot \mathbf{E} + \langle \hat{\mathbf{G}}^{ME} \tilde{\mu}_{\mathrm{a}} \rangle_{V_\mu} \cdot \mathbf{H}, \\ \mathbf{H} &\approx \mathbf{H}^{\mathrm{b}} + \langle \hat{\mathbf{G}}^{JH} \tilde{\sigma}_{\mathrm{a}} \rangle_{V_\sigma} \cdot \mathbf{E} + \langle \hat{\mathbf{G}}^{MH} \tilde{\mu}_{\mathrm{a}} \rangle_{V_\mu} \cdot \mathbf{H}. \end{aligned}$$

Introducing new compact notations and placing the background field $\mathbf{E}^{\mathrm{b}}$, $\mathbf{H}^{\mathrm{b}}$ into the right-hand side of the expressions, we obtain the following system of equations:

$$\begin{aligned}&\mathbf{E}-\hat{\mathbf{\Gamma}}^{\mathrm{e}}\hat{\mathbf{\Pi}}^{ME}\cdot\mathbf{H}\approx\hat{\mathbf{\Gamma}}^{\mathrm{e}}\cdot\mathbf{E}^{\mathrm{b}},\\&\hat{\mathbf{\Gamma}}^{\mathrm{m}}\hat{\mathbf{\Pi}}^{JH}\cdot\mathbf{E}-\mathbf{H}\approx-\hat{\mathbf{\Gamma}}^{\mathrm{m}}\cdot\mathbf{H}^{\mathrm{b}},\end{aligned}\tag{5.11}$$

where dimensionless scattering tensors are

$$\hat{\mathbf{\Gamma}}^{\mathrm{e}}=[\hat{\mathbf{I}}-\hat{\mathbf{\Pi}}^{JE}]^{-1},\qquad\hat{\mathbf{\Gamma}}^{\mathrm{m}}=[\hat{\mathbf{I}}-\hat{\mathbf{\Pi}}^{MH}]^{-1}$$

and the integrals of the Green's dyadic functions are

$$\begin{aligned}&\hat{\mathbf{\Pi}}^{JE}=\langle\hat{\mathbf{G}}^{JE}\tilde{\sigma}_{\mathrm{a}}\rangle_{V_\sigma},\qquad\hat{\mathbf{\Pi}}^{ME}=\langle\hat{\mathbf{G}}^{ME}\tilde{\mu}_{\mathrm{a}}\rangle_{V_\mu},\\&\hat{\mathbf{\Pi}}^{JH}=\langle\hat{\mathbf{G}}^{JH}\tilde{\sigma}_{\mathrm{a}}\rangle_{V_\sigma},\qquad\hat{\mathbf{\Pi}}^{HM}=\langle\hat{\mathbf{G}}^{MH}\tilde{\mu}_{\mathrm{a}}\rangle_{V_\mu}.\end{aligned}$$

Applying simple linear operations to the equations in Formula (5.11), we express the anomalous fields in the form

$$\left.\begin{aligned}&\mathbf{E}^{\mathrm{LN}}=\hat{\mathbf{\Gamma}}^{JE}\mathbf{E}^{\mathrm{b}}+\hat{\mathbf{\Gamma}}^{ME}\mathbf{H}^{\mathrm{b}},\\&\mathbf{H}^{\mathrm{LN}}=\hat{\mathbf{\Gamma}}^{JH}\mathbf{E}^{\mathrm{b}}+\hat{\mathbf{\Gamma}}^{MH}\mathbf{H}^{\mathrm{b}},\end{aligned}\right.\qquad\mathbf{r}\in V_\sigma\cup V_\mu.\tag{5.12}$$

where $\hat{\mathbf{\Gamma}}^{JE}$, $\hat{\mathbf{\Gamma}}^{ME}$, $\hat{\mathbf{\Gamma}}^{JH}$ and $\hat{\mathbf{\Gamma}}^{MH}$ are new EM scattering tensors as introduced in (Murray et al., 1999; Cheryauka and Sato, 1999):

$$\begin{aligned}&\hat{\mathbf{\Gamma}}^{JF}=\hat{\mathbf{\Gamma}}_{\mathrm{e}}^{\mathrm{det}}\hat{\mathbf{\Gamma}}^{\mathrm{e}},\qquad\hat{\mathbf{\Gamma}}^{ME}=\hat{\mathbf{\Gamma}}^{JE}\hat{\mathbf{\Pi}}^{ME}\hat{\mathbf{\Gamma}}^{\mathrm{m}},\\&\hat{\mathbf{\Gamma}}^{MH}=\hat{\mathbf{\Gamma}}_{\mathrm{h}}^{\mathrm{det}}\hat{\mathbf{\Gamma}}^{\mathrm{m}},\qquad\hat{\mathbf{\Gamma}}^{JH}=\hat{\mathbf{\Gamma}}^{MH}\hat{\mathbf{\Gamma}}^{\mathrm{e}}\hat{\mathbf{\Pi}}^{ME},\end{aligned}$$

and

$$\begin{aligned}&\hat{\mathbf{\Gamma}}_{\mathrm{e}}^{\mathrm{det}}=[\hat{\mathbf{I}}-\hat{\mathbf{\Gamma}}^{\mathrm{e}}\hat{\mathbf{\Pi}}^{ME}\hat{\mathbf{\Gamma}}^{\mathrm{m}}\hat{\mathbf{\Pi}}^{JH}]^{-1},\\&\hat{\mathbf{\Gamma}}_{\mathrm{h}}^{\mathrm{det}}=[\hat{\mathbf{I}}-\hat{\mathbf{\Gamma}}^{\mathrm{m}}\hat{\mathbf{\Pi}}^{JH}\hat{\mathbf{\Gamma}}^{\mathrm{e}}\hat{\mathbf{\Pi}}^{ME}]^{-1}.\end{aligned}$$

The fields $\mathbf{E}^{\mathrm{LN}}$, $\mathbf{H}^{\mathrm{LN}}$ outside the perturbed areas are

$$\begin{aligned}&\mathbf{E}^{\mathrm{LN}}=\langle\hat{\mathbf{G}}^{JE}\mathbf{J}^{\mathrm{LN}}\rangle_{V_\sigma}+\langle\hat{\mathbf{G}}^{ME}\mathbf{M}^{\mathrm{LN}}\rangle_{V_\mu},\\&\mathbf{H}^{\mathrm{LN}}=\langle\hat{\mathbf{G}}^{JH}\mathbf{J}^{\mathrm{LN}}\rangle_{V_\sigma}+\langle\hat{\mathbf{G}}^{MH}\mathbf{M}^{\mathrm{LN}}\rangle_{V_\mu},\end{aligned}\tag{5.13}$$

where the LN current densities $\mathbf{J}^{\mathrm{LN}}, \mathbf{M}^{\mathrm{LN}}$ are

$$\mathbf{J}^{LN}=\tilde{\sigma}_{\mathrm{a}}(\hat{\mathbf{\Gamma}}^{JE}\mathbf{E}^{\mathrm{b}}+\hat{\mathbf{\Gamma}}^{ME}\mathbf{H}^{\mathrm{b}}),\qquad\mathbf{M}^{LN}=\tilde{\mu}_{\mathrm{a}}(\hat{\mathbf{\Gamma}}^{JH}\mathbf{E}^{\mathrm{b}}+\hat{\mathbf{\Gamma}}^{MH}\mathbf{H}^{\mathrm{b}}).\tag{0.1}$$

Note that the structures of the scattering tensors $\hat{\mathbf{\Gamma}}^{JE}$, $\hat{\mathbf{\Gamma}}^{ME}$, $\hat{\mathbf{\Gamma}}^{JH}$, $\hat{\mathbf{\Gamma}}^{MH}$ can be simplified if the locations of electrical and magnetic inhomogeneities do not intersect, or the contributions of anomalous conductivity $\tilde{\sigma}_{\mathrm{a}}$ or anomalous magnetic permeability $\tilde{\mu}_{\mathrm{a}}$ dominate the other.

5. QUASI-LINEAR APPROXIMATION

Zhdanov and Fang (1996, 1997) developed a quasi-linear (QL) approximation, based on a linear relationship between anomalous and background fields inside an inhomogeneous domain, expressed by an electrical reflectivity tensor $\hat{\lambda}^{e}$:

$$\mathbf{E}^{\mathrm{QL}}(\mathbf{r}) \approx \hat{\lambda}^{e}(\mathbf{r})\mathbf{E}^{b}(\mathbf{r}), \quad \mathbf{r} \in V_{\sigma}. \tag{5.14}$$

This tensor can be effectively approximated by a system of smooth basis functions on a coarse spatial grid because of a smooth variation of the field inside the inhomogeneity. Using a similar approach, we can introduce a 'magnetic reflectivity tensor' $\hat{\lambda}^{m}$:

$$\mathbf{H}^{\mathrm{QL}}(\mathbf{r}) \approx \hat{\lambda}^{m}(\mathbf{r})\mathbf{H}^{b}(\mathbf{r}), \quad \mathbf{r} \in V_{\mu}. \tag{5.15}$$

According to Equations (5.7), (5.14) and (5.15), the QL approximations of the anomalous fields outside the perturbed areas are expressed by the formulae

$$\begin{aligned} \mathbf{E}^{\mathrm{QL}} &= \left\langle \hat{\mathbf{G}}^{JE}\mathbf{J}^{\mathrm{QL}} \right\rangle_{V_{\sigma}} + \left\langle \hat{\mathbf{G}}^{ME}\mathbf{M}^{\mathrm{QL}} \right\rangle_{V_{\mu}}, \\ \mathbf{H}^{\mathrm{QL}} &= \left\langle \hat{\mathbf{G}}^{JH}\mathbf{J}^{\mathrm{QL}} \right\rangle_{V_{\sigma}} + \left\langle \hat{\mathbf{G}}^{MH}\mathbf{M}^{\mathrm{QL}} \right\rangle_{V_{\mu}}, \end{aligned} \tag{5.16}$$

where appropriate current densities $\mathbf{J}^{\mathrm{QL}}, \mathbf{M}^{\mathrm{QL}}$ are:

$$\mathbf{J}^{\mathrm{QL}} = \tilde{\sigma}_{a}(\hat{\mathbf{I}} + \hat{\lambda}^{e})\mathbf{E}^{b}, \qquad \mathbf{M}^{\mathrm{QL}} = \tilde{\mu}_{a}(\hat{\mathbf{I}} + \hat{\lambda}^{m})\mathbf{H}^{b}.$$

The electrical and magnetic reflectivity tensors, $\hat{\lambda}^{e}$ and $\hat{\lambda}^{m}$, are determined by a minimization technique applied to the corresponding areas of anomalous conductivity support, V_{σ}, and anomalous magnetic permeability support, V_{μ}, according to the following equations.

(1) Within the joint area of the domains V_{σ} and V_{μ}, $\mathbf{r} \in V = V_{\sigma} \bigcap V_{\mu}$

$$\left\| \begin{matrix} \hat{\lambda}^{e}\mathbf{E}^{b} - \left\langle \hat{\mathbf{G}}^{JE}\tilde{\sigma}_{a}(\mathbf{I} + \hat{\lambda}^{e})\mathbf{E}^{b} + \hat{\mathbf{G}}^{ME}\tilde{\mu}_{a}(\hat{\mathbf{I}} + \hat{\lambda}^{m})\mathbf{H}^{b} \right\rangle_{V} \\ \hat{\lambda}^{m}\mathbf{H}^{b} - \left\langle \hat{\mathbf{G}}^{JH}\tilde{\sigma}_{a}(\mathbf{I} + \hat{\lambda}^{e})\mathbf{E}^{b} + \hat{\mathbf{G}}^{MH}\tilde{\mu}_{a}(\hat{\mathbf{I}} + \hat{\lambda}^{m})\mathbf{H}^{b} \right\rangle_{V} \end{matrix} \right\| = \min. \tag{5.17}$$

(2) Within the area outside V_{μ} but inside V_{σ}, $\mathbf{r} \in V = V_{\sigma} \setminus V_{\mu}$

$$\left\| \hat{\lambda}^{e}\mathbf{E}^{b} - \left\langle \hat{\mathbf{G}}^{JE}\tilde{\sigma}_{a}(\hat{\mathbf{I}} + \hat{\lambda}^{e})\mathbf{E}^{b} \right\rangle_{V} \right\| = \min. \tag{5.18}$$

(3) Within the area outside V_{σ} but inside V_{μ}, $\mathbf{r} \in V = V_{\mu} \setminus V_{\sigma}$

$$\left\| \hat{\lambda}^{m}\mathbf{H}^{b} - \left\langle \hat{\mathbf{G}}^{MH}\tilde{\mu}_{a}(\hat{\mathbf{I}} + \hat{\lambda}^{m})\mathbf{H}^{b} \right\rangle_{V} \right\| = \min. \tag{5.19}$$

Note that the solution of Equations (5.17)–(5.19) is nonlinear with respect to $\tilde{\sigma}_{a}$ and $\tilde{\mu}_{a}$, because $\hat{\lambda}^{e}$ and $\hat{\lambda}^{m}$ are nonlinear functions of $\tilde{\sigma}_{a}$ and $\tilde{\mu}_{a}$.

6. QUASI-ANALYTICAL APPROXIMATION

The Quasi-Analytical (QA) approximation (Zhdanov et al., 2000) is based on the same assumption as the QL approximation that the anomalous fields inside an inhomogeneous domain are linearly proportional to the background EM fields through the reflectivity tensors $\hat{\lambda}^{\mathrm{e}}$ and $\hat{\lambda}^{\mathrm{m}}$ (Equations (5.14) and (5.15)). The main difference is that, using an analytic technique in the QA approximation, Zhdanov et al. (2000) obtained the reflectivity tensor $\hat{\lambda}^{\mathrm{e}}$ in explicit form. Here we extend the QA approach for media with joint electrical and magnetic inhomogeneities.

EM fields at inner points can be expressed using Equations (5.14) and (5.15) as

$$\hat{\lambda}^{\mathrm{e}}\mathbf{E}^{\mathrm{b}} = \mathbf{E}^{\mathrm{B}} + \langle \hat{\mathbf{G}}^{JE}\tilde{\sigma}_{\mathrm{a}}\hat{\lambda}^{\mathrm{e}}\mathbf{E}^{\mathrm{b}}\rangle_{V_\sigma} + \langle \hat{\mathbf{G}}^{ME}\tilde{\mu}_{\mathrm{a}}\hat{\lambda}^{\mathrm{m}}\mathbf{H}^{\mathrm{b}}\rangle_{V_\mu},$$
$$\hat{\lambda}^{\mathrm{m}}\mathbf{H}^{\mathrm{b}} = \mathbf{H}^{\mathrm{B}} + \langle \hat{\mathbf{G}}^{JH}\tilde{\sigma}_{\mathrm{a}}\hat{\lambda}^{\mathrm{e}}\mathbf{E}^{\mathrm{b}}\rangle_{V_\sigma} + \langle \hat{\mathbf{G}}^{MH}\tilde{\mu}_{\mathrm{a}}\hat{\lambda}^{\mathrm{m}}\mathbf{H}^{\mathrm{b}}\rangle_{V_\mu}, \tag{5.20}$$

where $\mathbf{E}^{\mathrm{B}}, \mathbf{H}^{\mathrm{B}}$ are anomalous Born fields:

$$\mathbf{E}^{\mathrm{B}} = \langle \hat{\mathbf{G}}^{JE}\tilde{\sigma}_{\mathrm{a}}\mathbf{E}^{\mathrm{b}}\rangle_{V_\sigma} + \langle \hat{\mathbf{G}}^{ME}\tilde{\mu}_{\mathrm{a}}\mathbf{H}^{\mathrm{b}}\rangle_{V_\mu} = \mathbf{E}^{\mathrm{B}\sigma} + \mathbf{E}^{\mathrm{B}\mu},$$
$$\mathbf{H}^{\mathrm{B}} = \langle \hat{\mathbf{G}}^{JH}\tilde{\sigma}_{\mathrm{a}}\mathbf{E}^{\mathrm{b}}\rangle_{V_\sigma} + \langle \hat{\mathbf{G}}^{MH}\tilde{\mu}_{\mathrm{a}}\mathbf{H}^{\mathrm{b}}\rangle_{V_\mu} = \mathbf{H}^{\mathrm{B}\sigma} + \mathbf{H}^{\mathrm{B}\mu}. \tag{5.21}$$

Subtracting weighted Born fields from Equation (5.20), we obtain

$$\begin{aligned}\hat{\lambda}^{\mathrm{e}}(\mathbf{E}^{\mathrm{b}} - \mathbf{E}^{\mathrm{B}\sigma}) - \hat{\lambda}^{\mathrm{m}}\mathbf{E}^{\mathrm{B}\mu} &= \mathbf{E}^{\mathrm{B}} + \langle (\hat{\mathbf{G}}^{JE}\tilde{\sigma}_{\mathrm{a}}\hat{\lambda}^{\mathrm{e}}(\mathbf{r}') - \hat{\lambda}^{\mathrm{e}}(\mathbf{r})\hat{\mathbf{G}}^{JE}\tilde{\sigma}_{\mathrm{a}})\mathbf{E}^{\mathrm{b}}\rangle_{V_\sigma}\\ &\quad + \langle (\hat{\mathbf{G}}^{ME}\tilde{\mu}_{\mathrm{a}}\hat{\lambda}^{\mathrm{m}}(\mathbf{r}') - \hat{\lambda}^{\mathrm{m}}(\mathbf{r})\hat{\mathbf{G}}^{ME}\tilde{\mu}_{\mathrm{a}})\mathbf{H}^{\mathrm{b}}\rangle_{V_\mu},\\ -\hat{\lambda}^{\mathrm{e}}\mathbf{H}^{\mathrm{B}\sigma} + \hat{\lambda}^{\mathrm{m}}(\mathbf{H}^{\mathrm{b}} - \mathbf{H}^{\mathrm{B}\mu}) &= \mathbf{H}^{\mathrm{B}} + \langle (\hat{\mathbf{G}}^{JH}\tilde{\sigma}_{\mathrm{a}}\hat{\lambda}^{\mathrm{e}}(\mathbf{r}') - \hat{\lambda}^{\mathrm{e}}(\mathbf{r})\hat{\mathbf{G}}^{JH}\tilde{\sigma}_{\mathrm{a}})\mathbf{E}^{\mathrm{b}}\rangle_{V_\sigma}\\ &\quad + \langle (\hat{\mathbf{G}}^{MH}\tilde{\mu}_{\mathrm{a}}\hat{\lambda}^{\mathrm{m}}(\mathbf{r}') - \hat{\lambda}^{\mathrm{m}}(\mathbf{r})\hat{\mathbf{G}}^{MH}\tilde{\mu}_{\mathrm{a}})\mathbf{H}^{\mathrm{b}}\rangle_{V_\mu}.\end{aligned} \tag{5.22}$$

Following Habashy et al. (1993) and Torres-Verdin and Habashy (1994), we can take into account that the Green's tensors $\hat{\mathbf{G}}^{JE}$, $\hat{\mathbf{G}}^{ME}$, $\hat{\mathbf{G}}^{JH}$ and $\hat{\mathbf{G}}^{MH}$ exhibit either a singularity or a peak at the point $\mathbf{r}_j = \mathbf{r}$. Therefore, one can expect that the dominant contribution to the integral in Equation (5.22) is from some vicinity of the point $\mathbf{r}_j = \mathbf{r}$. Assuming also that $\hat{\lambda}^{\mathrm{e}}$ and $\hat{\lambda}^{\mathrm{m}}$ are the slowly varying functions within domains V_σ, V_μ one can rewrite Equation (5.22) in the form

$$\hat{\lambda}^{\mathrm{e}}(\mathbf{E}^{\mathrm{b}} - \mathbf{E}^{\mathrm{B}\sigma}) - \hat{\lambda}^{\mathrm{m}}\mathbf{E}^{\mathrm{B}\mu} \approx \mathbf{E}^{\mathrm{B}}, \tag{5.23}$$

$$-\hat{\lambda}^{\mathrm{e}}\mathbf{H}^{\mathrm{B}\sigma} + \hat{\lambda}^{\mathrm{m}}(\mathbf{H}^{\mathrm{b}} - \mathbf{H}^{\mathrm{B}\mu}) \approx \mathbf{H}^{\mathrm{B}}. \tag{5.24}$$

Note that the system of Equations (5.23) and (5.24) is, in general cases, underdetermined, because we have two vector equations for two unknown tensors $\hat{\lambda}^{\mathrm{e}}$ and $\hat{\lambda}^{\mathrm{m}}$. Let us consider now two special cases with scalar and diagonal reflectivity tensors.

In the case of scalar reflectivity tensors ($\hat{\lambda}^{\mathrm{e}} = \lambda^{\mathrm{e}}\hat{\mathbf{I}}$, $\hat{\lambda}^{\mathrm{m}} = \lambda^{\mathrm{m}}\hat{\mathbf{I}}$, where $\hat{\mathbf{I}}$ is a unit tensor), the linear system of Equations (5.23) and (5.24) is overdetermined. We can obtain two scalar equations for λ^{e} and λ^{m} by choosing the specific type of the multipliers. In particular, we calculate first, the dot product of both sides of Equation (5.23) and the complex conjugate electric field $\mathbf{E}^{\mathrm{b}*}$, and, second, calculate the dot product of both sides of Equation (5.24), the complex conjugate magnetic field $\mathbf{H}^{\mathrm{b}*}$:

$$\lambda^{\mathrm{e}}(\mathbf{E}^{\mathrm{b}} \cdot \mathbf{E}^{\mathrm{b}*} - \mathbf{E}^{\mathrm{B}\sigma} \cdot \mathbf{E}^{\mathrm{b}*}) - \lambda^{\mathrm{m}}\mathbf{E}^{\mathrm{B}\mu} \cdot \mathbf{E}^{\mathrm{b}*} \approx \mathbf{E}^{\mathrm{B}} \cdot \mathbf{E}^{\mathrm{b}*}, \tag{5.25}$$

$$-\lambda^{\mathrm{e}}\mathbf{H}^{\mathrm{B}\sigma} \cdot \mathbf{H}^{\mathrm{b}*} + \lambda^{\mathrm{m}}(\mathbf{H}^{\mathrm{b}} \cdot \mathbf{H}^{\mathrm{b}*} - \mathbf{H}^{\mathrm{B}\mu} \cdot \mathbf{H}^{\mathrm{b}*}) \approx \mathbf{H}^{\mathrm{B}} \cdot \mathbf{H}^{\mathrm{b}*}, \tag{5.26}$$

where '*' denotes the complex conjugate vectors.

As a result, we obtain a system of two linear equations with respect to $\hat{\lambda}^{\mathrm{e}}$ and $\hat{\lambda}^{\mathrm{m}}$:

$$\begin{pmatrix} \mathbf{E}^{\mathrm{b}}\cdot\mathbf{E}^{\mathrm{b}*}-\mathbf{E}^{\mathrm{B}\sigma}\cdot\mathbf{E}^{\mathrm{b}*} & -\mathbf{E}^{\mathrm{B}\mu}\cdot\mathbf{E}^{\mathrm{b}*} \\ -\mathbf{H}^{\mathrm{B}\sigma}\cdot\mathbf{H}^{\mathrm{b}*} & \mathbf{H}^{\mathrm{b}}\cdot\mathbf{H}^{\mathrm{b}*}-\mathbf{H}^{\mathrm{B}\mu}\cdot\mathbf{H}^{\mathrm{b}*} \end{pmatrix} \begin{pmatrix} \lambda^{\mathrm{e}} \\ \lambda^{\mathrm{m}} \end{pmatrix} = \begin{pmatrix} \mathbf{E}^{\mathrm{B}}\cdot\mathbf{E}^{\mathrm{b}*} \\ \mathbf{H}^{\mathrm{B}}\cdot\mathbf{H}^{\mathrm{b}*} \end{pmatrix}, \tag{5.27}$$

which can be easily resolved as follows:

$$\lambda^{\mathrm{e}} = D^{-1}\left[\left(\mathbf{E}^{\mathrm{B}}\cdot\mathbf{E}^{\mathrm{b}*}\right)\left(\mathbf{H}^{\mathrm{b}}\cdot\mathbf{H}^{\mathrm{b}*}-\mathbf{H}^{\mathrm{B}\mu}\cdot\mathbf{H}^{\mathrm{b}*}\right)+\left(\mathbf{E}^{\mathrm{B}\mu}\cdot\mathbf{E}^{\mathrm{b}*}\right)\left(\mathbf{H}^{\mathrm{B}}\cdot\mathbf{H}^{\mathrm{b}*}\right)\right],$$

$$\lambda^{\mathrm{m}} = D^{-1}\left[\left(\mathbf{H}^{\mathrm{B}}\cdot\mathbf{H}^{\mathrm{b}*}\right)\left(\mathbf{E}^{\mathrm{b}}\cdot\mathbf{E}^{\mathrm{b}*}-\mathbf{E}^{\mathrm{B}\sigma}\cdot\mathbf{E}^{\mathrm{b}*}\right)+\left(\mathbf{H}^{\mathrm{B}\sigma}\cdot\mathbf{H}^{\mathrm{b}*}\right)\left(\mathbf{E}^{\mathrm{B}}\cdot\mathbf{E}^{\mathrm{b}*}\right)\right], \tag{5.28}$$

assuming that the determinant, D, of the matrix of the linear Equation (5.27) is not equal to zero:

$$D = \left(\mathbf{E}^{\mathrm{b}}\cdot\mathbf{E}^{\mathrm{b}*}-\mathbf{E}^{\mathrm{B}\sigma}\cdot\mathbf{E}^{\mathrm{b}*}\right)\left(\mathbf{H}^{\mathrm{b}}\cdot\mathbf{H}^{\mathrm{b}*}-\mathbf{H}^{\mathrm{B}\mu}\cdot\mathbf{H}^{\mathrm{b}*}\right)-\left(\mathbf{E}^{\mathrm{B}\mu}\cdot\mathbf{E}^{\mathrm{b}*}\right)\left(\mathbf{H}^{\mathrm{B}\sigma}\cdot\mathbf{H}^{\mathrm{b}*}\right)\neq 0.$$

For example, in the case of the purely electrical inhomogeneities,

$$\mathbf{E}^{\mathrm{B}\mu}=0, \qquad \mathbf{H}^{\mathrm{B}\mu}=0,$$
$$\mathbf{E}^{\mathrm{B}}=\mathbf{E}^{\mathrm{B}\sigma}, \qquad \mathbf{H}^{\mathrm{B}}=\mathbf{H}^{\mathrm{B}\sigma},$$

and the formula for the electrical reflectivity coefficient becomes

$$\lambda^{\mathrm{e}} = \frac{\left(\mathbf{E}^{\mathrm{B}\sigma}\cdot\mathbf{E}^{\mathrm{b}*}\right)}{\left(\mathbf{E}^{\mathrm{b}}\cdot\mathbf{E}^{\mathrm{b}*}-\mathbf{E}^{\mathrm{B}\sigma}\cdot\mathbf{E}^{\mathrm{b}*}\right)}. \tag{5.29}$$

Formula (5.29) is equivalent to the one developed in Zhdanov et al. (2000). Note that by choosing different multipliers in Equations (5.25) and (5.26) one can select various situations for reflectivity coefficients.

In the case of diagonal reflectivity tensors $\hat{\lambda}^{\mathrm{e}}$, $\hat{\lambda}^{\mathrm{m}}$, we obtain from Equations (5.23) and (5.24) a 6×6 system of linear equations with respect to the six unknown diagonal components of tensors $\lambda^{\mathrm{e}}_{ii}, \lambda^{\mathrm{m}}_{ii}, i=1,2,3$:

$$\begin{pmatrix} E_1^b-E_1^{B\sigma} & 0 & 0 & -E_1^{B\mu} & 0 & 0 \\ 0 & E_2^b-E_2^{B\sigma} & 0 & 0 & -E_2^{B\mu} & 0 \\ 0 & 0 & E_3^b-E_3^{B\sigma} & 0 & 0 & -E_3^{B\mu} \\ -H_1^{B\sigma} & 0 & 0 & H_1^b-H_1^{B\mu} & 0 & 0 \\ 0 & -H_2^{B\sigma} & 0 & 0 & H_2^b-H_2^{B\mu} & 0 \\ 0 & 0 & -H_3^{B\sigma} & 0 & 0 & H_3^b-H_3^{B\mu} \end{pmatrix}$$

$$\times \begin{pmatrix} \lambda^{\mathrm{e}}_{11} \\ \lambda^{\mathrm{e}}_{22} \\ \lambda^{\mathrm{e}}_{33} \\ \lambda^{\mathrm{m}}_{11} \\ \lambda^{\mathrm{m}}_{22} \\ \lambda^{\mathrm{m}}_{33} \end{pmatrix} = \begin{pmatrix} E_1^{\mathrm{B}} \\ E_2^{\mathrm{B}} \\ E_3^{\mathrm{B}} \\ H_1^{\mathrm{B}} \\ H_2^{\mathrm{B}} \\ H_3^{\mathrm{B}} \end{pmatrix}. \tag{5.30}$$

To solve this sparse problem we consider separately three pairs of equations: the 1st and the 4th, the 2nd and the 5th, and the 3rd and the 6th:

$$\begin{cases} (E_i^{\mathrm{b}}-E_i^{\mathrm{B}\sigma})\lambda^{\mathrm{e}}_{ii}-E_i^{\mathrm{B}\mu}\lambda^{\mathrm{m}}_{ii}=E_i^{\mathrm{B}} \\ -H_i^{\mathrm{B}\sigma}\lambda^{\mathrm{e}}_{ii}+(H_i^{\mathrm{b}}-H_i^{\mathrm{B}\mu})\lambda^{\mathrm{m}}_{ii}=H_i^{\mathrm{B}} \end{cases} \quad i=1,2,3. \tag{5.31}$$

Solving each of the 2×2 equations from the system (5.31) under the assumption that the determinant, D_i, of the corresponding matrix of the linear equations (5.31) is not equal to zero,

$$D_i = (E_i^{\mathrm{b}} - E_i^{\mathrm{B}\sigma})(H_i^{\mathrm{b}} - H_i^{\mathrm{B}\mu}) - H_i^{\mathrm{B}\sigma} E_i^{\mathrm{B}\mu} \neq 0,$$

we obtain the values of the diagonal tensor components $\lambda_{ii}^{\mathrm{e}}$ and $\lambda_{ii}^{\mathrm{m}}$

$$\begin{aligned} \lambda_{ii}^{\mathrm{e}} &= D_i^{-1}\left[E_i^{\mathrm{B}}(H_i^{\mathrm{b}} - H_i^{\mathrm{B}\mu}) + H_i^{\mathrm{B}} E_i^{\mathrm{B}\mu}\right], \\ \lambda_{ii}^{\mathrm{m}} &= D_i^{-1}\left[H_i^{\mathrm{B}}(E_i^{\mathrm{b}} - E_i^{\mathrm{B}\sigma}) + E_i^{\mathrm{B}} H_i^{\mathrm{B}\sigma}\right]. \end{aligned} \qquad i = 1,2,3. \tag{5.32}$$

Finally, for purely electrical inhomogeneities, we find

$$\lambda_{ii}^{\mathrm{e}} = \frac{E_i^{\mathrm{B}\sigma}}{E_i^{\mathrm{b}} - E_i^{\mathrm{B}\sigma}}, \quad i = 1,2,3, \tag{5.33}$$

while for purely magnetic inhomogeneities we have

$$\lambda_{ii}^{\mathrm{m}} = \frac{H_i^{\mathrm{B}\mu}}{H_i^{\mathrm{b}} - H_i^{\mathrm{B}\mu}}, \quad i = 1,2,3. \tag{5.34}$$

7. NUMERICAL RESULTS AND VALIDATION

Let us consider now several numerical applications of the developed approximations.

In the first set of numerical experiments, we check the validity of the linear and nonlinear approximations in a one-dimensional cylindrically layered model, excited by a vertical magnetic dipole (Figure 2). The EM induction response in the cylindrically layered model can be calculated using an integral equation method. In Appendix A we

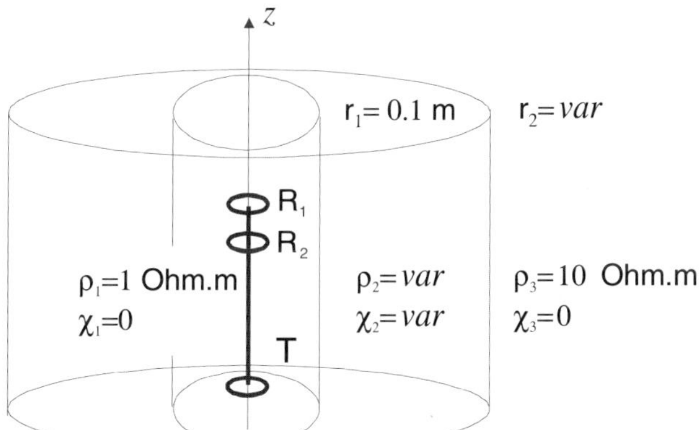

Figure 2. One-dimensional cylindrically layered model and induction array. The characteristics of the three-coil induction tool $TxRx_1Rx_2$ are as follows: $f = 50$ kHz; $L_1 = 1.5$ m, $L_2 = 1$ m; $M_1 = 1$ A m^2; $M_2 = -0.2963$ A m^2, where f is the frequency, L_1 and L_2 are the tool spacings, and M_1 and M_2 are the receiver moments.

implement a quasi-analytical approach to computing the EM field in the model with axial symmetry and present the corresponding formulae. In our simple computational test we consider the second layer in the three-layered model as an anomalous domain (scatterer) with respect to the two-layer model 'borehole space–background formation', chosen as a background model. This model study simulates an invaded zone effect in a thick formation, which is a typical problem in induction logging.

At the same time, electromagnetic fields in cylindrically layered models with a radial piecewise distribution of electromagnetic parameters can be expressed in closed integral form and can be calculated with arbitrary accuracy. The mathematical description and the solution for the vertical component of the magnetic field can be found, for instance, in Augustin et al. (1989). We treat the result obtained by using this calculation technique as an 'exact' solution. We analyze a synthetic voltage signal of a borehole compensated array with characteristics close to practical ones. The voltage signal of the conventional three-coil differential induction tool is represented as a linear combination of the two-coil tool's responses (Kaufman and Keller, 1989):

$$V = V_1 + V_2 = \tilde{\mu}_1 \left[M_1 H_{zz}^0(L_1) + M_2 H_{zz}^0(L_2) \right] + \frac{\tilde{\mu}_1}{2\pi^2} \int_0^\infty p_1^2 \nu_1(\lambda)(M_1 \cos\lambda L_1 + M_2 \cos\lambda L_2)\,\mathrm{d}\lambda,$$

where $\{V_1, M_1, L_1\}$, $\{V_2, M_2, L_2\}$ are the voltage signals, the coil moments, and the spacings of the 1st and the 2nd two-coil induction subarrays; $\nu_1(\lambda)$ is the so-called 'layered function' within the borehole space and H_{zz}^0 is a primary field in the homogeneous medium with the parameters of the borehole space $k_1^2 = \tilde{\mu}_1 \tilde{\sigma}_1$, $p_1^2 = \lambda^2 - k_1^2$.

Figure 3 shows the real and imaginary parts of the voltage (R and X signals in logging terminology), calculated by using the exact data (the closed form solution: real part, solid line; imaginary part, dotted line) and the combined approximation (real part, '+' symbols; imaginary part, 'o' symbols). As one can find from Equations (5.6) and (5.7), the anomalous field components are determined by the superposition of electric and magnetic scattering currents depending on fluctuations of the material properties. We can separately choose a form of the approximation for these currents and compose a hybrid type approximation. Here we implement the quasi-analytical approximation for the anomalous electrical resistivity ϱ and the Born approximation for the anomalous magnetic susceptibility χ.

We compute three-coil induction tool responses for three cases:

(1) a variable resistivity ρ_2 of the second cylindrical layer with a fixed magnetic susceptibility $\chi_2 = 0.01$ and the radius of the outer boundary $r_2 = 1.5$ m (Figure 3a);

(2) a variable susceptibility χ_2 with a fixed resistivity $\rho_2 = 2$ Ohm m and the radius of the outer boundary $r_2 = 1.5$ m (Figure 3b); and

(3) a variable radius of the outer boundary r_2 of the second cylindrical layer with a fixed resistivity $\rho_2 = 2$ Ohm m and magnetic susceptibility $\chi_2 = 0.01$ (Figure 3c).

Our calculations demonstrate that the Born approximation provides a reasonable response estimation in a model with variable magnetic susceptibility, because the range of susceptibility variations in the invaded formations is small. At the same time, the electrical resistivity can change by a factor up four orders of magnitude, and one has

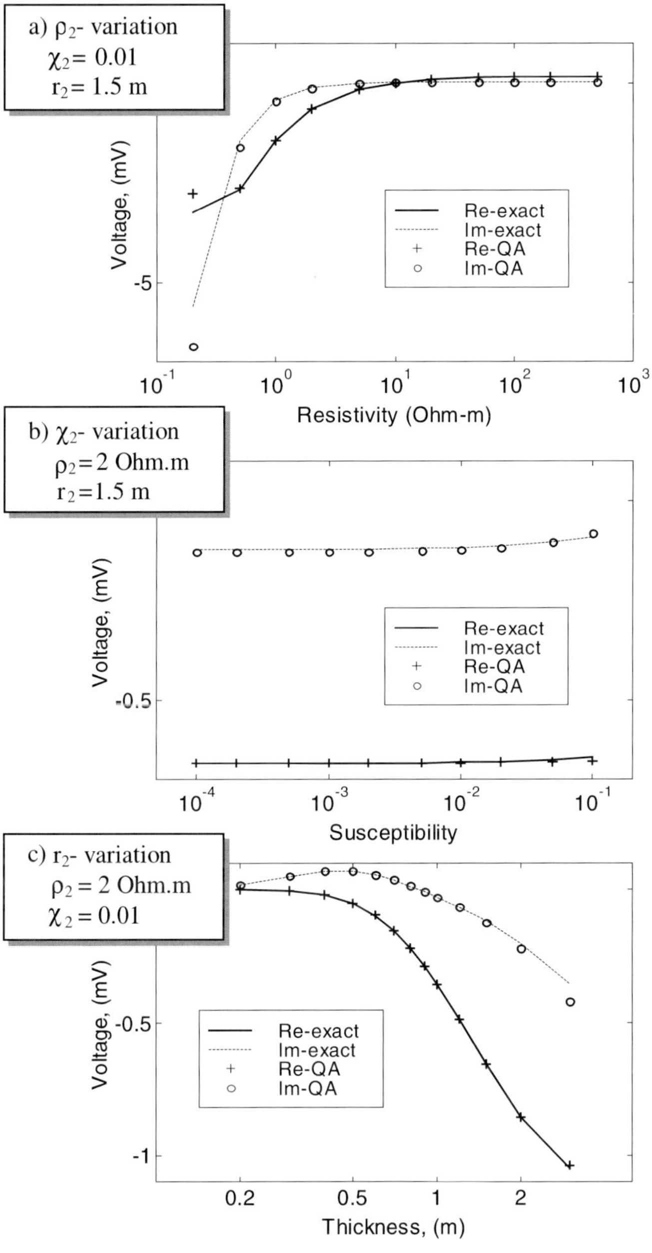

Figure 3. The comparison of the synthetic voltage signals. The responses were calculated using the closed form solution (real part, solid line; imaginary part, dashed line) and the combined QA approximation for the electrical resistivity ϱ_2 and Born approximation for the magnetic susceptibility χ_2 (real part, '+' symbols; imaginary part, 'o' symbols). (a) Variation of the layer electrical resistivity. (b) Variation of the layer magnetic susceptibility. (c) Variation of the layer outer boundary.

to apply the nonlinear approximations (in our case we use the QA approximation) for high-contrast and large-size anomalous areas.

In the second set of numerical simulations, we study the ability of the nonlinear methods to simulate an EM field in borehole models with casing. These models have extremely high contrasts in electrical conductivity and magnetic susceptibility parameters. For instance, the mild steel, which is widely used for borehole casing, has electrical resistivity in the order of 10^{-6} Ohm m and magnetic susceptibility in the order of 10^3 (Balasnis, 1989).

Figure 4 demonstrates the results of the comparison between exact, localized and quasi-analytical solutions. The approximate data produced by both nonlinear methods for anomalous fields have good accuracy and graphically fit the exact data. At the same time, the quasi-analytical formulation gives the better approximation, because it takes into account the contribution from scattering currents depending on a primary source location.

For this model we study also the validity of casing approximation using well known electrical (S) and magnetic (M) thin sheets models (Figures 5 and 6). In the general case of electrical and magnetic inhomogeneities and arbitrary polarization of a primary source the EM fields are functions not only of wave numbers k_i of a medium, but they also depend on ratios (contrasts) of electrical and magnetic properties (Felsen and Marcuvitz, 1994). Thus, we plan to consider the quality of the approximate solutions and the effects of anomalous electrical conductivity and magnetic susceptibility separately.

We have found that the casing can be considered as an S thin sheet, which is characterized by a specific conductance (Figure 5):

$$S = \frac{1}{\rho_2}\,\mathrm{d}r = \text{const.}$$

and low-magnetic properties only. The real thickness of the casing, $\mathrm{d}r$, may vary from 0.001 m to 0.05 m (with the corresponding change of the conductivity $1/\rho_2$ to keep the conductance constant) without significant effect in the induction tool response for materials of low-magnetic susceptibility value, $\chi_2 \leq 1$, (upper panel, Figure 5). Highly magnetized casing with $\chi_2 = 10$ to 10^3 (Figure 5, middle and bottom panels) cannot be satisfactory simulated by S thin-sheet approximation.

A similar effect is observed for an M thin sheet with a resistivity $\rho_2 = 10^{-1}$ Ohm m and with a constant integrated magnetic susceptibility (Figure 6, upper panel):

$$M = (1 + \chi_2)\,\mathrm{d}r = \text{const.}$$

However, this equivalence is not perfect for a lower resistivity of 10^{-3} Ohm m (Figure 6, middle panel); and one cannot neglect the casing thickness for a highly conductive casing with $\rho_2 = 10^{-6}$ Ohm m (Figure 6, bottom panel).

8. CONCLUSIONS

In this paper we have introduced a family of nonlinear approximations of electromagnetic field in models with joint electrical and magnetic inhomogeneities. These approximations are presented in the form of tensor integrals over a domain with anomalous

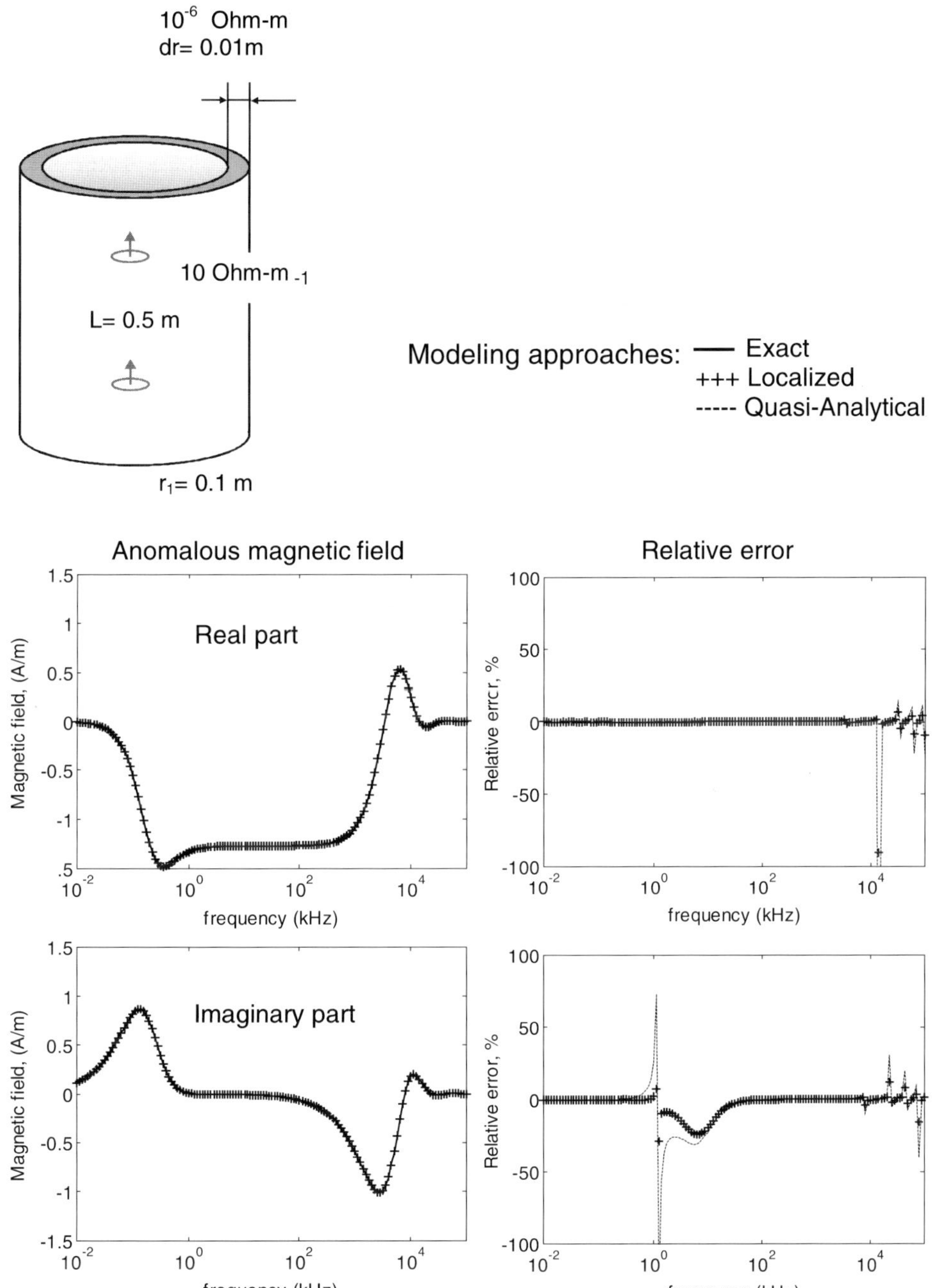

Figure 4. The comparison of the exact, LN, and QA solutions in the casing model.

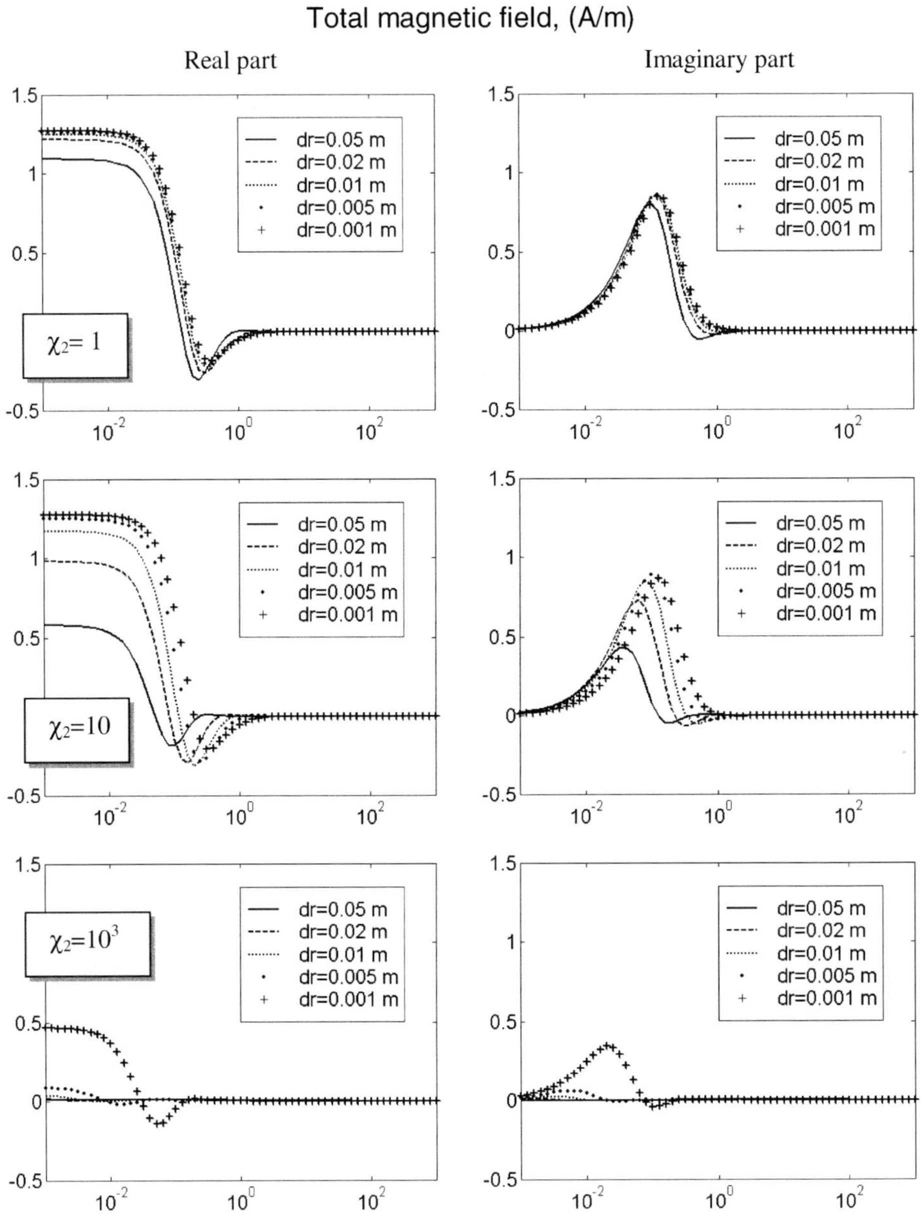

Figure 5. *S*-**equivalence in the casing model,** $\rho_2^{-1}\,\mathbf{d}r = 10^4\,S$.

parameters. The developed approximations combine linear and nonlinear estimations depending on the range of the complex conductivity and magnetic susceptibility perturbations. The introduced family of nonlinear approximations could form a basis for

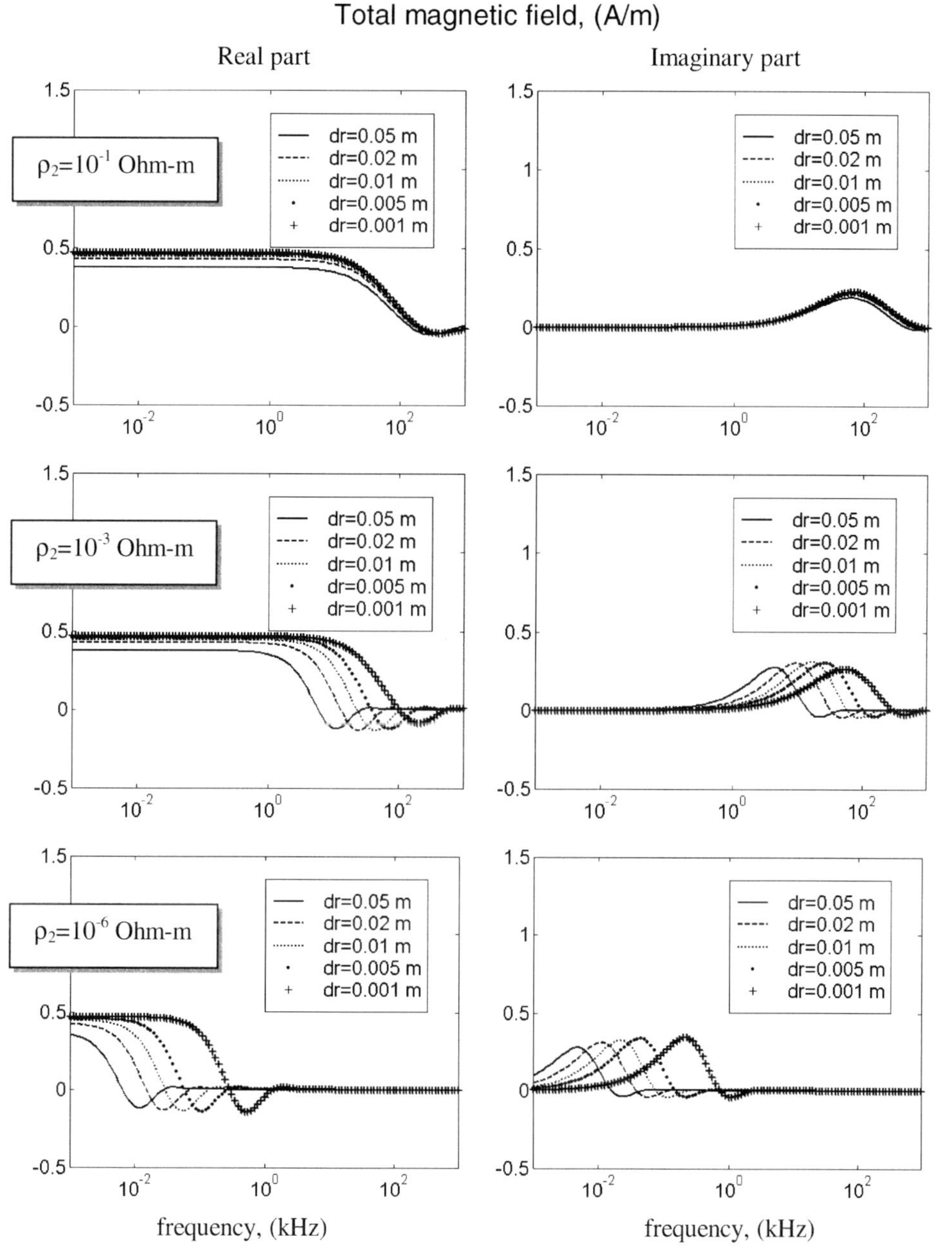

Figure 6. ***M*-equivalence in the casing model, $(1+\chi_2)\,dr = 1M$.**

fast EM modeling and imaging in the multi-dimensional environment where the joint electrical and magnetic inhomogeneities are the essential feature of the model. The numerical tests carried out for one-dimensional electromagnetic logging application have demonstrated the validity of the theory and the effectiveness of the proposed approximations.

ACKNOWLEDGEMENTS

The financial support for this work was provided by the National Science Foundation under the grant No. ECS-998 7779. The authors also acknowledge the support of the University of Utah Consortium for Electromagnetic Modeling and Inversion (CEMI), which includes Advanced Power Technologies Inc., AGIP, Baker Atlas Logging Services, BHP Minerals, ExxonMobil Upstream Research Company, INCO Exploration, Japan National Oil Corporation, MINDECO, Naval Research Laboratory, Rio Tinto, 3JTech Corporation, and Zonge Engineering. One of the authors (A. Cheryauka) received partial support for these studies from the Japan Ministry of Education, Culture and Sport (Grant-in Aid for Scientific Research no. 08555253 and 10044122), while he was visiting the Center for Northeast Asian Studies, Tohoku University. The authors are also grateful to K.H. Lee and C. Farquharson for their critical reviews, which have improved the manuscript.

Appendix A. QUASI-ANALYTICAL ONE-DIMENSIONAL SOLUTION

Consider a one-dimensional cylindrically layered model with the cylindrical coordinate system $\{\rho, \phi, z\}$ and the axial distribution of electrical conductivity $\tilde{\sigma}(\rho)$ and magnetic permeability $\tilde{\mu}(\rho)$. The incident field is excited by the vertical magnetic dipole of a unit moment located at the axis of symmetry at the point z_p . We formulate a simple axial symmetrical one-dimensional boundary value problem, where the model properties do not depend on the ϕ and z coordinates and the EM field is also azimuthally uniform.

Applying this Fourier transform with respect to the variable z to the initial Equations (5.6) and (5.7) replacing all the functions $\mathbf{E}, \mathbf{H}, \hat{\mathbf{G}}$ with their appropriate spectra $\mathbf{e}, \mathbf{h}, \hat{\mathbf{g}}$, we obtain

$$\mathbf{e}^{\mathrm{a}} = \langle \hat{\mathbf{g}}^{JE} \tilde{\sigma}_{\mathrm{a}} (\mathbf{e}^{\mathrm{b}} + \mathbf{e}^{\mathrm{a}}) \rangle_{V_\sigma(\rho',\phi')} + \langle \hat{\mathbf{g}}^{ME} \tilde{\mu}_{\mathrm{a}} (\mathbf{h}^{\mathrm{b}} + \mathbf{h}^{\mathrm{a}}) \rangle_{V_\mu(\rho',\phi')},$$
$$\mathbf{h}^{\mathrm{a}} = \langle \hat{\mathbf{g}}^{JH} \tilde{\sigma}_{\mathrm{a}} (\mathbf{e}^{\mathrm{b}} + \mathbf{e}^{\mathrm{a}}) \rangle_{V_\sigma(\rho',\phi')} + \langle \hat{\mathbf{g}}^{MH} \tilde{\mu}_{\mathrm{a}} (\mathbf{h}^{\mathrm{b}} + \mathbf{h}^{\mathrm{a}}) \rangle_{V_\mu(\rho',\phi')}. \tag{5.A1}$$

Defining Green's tensor functions, $\check{\mathbf{G}}$, of the circular source by integration of the Green's tensor functions $\hat{\mathbf{G}}$ of the point sources as

$$\check{\mathbf{G}} = \int_0^{2\pi} \hat{\mathbf{G}}(\phi') d\phi', \tag{5.A2}$$

and substituting their spectra $\check{\mathbf{g}}^{\alpha\beta}$ in Equation (5.A1), we obtain

$$\mathbf{e}^{\mathrm{a}} = \langle \check{\mathbf{g}}^{DE} \tilde{\sigma}_{\mathrm{a}} (\mathbf{e}^{\mathrm{b}} + \mathbf{e}^{\mathrm{a}}) \rangle_{V_\sigma(\rho')} + \langle \check{\mathbf{g}}^{ME} \tilde{\mu}_{\mathrm{a}} (\mathbf{h}^{\mathrm{b}} + \mathbf{h}^{\mathrm{a}}) \rangle_{V_\mu(\rho')},$$
$$\mathbf{h}^{\mathrm{a}} = \langle \check{\mathbf{g}}^{JH} \tilde{\sigma}_{\mathrm{a}} (\mathbf{e}^{\mathrm{b}} + \mathbf{e}^{\mathrm{a}}) \rangle_{V_\sigma(\rho')} + \langle \check{\mathbf{g}}^{MH} \tilde{\mu}_{\mathrm{a}} (\mathbf{h}^{\mathrm{b}} + \mathbf{h}^{\mathrm{a}}) \rangle_{V_\mu(\rho')}. \tag{5.A3}$$

According to Equation (5.5), the tensors $\breve{\mathbf{g}}^{JE}$, $\breve{\mathbf{g}}^{JH}$, $\breve{\mathbf{g}}^{MH}$, $\breve{\mathbf{g}}^{ME}$ satisfy the following equations:

$$
\begin{aligned}
&(\nabla^* \times \frac{1}{\tilde{\mu}_b(\rho)}\nabla^* \times -\tilde{\sigma}_b(\rho))\breve{\mathbf{g}}^{JE}(\rho'|\rho) = \hat{\mathbf{I}}\frac{\delta(\rho-\rho')}{\rho},\\
&\breve{\mathbf{g}}^{JH}(\rho'|\rho) = \frac{1}{\tilde{\mu}_b(\rho')}\nabla^* \times \breve{\mathbf{g}}^{JE}(\rho'|\rho),\\
&(\nabla^* \times \frac{1}{\tilde{\sigma}_b(\rho)}\nabla^* \times -\tilde{\mu}_b(\rho))\breve{\mathbf{g}}^{MH}(\rho'|\rho) = \hat{\mathbf{I}}\frac{\delta(\rho-\rho')}{\rho},\\
&\breve{\mathbf{g}}^{ME}(\rho'|\rho) = \frac{0}{\tilde{\sigma}_b(\rho')}\nabla^* \times \breve{\mathbf{g}}^{MH}(\rho'|\rho),
\end{aligned}
\tag{5.A4}
$$

where $\nabla^* = \frac{\partial}{\partial\rho}\mathbf{i}_\rho + i\lambda\mathbf{i}_z$ is a Fourier spectrum of the axial symmetrical vector-operator ∇.

Equations (5.A4) for the tensor functions $\breve{\mathbf{g}}^{\alpha\beta}$ in the cylindrically layered model can be solved in close analytical form. However, we do not present here the general analytical solutions because of the tremendous length and complexity of these expressions. We restrict our study to the simplest case of the homogeneous model only. In this case, the solutions of Equations (5.A4) can be expressed in the compact form

$$
\begin{aligned}
&\breve{\mathbf{g}}^{JE}(\rho'|\rho) = \tilde{\mu}_b\breve{\mathbf{g}}(\rho'|\rho)\\
&\breve{\mathbf{g}}^{JH}(\rho'|\rho) = \breve{\mathbf{g}}'(\rho'|\rho)\\
&\breve{\mathbf{g}}^{MH}(\rho'|\rho) = \tilde{\sigma}_b\breve{\mathbf{g}}(\rho'|\rho)\\
&\breve{\mathbf{g}}^{ME}(\rho'|\rho) = \breve{\mathbf{g}}'(\rho'|\rho)
\end{aligned}
, \quad \breve{\mathbf{g}}' = \nabla^* \times \breve{\mathbf{g}},
\tag{5.A5}
$$

where the function $\breve{\mathbf{g}}(\rho'|\rho)$ is generated by the diagonal Green's tensor function $\hat{\mathbf{g}}(\rho'|\rho)$ for the homogeneous full space:

$$
\breve{\mathbf{g}} = \hat{\mathbf{g}} + k^{-2}\nabla^*(\nabla^* \cdot \hat{\mathbf{g}}), \qquad k^2 = \tilde{\mu}_b\tilde{\sigma}_b \neq 0.
\tag{5.A6}
$$

The Green's tensor function $\hat{\mathbf{g}}(\rho'|\rho)$ satisfies the tensor Helmholtz equation:

$$
\nabla^{*2}\hat{\mathbf{g}} + k^2\hat{\mathbf{g}} = -\hat{\mathbf{I}}\frac{\delta(\rho-\rho')}{\rho}.
\tag{5.A7}
$$

In a cylindrical coordinate system it is determined by a differential matrix operator applied to the scalar function $g(\rho'|\rho)$:

$$
\hat{\mathbf{g}} = \begin{pmatrix}
-\frac{1}{\alpha^2}\frac{\partial^2}{\partial\rho\partial\rho'} & 0 & 0\\
0 & -\frac{1}{\alpha^2}\frac{\partial^2}{\partial\rho\partial\rho'} & 0\\
0 & 0 & 1
\end{pmatrix} g,
\tag{5.A8}
$$

$$
g = \begin{cases}
\mathbf{K}_0(\alpha\rho')\mathbf{I}_0(\alpha\rho), & \rho \le \rho'\\
\mathbf{I}_0(\alpha\rho')\mathbf{K}_0(\alpha\rho), & \rho > \rho'
\end{cases}, \qquad \alpha = \sqrt{\lambda^2 - k^2}, \quad \mathrm{Re}(\alpha) > 0.
\tag{5.A9}
$$

Here $\mathbf{I}_0$, $\mathbf{K}_0$ are the modified Bessel and MacDonald functions of the zero order.

Using Equations (5.A5), (5.A8) and (5.A9) and the similar notations as in Equations (5.5)–(5.9), we can calculate the spectra of the background fields,

$$\mathbf{e}^{\mathrm{b}} = \tilde{\mu}_{\mathrm{b}}\rho'\mathbf{i}_z \cdot \check{\mathbf{g}}', \qquad \mathbf{h}^{\mathrm{b}} = k^2\rho'\mathbf{i}_z \cdot \check{\mathbf{g}}, \tag{5.A10}$$

and the spectra of the Born approximations are

$$\mathbf{e}^{\mathrm{B}} = \mathbf{e}^{\mathrm{B}\sigma} + \mathbf{e}^{\mathrm{B}\mu} = \langle \tilde{\mu}_{\mathrm{b}}\check{\mathbf{g}}\tilde{\sigma}_{\mathrm{a}}\mathbf{e}^{\mathrm{b}}\rangle_{V_\sigma(\rho)} + \langle \check{\mathbf{g}}'\tilde{\mu}_{\mathrm{a}}\mathbf{h}^{\mathrm{b}}\rangle_{V_\mu(\rho)},$$

$$\mathbf{h}^{\mathrm{B}} = \mathbf{h}^{\mathrm{B}\sigma} + \mathbf{h}^{\mathrm{B}\mu} = \langle \check{\mathbf{g}}'\tilde{\sigma}_{\mathrm{a}}\mathbf{e}^{\mathrm{b}}\rangle_{V_\sigma(\rho)} + \langle \tilde{\sigma}_{\mathrm{b}}\check{\mathbf{g}}\tilde{\mu}_{\mathrm{a}}\mathbf{h}^{\mathrm{b}}\rangle_{V_\mu(\rho)}. \tag{5.A11}$$

The diagonal reflectivity tensors, $\hat{\lambda}^{\mathrm{e}}$, $\hat{\lambda}^{\mathrm{m}}$, are determined as the solutions of the linear system of equations obtained by the Fourier transform of Equations (5.22) in diagonal tensor form. In the case of an axial symmetrical model excited by the vertical magnetic dipole, the electrical reflectivity coefficient λ^{e} can be found from a scalar Equation (5.27). In particular, in the model with high-contrast conductivity and low-contrast magnetic susceptibility distributions (Figure 3), we can use the analytical expression for electrical reflectivity coefficient λ^{e}, based on Formula (5.33):

$$\lambda^{\mathrm{e}}(\rho) = \frac{e_\phi^{\mathrm{B}\sigma}(\rho)}{e_\phi^{\mathrm{b}}(\rho) - e_\phi^{\mathrm{B}\sigma}(\rho)}. \tag{5.A12}$$

Note that in this model we use the conventional Born approximation with respect to the magnetic anomaly and do not introduce the magnetic reflectivity coefficient λ^{m}. Using Equation (5.A12), we write

$$\left(1+\lambda^{\mathrm{e}}(\rho)\right)^{-1} = 1 - \frac{e_\phi^{B\sigma}(\rho)}{e_\phi^{\mathrm{b}}(\rho)}$$

$$= 1 - \tilde{\mu}_{\mathrm{b}}\tilde{\sigma}_{\mathrm{a}}\left(\int_{\rho_1}^{\rho} \mathbf{I}_1(\alpha\rho')\mathbf{K}_1(\alpha\rho')\rho'\,\mathrm{d}\rho' + \frac{\mathbf{I}_1(\alpha\rho)}{\mathbf{K}_1(\alpha\rho)}\int_{\rho}^{\rho_2} \mathbf{K}_1^2(\alpha\rho')\rho'\,\mathrm{d}\rho'\right), \tag{5.A13}$$

where $\alpha = \sqrt{\lambda^2 - k^2}$. The integrals in Formula (5.A13) can be evaluated in explicit form using the tabular integral (Gradstein and Ryzhik, 1994, p. 661, 5.54)

$$\int x\mathbf{U}_p(\alpha x)\mathbf{V}_p(\beta x)\,\mathrm{d}x = \frac{\beta x\mathbf{U}_p(\alpha x)\mathbf{V}_{p-1}(\beta x) - \alpha x\mathbf{U}_{p-1}(\alpha x)\mathbf{V}_p(\beta x)}{\alpha^2 - \beta^2},$$

where $\mathbf{U}_p(x)$and $\mathbf{V}_p(x)$ are arbitrary cylindrical functions of p order, and the interval $\rho' \in [\rho_1, \rho_2]$ is the thickness of the cylindrical layer with the anomalous conductivity $\tilde{\sigma}_{\mathrm{a}}$.

Finally, the vertical magnetic field H_z measured at the axis of symmetry at the point z_q is derived from the expressions

$$H_z = H_z^{\mathrm{b}} + \frac{1}{\pi}\int_0^\infty h_z(\lambda)\cos\lambda L\,\mathrm{d}\lambda,$$

$$H_z^{\mathrm{b}} = \frac{\mathrm{e}^{ikL}}{2\pi L^3}(1 - ikL),$$

$$h_z = \left\langle \mathbf{i}_z \cdot \hat{\mathbf{g}}'\tilde{\sigma}_{\mathrm{a}}(1+\lambda^{\mathrm{e}})\mathbf{e}^{\mathrm{b}}\right\rangle_{V_\sigma(\rho)} + \left\langle \mathbf{i}_z \cdot \tilde{\sigma}_{\mathrm{b}}\hat{\mathbf{g}}\tilde{\mu}_{\mathrm{a}}\mathbf{h}^{\mathrm{b}}\right\rangle_{V_\mu(\rho)}, \tag{5.A14}$$

where $L = |z_p - z_q|$ is the distance between the position of the transmitting magnetic dipole z_p and the position of the receiver z_q and k is the wave number of background homogeneous space.

REFERENCES

Augustin, A.M., Kennedy, W.D., Morrison, H.F. and Lee, K.H., 1989. A theoretical study of surface-to-borehole electromagnetic logging in cased holes. Geophysics, 54, 90–99.

Balasnis, C.A., 1989. Advanced Engineering Electromagnetics. Wiley New York, NY, 982 pp.

Cheryauka, A.B. and Sato, M., 1999. Nonlinear approximation for EM scattering problem in jointly inhomogeneous medium. Proc. 3DEM-2 Int. Symp., Salt Lake City, pp. 95–98.

Felsen, L.B. and Marcuvitz, N., 1994. Radiation and Scattering of Waves. Prentice-Hall, Englewood Cliffs, NJ, 888 pp.

Gradstein, I.S. and Ryzhik, I.M., 1994. In: A. Jeffery (Ed.), Table of Integrals, Series and Products. Academic Press, New York, NY, 1204 pp.

Habashy, T.M., Groom, R.W. and Spies, B.R., 1993. Beyond the Born and Rytov approximations: a nonlinear approach to electromagnetic scattering. J. Geophys. Res., 98(B2) 1759–1777.

Hohmann, G.W., 1975. Three-dimensional induced polarization and electromagnetic modeling. Geophysics, 40, 389–324.

Kaufman, A.A. and Keller, G., 1989. Induction Logging. Elsevier, Amsterdam, 685 pp.

Murray, I.R., Alvarez, I. and Groom, R.W., 1999. Modeling of complex electromagnetic targets using advanced non-linear approximator techniques. 69th SEG Annu. Int. Meet., Houston, TX, pp. 271–274.

Tabarovsky, L.A., 1975. Integral Equation Method in Geoelectrics Problems (in Russian). Nauka, Novosibirsk, 175 pp.

Tai, C.T., 1979. Dyadic Green's Functions in Electromagnetic Theory. Intext Educational Publishers, Scranton, PA, 235 pp.

Torres-Verdin, C. and Habashy, T.M., 1994. Rapid 2.5-dimensional forward modeling and inversion via a new nonlinear scattering approximation. Radio Sci., 29(4), 1084–1079.

Weidelt, P., 1975. Electromagnetic induction in three-dimensional structures. Geophysics, 41, 85–109.

Zhdanov, M.S. and Fang, S., 1996. Quasi-linear approximation in 3-D electromagnetic modeling. Geophysics, 61, 646–665.

Zhdanov, M.S. and Fang, S., 1997. Quasi-linear series in three-dimensional electromagnetic modeling. Radio Sci., 31(6), 2167–2188.

Zhdanov, M.S., Dmitriev, V.I., Fang, S. and Hursan, G., 2000. Quasi-analytical approximations and series in electromagnetic modeling. Geophysics, 65, 1746–1757.

Chapter 6

THREE-DIMENSIONAL MODELING CONSIDERING THE TOPOGRAPHY FOR THE CASE OF THE TIME-DOMAIN ELECTROMAGNETIC METHOD

Masashi Endo [a] and Koji Noguchi [b]

[a] *Sumitomo Metal Mining Co., Ltd., Tokyo, Japan*
[b] *School of Science and Engineering, Waseda University, Tokyo, Japan*

Abstract: A finite-difference, time-domain solution to Maxwell's equation has been developed for a three-dimensional subsurface and topography. We use the staggered-grid scheme combined with a modified version of the DuFort–Frankel scheme to discretize the quasi-static Maxwell's equations. The algorithm allows not only arbitrary electrical conductivity variations but also arbitrary topography within a model. We use the boundary-fitted coordinate method to calculate the response of the topographic model. This method is used popularly in the field of fluid dynamics, and also in the field of electromagnetic propagation. In this method, calculation is accomplished in the computational domain, which is transformed from the physical domain by using coordinate-transformation coefficients. Numerical checks against the analytical solution for a homogeneous half-space, the numerical solution for a 3-D model, and an inclined-surface model showed that the solution provides accurate results. The TDEM responses of topographic models are shown as numerical examples, and it is indicated that the three-dimensional modeling considering the topography is required to interpret the data obtained in mountainous areas.

1. INTRODUCTION

Recently, the time-domain electromagnetic method (TDEM) has been widely used for the exploration of not only mineral resources such as metals, petroleum, and geothermal ones, but also for civil, base rock and underground water research for the prevention of disasters etc. Because these explorations are often carried out in mountainous regions, the topographic effect is considered not to be negligible. Also, the target of the explorations is often complicated and/or increasingly buried. Therefore, high quality measurement equipment has been developed to acquire weaker responses with high accuracy. In these circumstances, the modeling technique has to be able to represent arbitrary three-dimensional subsurface and topography with high accuracy to increase reliability of the exploration method.

To solve these problems, the authors have developed a finite-difference, time-domain solution to Maxwell's equation for three-dimensional subsurface, based on Wang and Hohmann (1993), and topography. We use a modified version of Yee's (1966) staggered-grid scheme combined with a modified version of the DuFort and Frankel (1953) scheme to discretize the quasi-static Maxwell's equations. And the boundary-fitted coordinate method is used to represent the topography. This method is used popularly in the field of fluid dynamics, and also in the field of electromagnetic propagation. In this method, calculation is accomplished in the computational domain, which is transformed from the physical domain by using coordinate-transformation coefficients.

2. THEORY

2.1. Governing equations

Like Wang and Hohmann (1993), we also use the following equations as governing equations for TDEM fields under the quasi-static approximation,

$$-\frac{\partial \mathbf{b}(\mathbf{r},t)}{\partial t} = \nabla \times \mathbf{e}(\mathbf{r},t), \tag{6.1}$$

$$\gamma \frac{\partial \mathbf{e}(\mathbf{r},t)}{\partial t} + \sigma(\mathbf{r})\mathbf{e}(\mathbf{r},t) = \frac{1}{\mu}\nabla \times \mathbf{b}(\mathbf{r},t), \tag{6.2}$$

$$\nabla \cdot \mathbf{b}(\mathbf{r},t) = 0, \tag{6.3}$$

$$\nabla \cdot \mathbf{j}(\mathbf{r},t) = 0, \tag{6.4}$$

where $\mathbf{b}(\mathbf{r},t)$ is the magnetic field, $\mathbf{e}(\mathbf{r},t)$ is the electric field, $\mathbf{j}(\mathbf{r},t)$ is the current density, μ is the magnetic permeability, $\sigma(r)$ is the electric conductivity, and γ is an artificial coefficient related to the phase velocity of the wave-like field.

2.2. Modeling technique for the topography

In this study, we use the boundary-fitted coordinate method to calculate the response of the topographic model. This method is used popularly in the field of fluid dynamics, and in the field of electromagnetic propagation. In this method, the physical domain, which is the real space for calculation, is discretized by the boundary-fitted grids (non-rectangular grids). The governing equations are solved numerically in the computational domain that is discretized by rectangular grids. The computational domain is related to the physical domain by coordinate-transformation coefficients. These coefficients are described in Appendix A.

2.3. Model discretization

The electric fields and the magnetic fields are discretized by using the staggered-grid shown in Figure 1. Electromagnetic fields are defined in the same direction as each axis of the physical domain. In both the physical and the computational domain, the electric field is sampled at the center of the prism edge, and the magnetic field is sampled at the center of the prism face. The problem of discontinuity of the electric field at the conductivity boundary can be avoided by using this grid. Correspondence of grids between the physical domain and the computational domain (Figure 2) is defined as follows: (1) the horizontal coordinates of grids in both domains are set to be equal

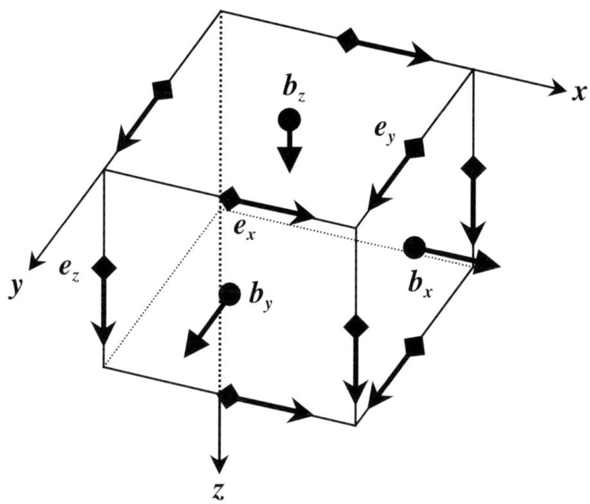

Figure 1. A staggered-grid. The electric field is sampled at the centers of the prism edges, and the magnetic field is sampled at the center of the prism faces.

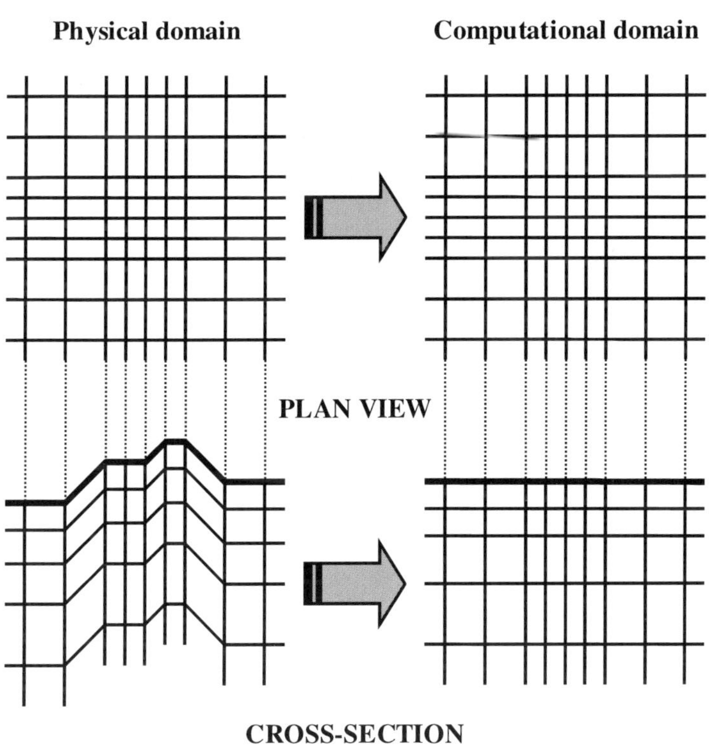

Figure 2. Correspondence of grids between the physical domain and the computational domain.

(horizontal grid spacings in both domains are equal to each other); (2) the vertical grid spacings in both domains are set to be equal (vertical coordinates of grids in both domains are not always equal to each other).

Thus, coordinate transformation coefficients at $(\xi(i), \eta(j), \zeta(k))$ in the computational domain are given as follows,

$$\begin{bmatrix} \dfrac{\partial x}{\partial \xi} & \dfrac{\partial y}{\partial \xi} & \dfrac{\partial z}{\partial \xi} \\ \dfrac{\partial x}{\partial \eta} & \dfrac{\partial y}{\partial \eta} & \dfrac{\partial z}{\partial \eta} \\ \dfrac{\partial x}{\partial \zeta} & \dfrac{\partial y}{\partial \zeta} & \dfrac{\partial z}{\partial \zeta} \end{bmatrix} = \begin{bmatrix} 1 & 0 & \dfrac{z(i+1,j,k)-z(i-1,j,k)}{\xi(i+1)-\xi(i-1)} \\ 0 & 1 & \dfrac{z(i,j+1,k)-z(i,j-1,k)}{\eta(j+1)-\eta(j-1)} \\ 0 & 0 & 1 \end{bmatrix}. \tag{6.5}$$

Following a time-staggering scheme (Yee, 1966), the electric field is defined at the integer indices of the time axis, and the magnetic field is defined at the center of the integer indices.

2.4. Finite-difference equations

In Appendix B, we describe derivation of the finite-difference equations for e_x, b_y, and b_z. Equations for other electromagnetic components can be derived in the same way. The notation $e_x^n(i,j,k)$ indicates the x-component electric field at $t = t_n$, and the other components are indicated in the same way.

2.5. Stability and reliability

The phase velocity of the wave-like field defined by Equation (6.2) can be written,

$$v_\mathrm{p} = \frac{1}{\sqrt{\mu\gamma}}. \tag{6.6}$$

The Courant–Friedrichs–Levy condition for a three-dimensional wave equation is given in the following inequality,

$$v_\mathrm{p} \leq \frac{\Delta}{\sqrt{3}\Delta t}, \tag{6.7}$$

where Δ is the grid spacing and Δt is the time step. Rearranging Equation (6.7) by substituting Equation (6.6), we obtain

$$\gamma \geq \frac{3}{\mu}\left(\frac{\Delta t}{\Delta}\right)^2. \tag{6.8}$$

This inequality is the necessary condition for the stability of the finite-difference equation. For the case that the grid spacing is not constant, we use the following condition,

$$\gamma \geq \frac{3}{\mu}\left(\frac{\Delta t}{\Delta_\mathrm{min}}\right)^2, \tag{6.9}$$

where Δ_min is the minimum grid spacing.

The term in Equation (6.2) acts as a displacement current. But its magnitude can be much larger than the real displacement current. To prevent the fictitious displacement current from being dominant over the diffusive EM field, that is to say, to let the solution for Equation (6.2) approximate sufficiently the real Maxwell equation, the following condition for the time-stepping is required (Oristaglio and Hohmann, 1984; Adhidjaja and Hohmann, 1989),

$$\Delta t \ll \left(\frac{\mu\sigma_{\min} t}{6}\right)^{1/2} \cdot \Delta_{\min}, \qquad (6.10)$$

where $\sigma_{\min}$ can be taken as the minimum conductivity in the model.

2.6. Boundary conditions

2.6.1. Air–earth boundary

The finite-difference equations (Appendix B) show that the EM fields in free space are required to calculate the EM field at the surface of the earth. But the time-step would be 0 by Equation (6.10), if the air were taken into account as the model. To avoid the problem, Oristaglio and Hohmann (1984) have implemented the upward-continuation boundary condition. Because this boundary condition is derived from the assumption that the surface is flat, we use the approximation based on the upward-continuation boundary condition for the air–earth boundary condition for the topographic model.

We assume that the upward-continuation boundary condition is established in the computational domain defined by the Cartesian coordinates. That is to say, we assume that the magnetic field in free space in the computational domain satisfies the following Laplacian equation,

$$\nabla^2 \mathbf{b} = 0. \qquad (6.11)$$

The horizontal components of **b** satisfying Equation (6.11) can be derived from its vertical component on the same horizontal plane (Nabighian, 1972, 1984; Macnae, 1984). Thus, we obtain,

$$B_\xi(u,v,\zeta=0) = -\frac{iu}{\sqrt{u^2+v^2}} B_\zeta(u,v,\zeta=0), \qquad (6.12)$$

$$B_\eta(u,v,\zeta=0) = -\frac{iv}{\sqrt{u^2+v^2}} B_\zeta(u,v,\zeta=0), \qquad (6.13)$$

where **B** is the Fourier transform of **b**; u and v are the wavenumber domain variables corresponding to ξ and η, respectively.

B_ξ and B_η can be upward continued to give their values in free-space,

$$B_\xi(u,v,\zeta=-h) = -\frac{iu}{\sqrt{u^2+v^2}} \exp\left(-h\sqrt{u^2+v^2}\right) B_\zeta(u,v,\zeta=0), \qquad (6.14)$$

$$B_\eta(u,v,\zeta=-h) = -\frac{iv}{\sqrt{u^2+v^2}} \exp\left(-h\sqrt{u^2+v^2}\right) B_\zeta(u,v,\zeta=0). \qquad (6.15)$$

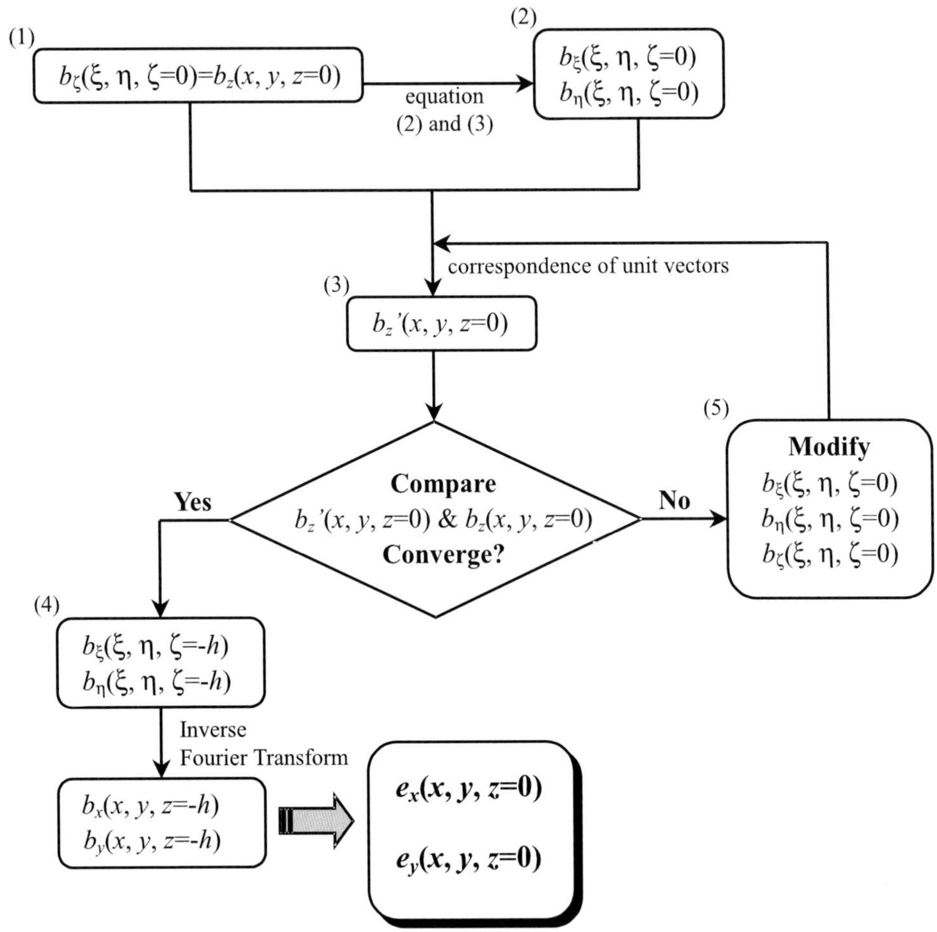

Figure 3. Calculation flow for approximation for the upward-continuation boundary condition at the air–earth boundary for a topographic model.

Solutions in computational domain are acquired by using inverse Fourier transform to solutions of Equations (6.14) and (6.15).

These solutions are established only for the case that the surface is flat or inclines equally to all directions in the Cartesian coordinate system. We use an approximation as shown in Figure 3 for air–earth boundary condition in arbitrary topographic model. Implementation of the approximation is as follows.

(1) The vertical magnetic field at the surface in the physical domain (known value), $b_z(x, y, z = 0)$, is used as the initial value of the unknown vertical magnetic field at the surface in computational domain $b_\zeta(\xi, \eta, \zeta = 0)$.

(2) Assuming that Equation (6.11) is established in the computational domain, horizontal magnetic fields $b_\xi(\xi, \eta, \zeta = 0)$ and $b_\eta(\xi, \eta, \zeta = 0)$ are calculated by using Equations (6.12), (6.13) and the inverse Fourier transform.

(3) Substitute magnetic fields at the surface in the computational domain (b_ξ, b_η, b_ζ) into unit vectors ($\mathbf{n}^{(\xi)}$, $\mathbf{n}^{(\eta)}$, $\mathbf{n}^{(\zeta)}$) in the following system equations according to unit vectors in the physical and computational domains and assume solution for b_ζ to the vertical magnetic field at the surface in the physical domain (denoted as b'_z),

$$\mathbf{n}^{(\xi)} = \frac{\nabla\xi}{|\nabla\xi|} = \frac{\frac{\partial\xi}{\partial x}\mathbf{i} + \frac{\partial\xi}{\partial y}\mathbf{j} + \frac{\partial\xi}{\partial z}\mathbf{k}}{\sqrt{\left(\frac{\partial\xi}{\partial x}\right)^2 + \left(\frac{\partial\xi}{\partial y}\right)^2 + \left(\frac{\partial\xi}{\partial z}\right)^2}}, \tag{6.16}$$

$$\mathbf{n}^{(\eta)} = \frac{\nabla\eta}{|\nabla\eta|} = \frac{\frac{\partial\eta}{\partial x}\mathbf{i} + \frac{\partial\eta}{\partial y}\mathbf{j} + \frac{\partial\eta}{\partial z}\mathbf{k}}{\sqrt{\left(\frac{\partial\eta}{\partial x}\right)^2 + \left(\frac{\partial\eta}{\partial y}\right)^2 + \left(\frac{\partial\eta}{\partial z}\right)^2}}, \tag{6.17}$$

$$\mathbf{n}^{(\zeta)} = \frac{\nabla\zeta}{|\nabla\zeta|} = \frac{\frac{\partial\zeta}{\partial x}\mathbf{i} + \frac{\partial\zeta}{\partial y}\mathbf{j} + \frac{\partial\zeta}{\partial z}\mathbf{k}}{\sqrt{\left(\frac{\partial\zeta}{\partial x}\right)^2 + \left(\frac{\partial\zeta}{\partial y}\right)^2 + \left(\frac{\partial\zeta}{\partial z}\right)^2}}. \tag{6.18}$$

(4) For the case that the difference between b_z and b'_z is negligible (less than 10^{-20}), the magnetic field in free-space (physical domain) is calculated by using Equations (6.16), (6.17), (6.18) and the inverse Fourier transform.

(5) For the case that the difference between b_z and b'_z is not negligible, all components of the magnetic field are modified by using the ratio of coefficients for each component in the following equation. Then, return to process (3).

$$\begin{aligned} k = \Bigg\{ & \left(\frac{\partial\zeta}{\partial x}\frac{\partial\eta}{\partial y} - \frac{\partial\eta}{\partial x}\frac{\partial\zeta}{\partial y}\right)\sqrt{\left(\frac{\partial\xi}{\partial x}\right)^2 + \left(\frac{\partial\xi}{\partial y}\right)^2 + \left(\frac{\partial\xi}{\partial z}\right)^2} \cdot \mathbf{n}^{(\xi)} \\ & + \left(\frac{\partial\xi}{\partial x}\frac{\partial\zeta}{\partial y} - \frac{\partial\zeta}{\partial x}\frac{\partial\xi}{\partial y}\right)\sqrt{\left(\frac{\partial\eta}{\partial x}\right)^2 + \left(\frac{\partial\eta}{\partial y}\right)^2 + \left(\frac{\partial\eta}{\partial z}\right)^2} \cdot \mathbf{n}^{(\eta)} \\ & + \left(\frac{\partial\xi}{\partial x}\frac{\partial\eta}{\partial y} - \frac{\partial\eta}{\partial x}\frac{\partial\xi}{\partial y}\right)\sqrt{\left(\frac{\partial\zeta}{\partial x}\right)^2 + \left(\frac{\partial\zeta}{\partial y}\right)^2 + \left(\frac{\partial\zeta}{\partial z}\right)^2} \cdot \mathbf{n}^{(\eta)} \Bigg\} \\ \Bigg/ \Bigg\{ & \left(\frac{\partial\eta}{\partial y}\frac{\partial\xi}{\partial z} - \frac{\partial\xi}{\partial y}\frac{\partial\eta}{\partial z}\right) \cdot \frac{\partial\zeta}{\partial x} + \left(\frac{\partial\eta}{\partial z}\frac{\partial\xi}{\partial x} - \frac{\partial\xi}{\partial z}\frac{\partial\eta}{\partial x}\right) \cdot \frac{\partial\zeta}{\partial y} + \left(\frac{\partial\eta}{\partial x}\frac{\partial\xi}{\partial y} - \frac{\partial\xi}{\partial x}\frac{\partial\eta}{\partial y}\right) \cdot \frac{\partial\zeta}{\partial z} \Bigg\}. \end{aligned} \tag{6.19}$$

2.6.2. Subsurface boundary

On the subsurface boundaries, we impose the homogeneous Dirichlet condition, i.e., the tangential electric fields on the subsurface boundaries are set to zero.

2.7. Initial condition

Commonly in a TDEM survey, the electromotive force (EMF) induced in the receiver coil is observed as the response excited by the transmitter current of which the wave form is the step function (step response). Thus, the EMF is equal to the time-derivative of the magnetic field (step response). On the other hand, the EMF is equal to the magnetic field (impulse response).

In this study, we have calculated the impulse response of **e** and **b**, and incorporated the impulse response of **e** as the initial condition. The wave form of the transmitter current is approximated by the Gaussian pulse defined as follows,

$$I(t) = \frac{1}{\sqrt{2\pi}a}\exp\left[-\frac{t^2}{2a^2}\right], \qquad (6.20)$$

where $I(t)$ is the transmitter current and a^2 is the variance.

3. NUMERICAL CHECKS

3.1. Homogeneous half-space

Consider a 100 m × 100 m loop source on a homogeneous half-space of 100 Ω m. The calculated responses at the center of the loop are shown in Figure 4. The vertical electromotive forces (EMF) correspond to the magnetic field measured with the horizontal unit coil. The calculated responses are in good agreement with the analytical responses (Ward and Hohmann, 1988). In this model, the execution time is approximately 50 min on an Intel Pentium II 300 MHz CPU.

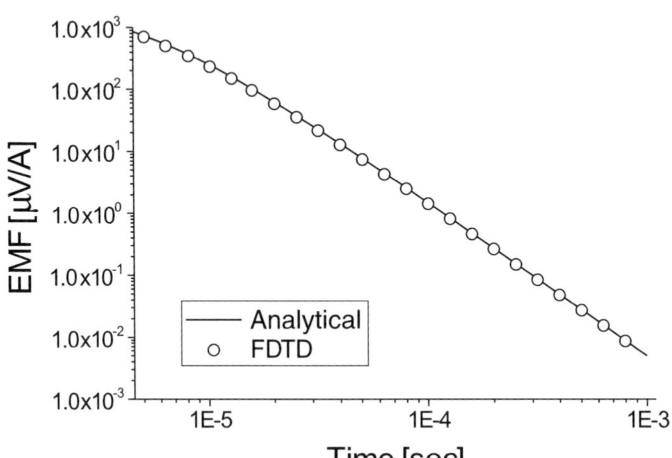

Figure 4. Comparison of analytical (Ward and Hohmann, 1988) and numerical solutions for vertical EMF at the center of the transmitter loop on a 100 Ω m homogeneous half-space.

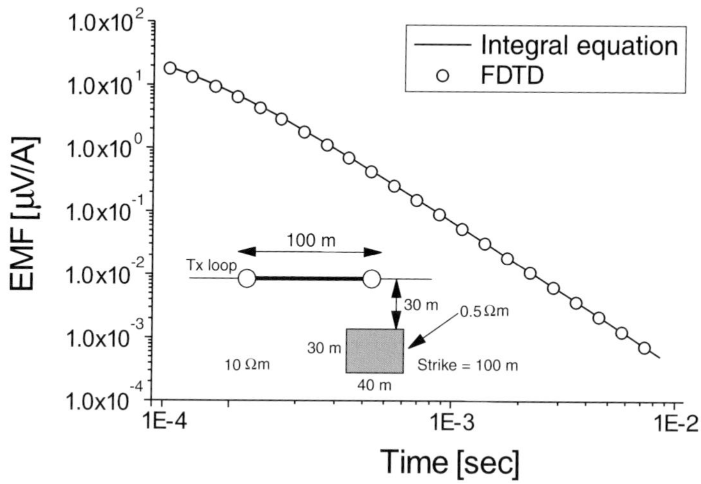

Figure 5. Comparison of the FDTD and integral-equation (Noguchi and Endo, 1995, 1996) solution for a central-loop survey over a 3-D conductor. The resistivity contrast is 20.

3.2. 3-D body in a homogeneous half-space

Wang and Hohmann (1993) computed the transient response of a 3-D model shown in Figure 5, using the FDTD technique. The 0.5 Ω m body is 100 m long, 40 m wide, and is 30 m deep. It is embedded in a 10 Ω m homogeneous half-space and excited by a 100 m $\times$ 100 m loop source on the surface. Figure 5 also shows the central-loop soundings from both the integral equation (Noguchi and Endo, 1995, 1996) and our solution. The two solutions are in good agreement with each other.

3.3. Inclined-surface model

To investigate the accuracy of the boundary-fitted coordinate system, we computed the model shown in Figure 6. In this model, consider a line source of 100 m (y-direction) on a homogeneous half-space of 100 Ω m and the angles of inclination θ are: (a) 26.6, (b) 31.0, (c) 38.7, (d) 45.0, (e) 50.2, and (f) 54.5 degrees. For the case of these models, the responses are acquired by the resultant vector of the responses of the flat-surface model. Figure 7 shows the responses (100 m apart in the x-direction from the center of the source) from both the solutions using the resultant vector and the solutions using the boundary-fitted coordinate method. The two solutions demonstrate an excellent agreement with each other for the case that the angle of inclination is below 45 degrees.

4. NUMERICAL EXAMPLES

Calculation models are shown in Figure 8. In all models, consider a 40 m $\times$ 40 m loop source on a homogeneous half-space of 100 Ω m. All receivers are located outside of

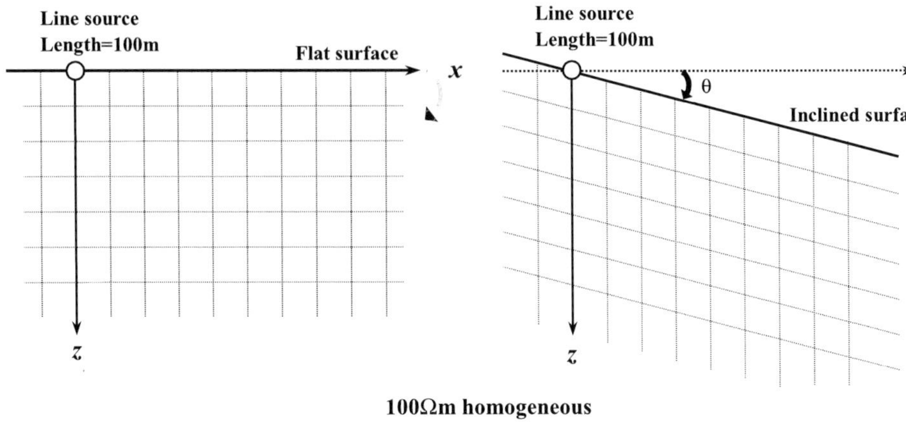

Figure 6. Flat-surface and inclined-surface models for investigation of the accuracy of the boundary-fitted coordinate method.

the transmitter loop. We have investigated the topographic effect by using the difference of the responses for different models. Using the relative error for the investigation causes the numerical dispersion because the response changes its sign for the case that a receiver is located outside of the transmitter loop.

Differences of the response between a 3-D valley model (Figure 8b; $\theta = 45$ degrees) and a flat-surface model are shown in Figure 9.

At all times, the difference of the response remarkably appears in the district to consider the point $(x, y) = (100, 0)$ as the center for the case of the x-component (Figure 9a–c). In addition, at a later time (795 μs; Figure 9c), the difference appears at the district where the surface inclines to the x-axis direction.

For the case of the z-component (Figure 9d–f), the same as in the case of the x-component, the difference of the response remarkably appears in the district to consider the point $(x, y) = (100, 0)$ as the center. But at a later time (795 μs; Figure 9f), the difference also appears at the district where the surface inclines to the y-axis direction.

Differences of responses at the surface between 3-D valley (Figure 8b; $\theta = 45$ degrees) and 2-D valley (Figure 8a; $\theta = 45$ degrees) models are shown in Figure 10.

Though the topography is exactly the same at $y = 0$ m in both models, the difference of the response remarkably appears near the line $x = 100$ m at an earlier time (8.07 and 79.5 μs), and at $500 \le x \le 700$ m at a later time (795 μs).

As mentioned above, the topographic effect on each component response appears in a different way by the direction of the inclination of surface. A 3-D topographic effect is shown, even if the topography is considered to be locally 2-D. Therefore, the surface topography has to be modeled in 3-D in practice.

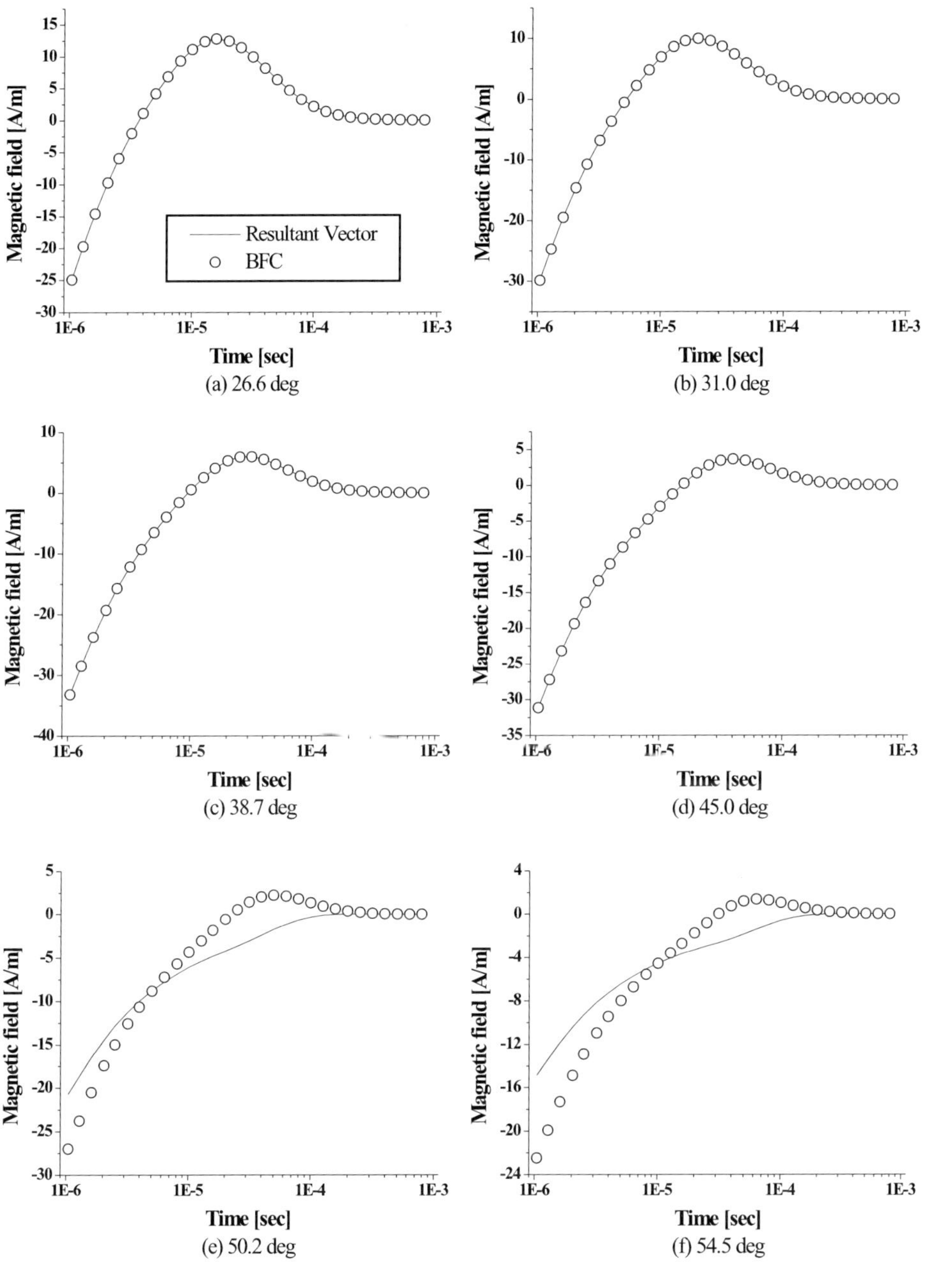

Figure 7. Comparison of the vertical magnetic field (impulse response) for inclined surface models to the field for a flat surface model.

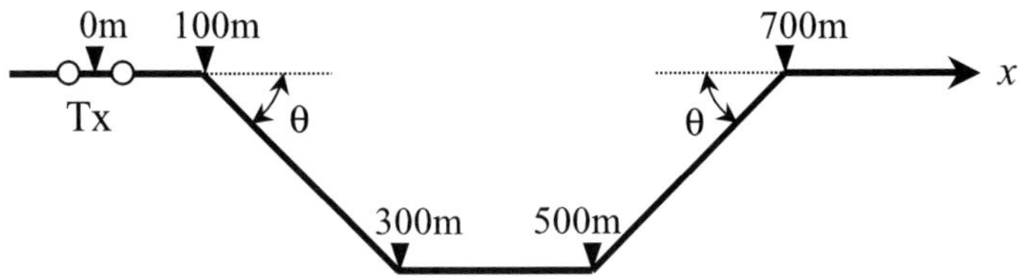

(a) 2-D topographic model

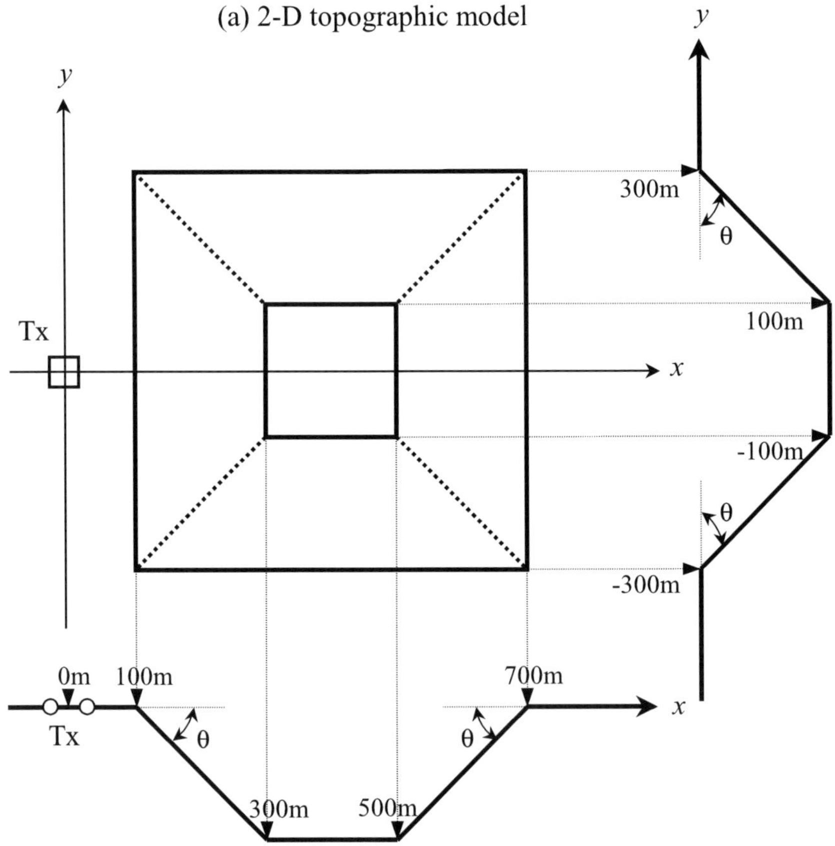

(b) 3-D topographic model

Figure 8. 2-D and 3-D topographic models. The source is a square (40 x 40 m), the angle of inclination (θ) is 45 degrees.

5. CONCLUSIONS

We have developed a modeling technique that allows not only arbitrary subsurface structure but also arbitrary topography by adapting the boundary-fitted coordinate method to the FDTD method.

The execution time for a typical model is about 50 min on an Intel Pentium II 300

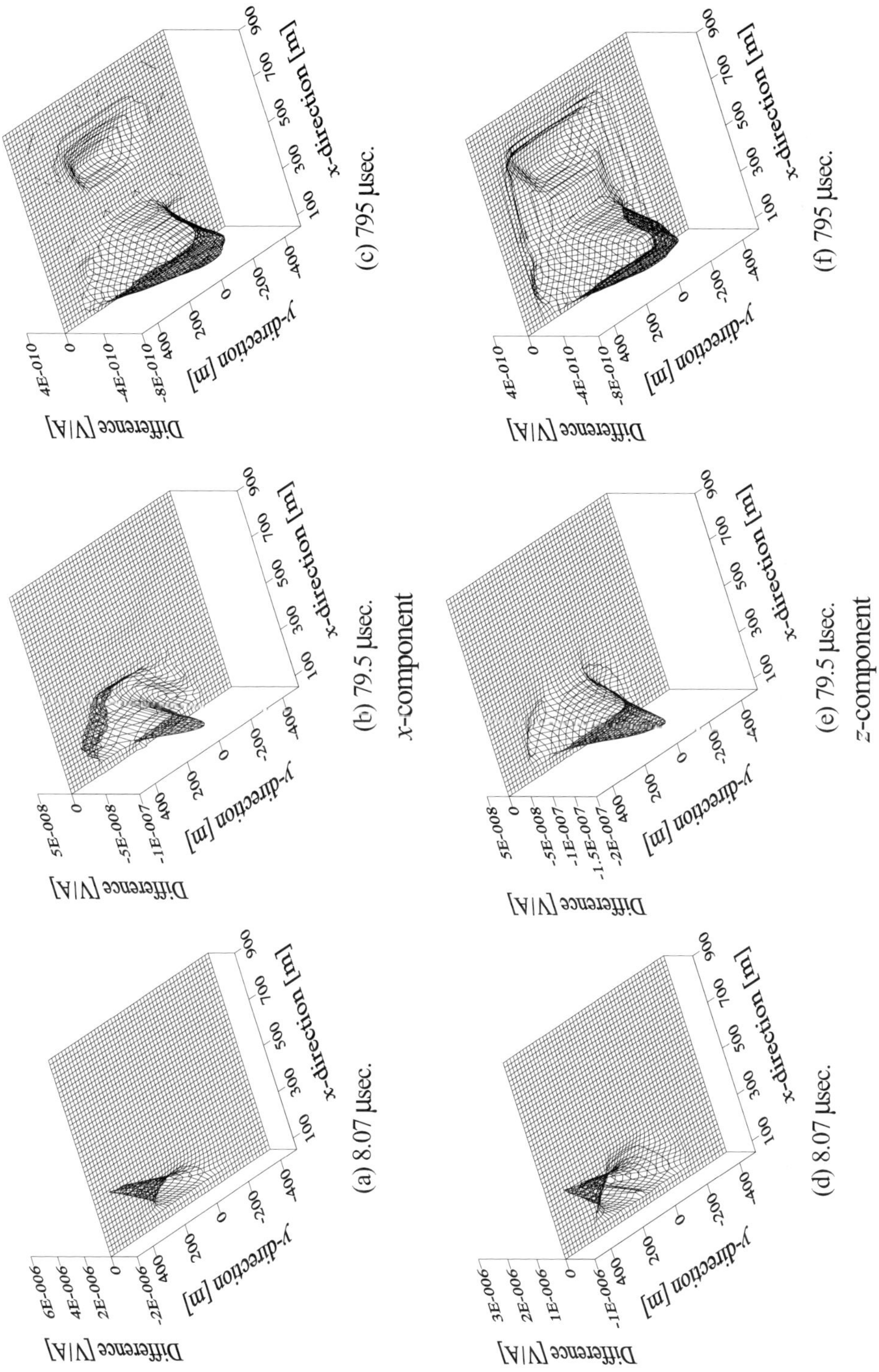

Figure 9. Difference of the EMF between 3-D valley and flat-surface models.

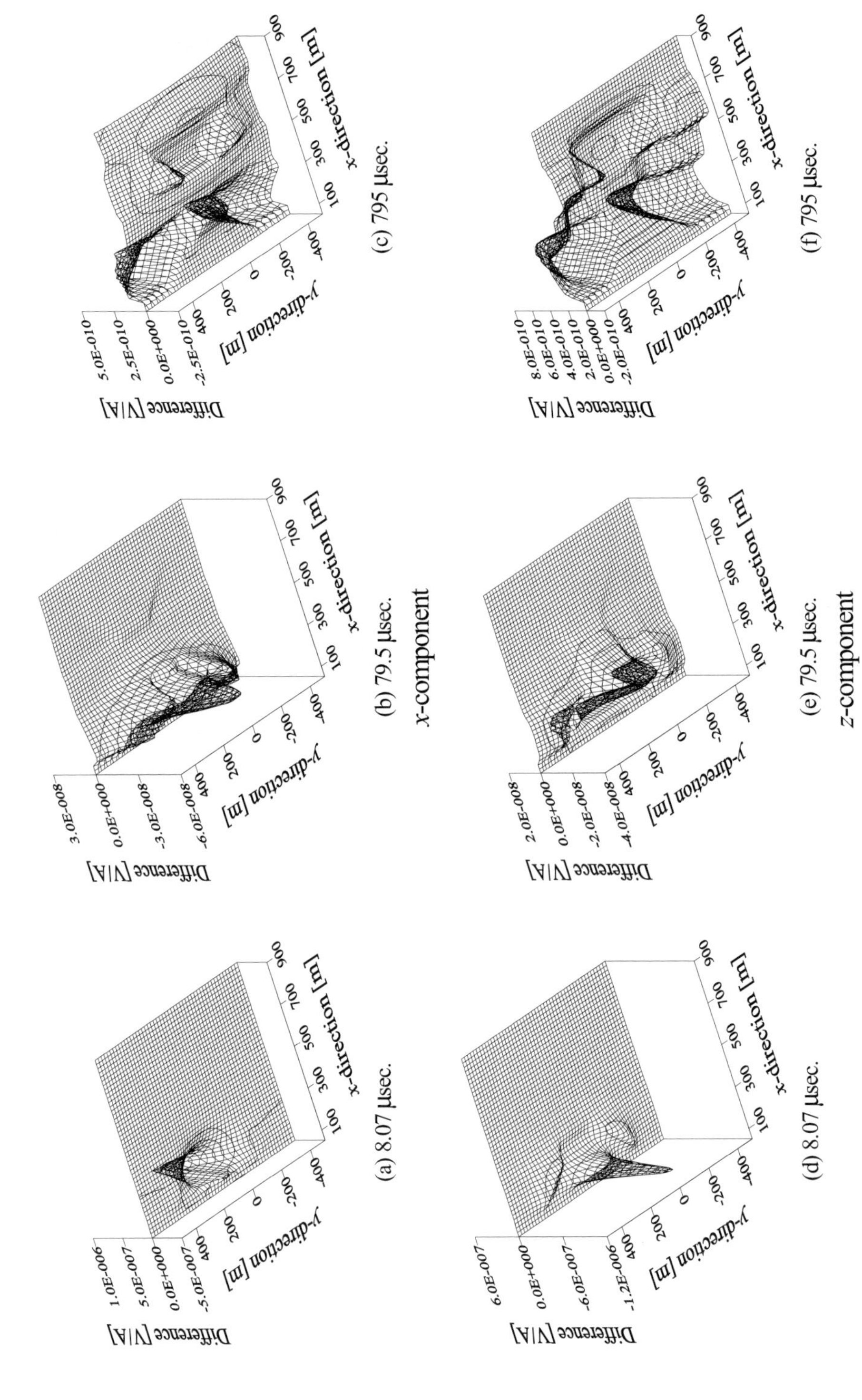

Figure 10. Difference of the EMF between 3-D valley and 2-D valley models.

MHz CPU to complete the field to 1 ms. The model contains $81 \times 81 \times 40$ grid points, with the smallest grid spacing of 10 m and the highest resistivity of 100 Ω m.

The TDEM responses of topographic models are shown as numerical examples. Finally, the topographic effect varies in a complicated manner using the model parameters so that the three-dimensional modeling considering the topography plays an important role in interpreting the data obtained in mountainous areas.

Appendix A. CORRESPONDENCE OF THE DIFFERENTIAL COEFFICIENTS BETWEEN THE PHYSICAL DOMAIN AND THE COMPUTATIONAL DOMAIN

Consider the coordinate transform from the physical domain defined by (x, y, z) to the computational domain defined by (ξ, η, ζ). Relationships of coordinates are given as follows,

$$\begin{cases} x = x(\xi, \eta, \zeta) \\ y = y(\xi, \eta, \zeta) \\ z = z(\xi, \eta, \zeta). \end{cases} \tag{6.A1}$$

Governing equations that are established in the physical domain have to be transformed to those in the computational domain. The governing equations are partial differential equations, so that the correspondence of the differential coefficients is required.

Following the chain rules, the equations are established as follows,

$$\begin{pmatrix} \frac{\partial}{\partial \xi} \\ \frac{\partial}{\partial \eta} \\ \frac{\partial}{\partial \zeta} \end{pmatrix} = \begin{bmatrix} \frac{\partial x}{\partial \xi} & \frac{\partial y}{\partial \xi} & \frac{\partial z}{\partial \xi} \\ \frac{\partial x}{\partial \eta} & \frac{\partial y}{\partial \eta} & \frac{\partial z}{\partial \eta} \\ \frac{\partial x}{\partial \zeta} & \frac{\partial y}{\partial \zeta} & \frac{\partial z}{\partial \zeta} \end{bmatrix} \cdot \begin{pmatrix} \frac{\partial}{\partial x} \\ \frac{\partial}{\partial y} \\ \frac{\partial}{\partial z} \end{pmatrix}, \tag{6.A2}$$

$$\begin{pmatrix} \frac{\partial}{\partial x} \\ \frac{\partial}{\partial y} \\ \frac{\partial}{\partial z} \end{pmatrix} = \begin{bmatrix} \frac{\partial \xi}{\partial x} & \frac{\partial \eta}{\partial x} & \frac{\partial \zeta}{\partial x} \\ \frac{\partial \xi}{\partial y} & \frac{\partial \eta}{\partial y} & \frac{\partial \zeta}{\partial y} \\ \frac{\partial \xi}{\partial z} & \frac{\partial \eta}{\partial z} & \frac{\partial \zeta}{\partial z} \end{bmatrix} \cdot \begin{pmatrix} \frac{\partial}{\partial \xi} \\ \frac{\partial}{\partial \eta} \\ \frac{\partial}{\partial \zeta} \end{pmatrix}, \tag{6.A3}$$

and using the next equation,

$$\begin{bmatrix} \frac{\partial x}{\partial \xi} & \frac{\partial y}{\partial \xi} & \frac{\partial z}{\partial \xi} \\ \frac{\partial x}{\partial \eta} & \frac{\partial y}{\partial \eta} & \frac{\partial z}{\partial \eta} \\ \frac{\partial x}{\partial \zeta} & \frac{\partial y}{\partial \zeta} & \frac{\partial z}{\partial \zeta} \end{bmatrix} \cdot \begin{bmatrix} \frac{\partial \xi}{\partial x} & \frac{\partial \eta}{\partial x} & \frac{\partial \zeta}{\partial x} \\ \frac{\partial \xi}{\partial y} & \frac{\partial \eta}{\partial y} & \frac{\partial \zeta}{\partial y} \\ \frac{\partial \xi}{\partial z} & \frac{\partial \eta}{\partial z} & \frac{\partial \zeta}{\partial z} \end{bmatrix} = \begin{bmatrix} 1 & 0 & 0 \\ 0 & 1 & 0 \\ 0 & 0 & 1 \end{bmatrix}, \tag{6.A4}$$

the coefficients for the coordinate transform are derived as follows,

$$\begin{cases} \dfrac{\partial \xi}{\partial x} = \left(\dfrac{\partial y}{\partial \eta}\dfrac{\partial z}{\partial \zeta} - \dfrac{\partial y}{\partial \zeta}\dfrac{\partial z}{\partial \eta}\right) \Big/ J \\ \dfrac{\partial \eta}{\partial x} = \left(\dfrac{\partial z}{\partial \xi}\dfrac{\partial y}{\partial \zeta} - \dfrac{\partial y}{\partial \xi}\dfrac{\partial z}{\partial \zeta}\right) \Big/ J \\ \dfrac{\partial \zeta}{\partial x} = \left(\dfrac{\partial y}{\partial \xi}\dfrac{\partial z}{\partial \eta} - \dfrac{\partial z}{\partial \xi}\dfrac{\partial y}{\partial \eta}\right) \Big/ J, \end{cases} \tag{6.A5}$$

$$\begin{cases} \dfrac{\partial \xi}{\partial y} = \left(\dfrac{\partial z}{\partial \eta}\dfrac{\partial x}{\partial \zeta} - \dfrac{\partial x}{\partial \eta}\dfrac{\partial z}{\partial \zeta}\right) \Big/ J \\ \dfrac{\partial \eta}{\partial y} = \left(\dfrac{\partial x}{\partial \xi}\dfrac{\partial z}{\partial \zeta} - \dfrac{\partial x}{\partial \zeta}\dfrac{\partial z}{\partial \xi}\right) \Big/ J \\ \dfrac{\partial \zeta}{\partial y} = \left(\dfrac{\partial x}{\partial \eta}\dfrac{\partial z}{\partial \xi} - \dfrac{\partial x}{\partial \xi}\dfrac{\partial z}{\partial \eta}\right) \Big/ J, \end{cases} \tag{6.A6}$$

$$\begin{cases} \dfrac{\partial \xi}{\partial z} = \left(\dfrac{\partial x}{\partial \eta}\dfrac{\partial y}{\partial \zeta} - \dfrac{\partial y}{\partial \eta}\dfrac{\partial x}{\partial \zeta}\right) \Big/ J \\ \dfrac{\partial \eta}{\partial z} = \left(\dfrac{\partial y}{\partial \xi}\dfrac{\partial x}{\partial \zeta} - \dfrac{\partial x}{\partial \xi}\dfrac{\partial y}{\partial \zeta}\right) \Big/ J \\ \dfrac{\partial \zeta}{\partial z} = \left(\dfrac{\partial x}{\partial \xi}\dfrac{\partial y}{\partial \eta} - \dfrac{\partial y}{\partial \xi}\dfrac{\partial x}{\partial \eta}\right) \Big/ J, \end{cases} \tag{6.A7}$$

where J is the jacobian of the 3-D transform, which is defined as follows,

$$J = \begin{vmatrix} \dfrac{\partial x}{\partial \xi} & \dfrac{\partial x}{\partial \eta} & \dfrac{\partial x}{\partial \zeta} \\ \dfrac{\partial y}{\partial \xi} & \dfrac{\partial y}{\partial \eta} & \dfrac{\partial y}{\partial \zeta} \\ \dfrac{\partial z}{\partial \xi} & \dfrac{\partial z}{\partial \eta} & \dfrac{\partial z}{\partial \zeta} \end{vmatrix}$$

$$= \frac{\partial x}{\partial \xi}\frac{\partial y}{\partial \eta}\frac{\partial z}{\partial \zeta} + \frac{\partial x}{\partial \eta}\frac{\partial y}{\partial \zeta}\frac{\partial z}{\partial \xi} + \frac{\partial x}{\partial \zeta}\frac{\partial y}{\partial \xi}\frac{\partial z}{\partial \eta} - \frac{\partial x}{\partial \xi}\frac{\partial y}{\partial \zeta}\frac{\partial z}{\partial \eta} - \frac{\partial x}{\partial \eta}\frac{\partial y}{\partial \xi}\frac{\partial z}{\partial \zeta} - \frac{\partial x}{\partial \zeta}\frac{\partial y}{\partial \eta}\frac{\partial z}{\partial \xi}. \tag{6.A8}$$

This jacobian corresponds to the volume of the block in the physical domain. Coefficients of the coordinate transform are acquired numerically. For example, using indices (i, j, k) in each direction of both domains, coefficients at $(\xi(i),\ \eta(j),\ \zeta(k))$ in the computational domain are acquired by the following approximation,

$$\frac{\partial x}{\partial \xi} = \frac{x(i+1, j, k) - x(i-1, j, k)}{\xi(i+1) - \xi(i-1)}.$$

Finally, relations of the differentiation (1st order) for function f are given as follows,

$$\frac{\partial f}{\partial x} = \left(\frac{\partial y}{\partial \eta}\frac{\partial z}{\partial \zeta} - \frac{\partial y}{\partial \zeta}\frac{\partial z}{\partial \eta}\right)\frac{\partial f}{\partial \xi}\bigg/ J + \left(\frac{\partial y}{\partial \zeta}\frac{\partial z}{\partial \xi} - \frac{\partial y}{\partial \xi}\frac{\partial z}{\partial \zeta}\right)\frac{\partial f}{\partial \eta}\bigg/ J$$
$$+\left(\frac{\partial y}{\partial \xi}\frac{\partial z}{\partial \eta} - \frac{\partial y}{\partial \eta}\frac{\partial z}{\partial \xi}\right)\frac{\partial f}{\partial \zeta}\bigg/ J, \quad (6.A9)$$

$$\frac{\partial f}{\partial y} = \left(\frac{\partial x}{\partial \zeta}\frac{\partial z}{\partial \eta} - \frac{\partial x}{\partial \eta}\frac{\partial z}{\partial \zeta}\right)\frac{\partial f}{\partial \xi}\bigg/ J + \left(\frac{\partial x}{\partial \xi}\frac{\partial z}{\partial \zeta} - \frac{\partial x}{\partial \zeta}\frac{\partial z}{\partial \xi}\right)\frac{\partial f}{\partial \eta}\bigg/ J$$
$$+\left(\frac{\partial x}{\partial \eta}\frac{\partial z}{\partial \xi} - \frac{\partial x}{\partial \xi}\frac{\partial z}{\partial \eta}\right)\frac{\partial f}{\partial \zeta}\bigg/ J, \quad (6.A10)$$

$$\frac{\partial f}{\partial z} = \left(\frac{\partial x}{\partial \eta}\frac{\partial y}{\partial \zeta} - \frac{\partial x}{\partial \zeta}\frac{\partial y}{\partial \eta}\right)\frac{\partial f}{\partial \xi}\bigg/ J + \left(\frac{\partial x}{\partial \zeta}\frac{\partial y}{\partial \xi} - \frac{\partial x}{\partial \xi}\frac{\partial y}{\partial \zeta}\right)\frac{\partial f}{\partial \eta}\bigg/ J$$
$$+\left(\frac{\partial x}{\partial \xi}\frac{\partial y}{\partial \eta} - \frac{\partial x}{\partial \eta}\frac{\partial y}{\partial \xi}\right)\frac{\partial f}{\partial \zeta}\bigg/ J. \quad (6.A11)$$

Appendix B. DERIVATION OF THE FINITE-DIFFERENCE EQUATIONS

B.1. x-Component electric field; e_x

The following equation is obtained by substituting Equations (6.A12), (6.A13) and (6.5) into Equation (6.2) for the x-component,

$$\gamma\left(\frac{\partial e_x(i+1/2,j,k)}{\partial t}\right)^{n+1/2} + \sigma e_x^{n+1/2}$$
$$= \left\{\frac{\partial b_z^{n+1/2}(i+1/2,j,k)}{\partial \eta} - \frac{z(i,j+1,k)-z(i,j-1,k)}{\eta(j+1)-\eta(j-1)}\frac{\partial b_z^{n+1/2}(i+1/2,j,k)}{\partial \zeta}\right.$$
$$\left. - \frac{\partial b_y^{n+1/2}(i+1/2,j,k)}{\partial \zeta}\right\}\bigg/ J. \quad (6.B1)$$

A central difference is used to approximate the time derivative in the left side of Equation (6.B1),

$$\left(\frac{\partial e_x(i+1/2,j,k)}{\partial t}\right)^{n+1/2} \approx \frac{e_x^{n+1}(i+1/2,j,k) - e_x^n(i+1/2,j,k)}{\Delta t_n}. \quad (6.B2)$$

The electric field at time index $[n+1/2]$ is obtained by the simple average,

$$e_x^{n+1/2}(i+1/2,j,k) \approx \frac{e_x^n(i+1/2,j,k) + e_x^{n+1}(i+1/2,j,k)}{2}. \quad (6.B3)$$

Spatial derivatives $\partial b_z/\partial \eta$ and $\partial b_y/\partial \zeta$, which appear in the right side of Equation

(6.B1), are obtained by the following approximations,

$$\frac{\partial b_z^{n+1/2}(i+1/2,j,k)}{\partial \eta} \approx \frac{b_z^{n+1/2}(i+1/2,j+1/2,k) - b_z^{n+1/2}(i+1/2,j-1/2,k)}{\{\eta(j+1)-\eta(j-1)\}/2}, \tag{6.B4}$$

$$\frac{\partial b_y^{n+1/2}(i+1/2,j,k)}{\partial \zeta} \approx \frac{b_y^{n+1/2}(i+1/2,j,k+1/2) - b_y^{n+1/2}(i+1/2,j,k-1/2)}{\{\zeta(k+1)-\zeta(k-1)\}/2}. \tag{6.B5}$$

As shown in Figure 11a, $\partial b_z/\partial\zeta$ is obtained by interpolation of four approximate values ($B_{z\zeta}^A, B_{z\zeta}^B, B_{z\zeta}^C, B_{z\zeta}^D$) around e_x. These values are given as follows,

$$B_{z\zeta}^{\mathrm{A}} = \frac{b_z^{n+1/2}(i+1/2,j-1/2,k) - b_z^{n+1/2}(i+1/2,j-1/2,k-1)}{\zeta(k)-\zeta(k-1)},$$

$$B_{z\zeta}^{\mathrm{B}} = \frac{b_z^{n+1/2}(i+1/2,j-1/2,k+1) - b_z^{n+1/2}(i+1/2,j-1/2,k)}{\zeta(k+1)-\zeta(k)},$$

$$B_{z\zeta}^{\mathrm{C}} = \frac{b_z^{n+1/2}(i+1/2,j+1/2,k) - b_z^{n+1/2}(i+1/2,j+1/2,k-1)}{\zeta(k)-\zeta(k-1)},$$

$$B_{z\zeta}^{\mathrm{D}} = \frac{b_z^{n+1/2}(i+1/2,j+1/2,k+1) - b_z^{n+1/2}(i+1/2,j+1/2,k)}{\zeta(k+1)-\zeta(k)}.$$

Thus we obtain

$$\frac{\partial b_z^{n+1/2}(i+1/2,j,k)}{\partial \zeta} = \frac{\Delta\eta_j\Delta\zeta_k B_{z\zeta}^{\mathrm{A}} + \Delta\eta_j\Delta\zeta_{k-1}B_{z\zeta}^{\mathrm{B}} + \Delta\eta_{j-1}\Delta\zeta_k B_{z\zeta}^{\mathrm{C}} + \Delta\eta_{j-1}\Delta\zeta_{k-1}B_{z\zeta}^{\mathrm{D}}}{(\Delta\eta_{j-1}+\Delta\eta_j)(\Delta\zeta_{k-1}+\Delta\zeta_k)}, \tag{6.B6}$$

where

$$\Delta\eta_j = \eta(j+1)-\eta(j),$$
$$\Delta\zeta_k = \zeta(k+1)-\zeta(k).$$

The difference equation for extrapolating e_x is obtained from rearranging Equation (6.B1) by substituting Equations (6.B2)–(6.B6).

$$e_x^{n+1}(i+1/2,j,k) = \frac{2\gamma-\Delta t_n\sigma}{2\gamma+\Delta t_n\sigma}e_x^n(i+1/2,j,k) + \frac{2\Delta t_n}{2\gamma+\Delta t_n\sigma}P. \tag{6.B7}$$

where P is used for the right side of Equation (6.B1).

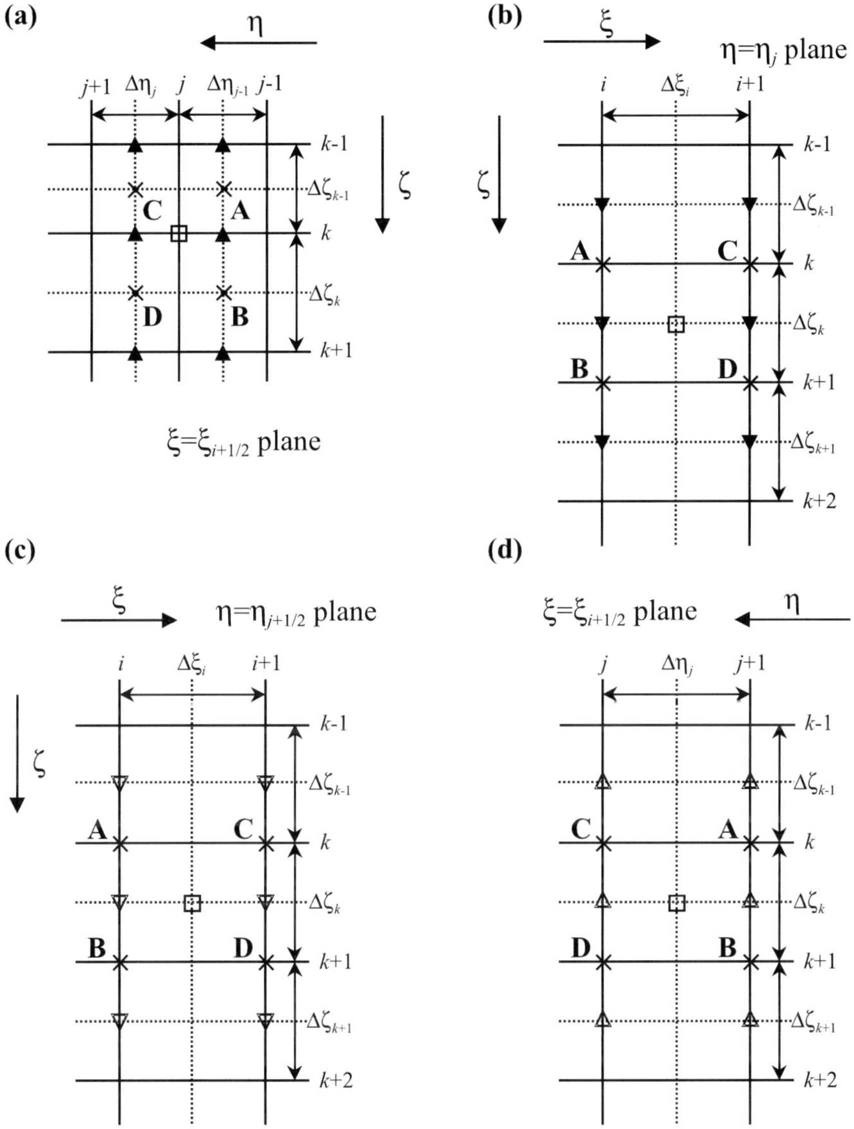

Figure 11. Interpolation points for differentiation of (a) b_z (▲), (b) e_z (▼), (c) b_x (▽) and (d) b_y (△). Pre-interpolation points and the desired point are indicated with × (A, B, C, D) and □, respectively.

B.2. y-Component magnetic field; b_y

The following equation is obtained by substituting Equations (6.A9), (6.A11) and (6.5) into Equation (6.1) for the y-component,

$$-\left(\frac{\partial b_y(i+1/2,j,k+1/2)}{\partial t}\right)^n = \left\{\frac{\partial e_x^n(i+1/2,j,k+1/2)}{\partial \zeta} - \frac{\partial e_z^n(i+1/2,j,k+1/2)}{\partial \xi} + \frac{z(i+1,j,k)-z(i-1,j,k)}{\xi(i+1)-\xi(i-1)}\frac{\partial e_z^n(i+1/2,j,k+1/2)}{\partial \zeta}\right\}. \tag{6.B8}$$

A central difference is used to approximate the time derivative in the left side,

$$\left(\frac{\partial b_y(i+1/2,j,k+1/2)}{\partial t}\right)^n \approx \frac{b_y^{n+1/2}(i+1/2,j,k+1/2)-b_y^{n-1/2}(i+1/2,j,k+1/2)}{(\Delta t_{n-1}+\Delta t_n)/2}. \tag{6.B9}$$

Spatial derivatives $\partial e_x/\partial\zeta$ and $\partial e_z/\partial\xi$ are obtained by the following approximations,

$$\frac{\partial e_x^n(i+1/2,j,k+1/2)}{\partial \zeta} \approx \frac{e_x^n(i+1/2,j,k+1)-e_x^n(i+1/2,j,k)}{\zeta(k+1)-\zeta(k)}, \tag{6.B10}$$

$$\frac{\partial e_z^n(i+1/2,j,k+1/2)}{\partial \xi} \approx \frac{e_z^n(i+1,j,k+1/2)-e_z^n(i,j,k+1/2)}{\xi(i+1)-\xi(i)}. \tag{6.B11}$$

$\partial e_z/\partial\zeta$ is obtained in the same way as $\partial b_z/\partial\zeta$ (Figure 11b),

$$E_{z\zeta}^{\mathrm{A}} = \frac{e_z^n(i,j,k+1/2)-e_z^n(i,j,k-1/2)}{\{\zeta(k+1)-\zeta(k-1)\}/2},$$

$$E_{z\zeta}^{\mathrm{B}} = \frac{e_z^n(i,j,k+3/2)-e_z^n(i,j,k+1/2)}{\{\zeta(k+2)-\zeta(k)\}/2},$$

$$E_{z\zeta}^{\mathrm{C}} = \frac{e_z^n(i+1,j,k+1/2)-e_z^n(i+1,j,k-1/2)}{\{\zeta(k+1)-\zeta(k-1)\}/2},$$

$$E_{z\zeta}^{\mathrm{D}} = \frac{e_z^n(i+1,j,k+3/2)-e_z^n(i+1,j,k+1/2)}{\{\zeta(k+2)-\zeta(k)\}/2}.$$

Thus we obtain

$$\frac{\partial e_z^n(i+1/2,j,k+1/2)}{\partial \zeta} \approx \frac{E_{z\zeta}^{\mathrm{A}}+E_{z\zeta}^{\mathrm{B}}+E_{z\zeta}^{\mathrm{C}}+E_{z\zeta}^{\mathrm{D}}}{4}. \tag{6.B12}$$

The difference equation for extrapolating is obtained by substituting Equations (6.B9)–(6.B12) into Equation (6.B8),

$$b_y^{n+1/2}(i+1/2,j,k+1/2) = b_y^{n-1/2}(i+1/2,j,k+1/2) - \frac{2Q}{\Delta t_{n-1}+\Delta t_n}, \tag{6.B13}$$

where Q is used for the right side of Equation (6.B8)

B.3. z-Component of the magnetic field; b_z

The following equation is obtained by substituting Equations (6.A9), (6.A10) and (6.5) into Equation (6.3) for the z-component,

$$\frac{\partial b_z^{n+1/2}(i+1/2,j+1/2,k+1/2)}{\partial\zeta}$$
$$= -\left\{\frac{\partial b_x^{n+1/2}(i+1/2,j+1/2,k+1/2)}{\partial\xi} - \frac{z(i+1,j,k)-z(i-1,j,k)}{\xi(i+1)-\xi(i-1)}\frac{\partial b_x^{n+1/2}(i+1/2,j+1/2,k+1/2)}{\partial\zeta}\right\}$$
$$-\left\{\frac{\partial b_y^{n+1/2}(i+1/2,j+1/2,k+1/2)}{\partial\eta} - \frac{z(i,j+1,k)-z(i,j-1,k)}{\eta(j+1)-\eta(j-1)}\frac{\partial b_y^{n+1/2}(i+1/2,j+1/2,k+1/2)}{\partial\zeta}\right\}. \quad (6.B14)$$

Spatial derivatives $\partial b_z/\partial\zeta$, $\partial b_x/\partial\xi$, and $\partial b_y/\partial\eta$ are obtained by the following approximations,

$$\frac{\partial b_z^{n+1/2}(i+1/2,j+1/2,k+1/2)}{\partial\zeta} \approx \frac{b_z^{n+1/2}(i+1/2,j+1/2,k+1)-b_z^{n+1/2}(i+1/2,j+1/2,k)}{\zeta(k+1)-\zeta(k)}, \quad (6.B15)$$

$$\frac{\partial b_x^{n+1/2}(i+1/2,j+1/2,k+1/2)}{\partial\xi} \approx \frac{b_x^{n+1/2}(i+1,j+1/2,k+1/2)-b_x^{n+1/2}(i,j+1/2,k+1/2)}{\xi(i+1)-\xi(i)}, \quad (6.B16)$$

$$\frac{\partial b_y^{n+1/2}(i+1/2,j+1/2,k+1/2)}{\partial\eta} \approx \frac{b_y^{n+1/2}(i+1/2,j+1,k+1/2)-b_y^{n+1/2}(i+1/2,j,k+1/2)}{\eta(j+1)-\eta(j)}. \quad (6.B17)$$

As shown in Figure 11c, $\partial b_x/\partial\zeta$ is obtained by using the following approximations,

$$B_{x\zeta}^{\mathrm{A}} = \frac{b_x^{n+1/2}(i,j+1/2,k+1/2)-b_x^{n+1/2}(i,j+1/2,k-1/2)}{\{\zeta(k+1)-\zeta(k-1)\}/2},$$

$$B_{x\zeta}^{\mathrm{B}} = \frac{b_x^{n+1/2}(i,j+1/2,k+3/2)-b_x^{n+1/2}(i,j+1/2,k+1/2)}{\{\zeta(k+2)-\zeta(k)\}/2},$$

$$B_{x\zeta}^{\mathrm{C}} = \frac{b_x^{n+1/2}(i+1, j+1/2, k+1/2) - b_x^{n+1/2}(i+1, j+1/2, k-1/2)}{\{\zeta(k+1) - \zeta(k-1)\}/2},$$

$$B_{x\zeta}^{\mathrm{D}} = \frac{b_x^{n+1/2}(i+1, j+1/2, k+3/2) - b_x^{n+1/2}(i+1, j+1/2, k+1/2)}{\{\zeta(k+2) - \zeta(k)\}/2}.$$

Thus we obtain

$$\frac{\partial b_x^{n+1/2}(i+1/2, j+1/2, k+1/2)}{\partial \zeta} \approx \frac{E_{x\zeta}^{\mathrm{A}} + E_{x\zeta}^{\mathrm{B}} + E_{x\zeta}^{\mathrm{C}} + E_{x\zeta}^{\mathrm{D}}}{4}. \tag{6.B18}$$

As shown in Figure 11d, $\partial b_y/\partial \zeta$ is obtained by using approximations as follows,

$$B_{y\zeta}^{\mathrm{A}} = \frac{b_y^{n+1/2}(i+1/2, j, k+1/2) - b_y^{n+1/2}(i+1/2, j, k-1/2)}{\{\zeta(k+1) - \zeta(k-1)\}/2},$$

$$B_{y\zeta}^{\mathrm{B}} = \frac{b_y^{n+1/2}(i+1/2, j, k+3/2) - b_y^{n+1/2}(i+1/2, j, k+1/2)}{\{\zeta(k+2) - \zeta(k)\}/2},$$

$$B_{y\zeta}^{\mathrm{C}} = \frac{b_y^{n+1/2}(i+1/2, j+1, k+1/2) - b_y^{n+1/2}(i+1/2, j+1, k-1/2)}{\{\zeta(k+1) - \zeta(k-1)\}/2},$$

$$B_{y\zeta}^{\mathrm{D}} = \frac{b_y^{n+1/2}(i+1/2, j+1, k+3/2) - b_y^{n+1/2}(i+1/2, j+1, k+1/2)}{\{\zeta(k+2) - \zeta(k)\}/2}.$$

Thus

$$\frac{\partial b_y^{n+1/2}(i+1/2, j+1/2, k+1/2)}{\partial \zeta} \approx \frac{B_{y\zeta}^{\mathrm{A}} + B_{y\zeta}^{\mathrm{B}} + B_{y\zeta}^{\mathrm{C}} + B_{y\zeta}^{\mathrm{D}}}{4}. \tag{6.B19}$$

The difference equation for extrapolating b_z is obtained from rearranging Equation (6.B14) by substituting Equations (6.B15)–(6.B19),

$$b_z^{n+1/2}(i+1/2, j+1/2, k) = b_z^{n+1/2}(i+1/2, j+1/2, k+1) - \Delta\zeta_k R. \tag{6.B20}$$

To solve Equation (6.B20) we start from the bottom subsurface boundary of the grid where $b_z = 0$ and step b_z upward.

REFERENCES

Adhidjaja, J.I. and Hohmann, G.W., 1989. A finite-difference algorithm for the transient electromagnetic response of a three-dimensional body. Geophys. J. Int., 98, 233–242.

DuFort, E.C. and Frankel, S.P., 1953. Stability conditions in the numerical treatment of parabolic differential equations. Math. Tables Other Aids Comput., 7, 135–152.

Macnae, J.C., 1984. Survey design for multicomponent electromagnetic systems. Geophysics, 49, 265–273.

Nabighian, M.N., 1972. The analytic signal of two-dimensional magnetic bodies with polygonal cross-section: its properties and use for automated interpretation. Geophysics, 37, 507–517.

Nabighian, M.N., 1984. Toward a three-dimensional automatic interpretation of potential field data via generalized Hilbert transforms: Fundamental relations. Geophysics, 49, 780–786.

Noguchi, K. and Endo, M., 1995. Three-dimensional TDEM modeling using integral-equation method, 93rd Annu. Meet., SEGJ, Expanded Abstracts, pp. 210–214.

Noguchi, K. and Endo, M., 1996. Three-dimensional TDEM modeling using integral-equation method (2). 94th Annu. Meet., SEGJ, Expanded Abstracts, pp. 245–249.

Oristaglio, M.L. and Hohmann, G.W., 1984. Diffusion of electromagnetic fields into a two-dimensional earth: a finite-difference approach. Geophysics, 49, 870–894.

Wang, T. and Hohmann, G.W., 1993. A finite-difference, time-domain solution for three-dimensional electromagnetic modeling. Geophysics, 58, 797–809.

Ward, S.H. and Hohmann, G.W., 1988. Electromagnetic theory for geophysical applications. In: M.N. Nabighian (Ed.), Electromagnetic Methods in Applied Geophysics, Vol. 1. Soc. Explor. Geophys., pp. 131–312.

Yee, K.S., 1966. Numerical solution of initial boundary problems involving Maxwell's equations in isotropic media. IEEE Trans. Ant. Prop., AP-14, 302–309.

Chapter 7

REDUCED-ORDER MODELING OF TRANSIENT DIFFUSIVE ELECTROMAGNETIC FIELDS

Rob F. Remis and Peter M. van den Berg

Laboratory of Electromagnetic Research, Center for Technical Geoscience, Delft University of Technology, Mekelweg 4, 2628 CD Delft, Netherlands

Abstract: The reduced-order modeling technique is applied to three-dimensional electromagnetic diffusion problems. The technique is based on the first-order Maxwell system. After a spatial discretization procedure that preserves the structure of Maxwell's equations, an algebraic system of equations is obtained in which the time coordinate appears as a continuous parameter. Approximations to the solution of this system are constructed with the help of a Lanczos algorithm for skew symmetric matrices. These approximations, called reduced-order models, give an accurate representation of the transient diffusive electromagnetic field on a certain bounded interval in time. The length of this interval can be extended by increasing the number of iterations of the underlying Lanczos algorithm. Some numerical examples for three-dimensional configurations illustrate the performance of the technique.

1. INTRODUCTION

The Spectral Lanczos Decomposition Method (SLDM) is a well known method for computing diffusive electromagnetic fields in inhomogeneous and possibly anisotropic media (see Druskin and Knizhnerman, 1994). The method is based on a second-order partial differential equation for either the electric or the magnetic field strength. After a spatial discretization procedure, an algebraic matrix equation is obtained in which the time coordinate appears as a continuous parameter. The solution of this equation is of the form $\exp(-At)\varphi$, where matrix A is a large, sparse, semi-positive definite, and symmetric matrix, t is the continuous time coordinate, and φ is a source vector. One way to evaluate the matrix exponential function, for any time instant t, is to diagonalize matrix A. However, this approach is impractical since the order of matrix A is simply too large. Druskin and Knizhnerman propose not to diagonalize matrix A, but to reduce it to a partial tridiagonal form with the help of a Lanczos algorithm for symmetric matrices. The algorithm takes the source vector φ as a starting vector and generates by means of a three-term recurrence relation a set of orthonormal vectors, known as Lanczos vectors, and a tridiagonal matrix containing the recurrence coefficients. The order of the tridiagonal matrix is equal to the number of iterations that have been carried out and is much smaller than the order of matrix A. The relation between the Lanczos vectors and the tridiagonal matrix is described by a single equation. This equation constitutes a partial reduction of matrix A to tridiagonal form. It is also known as a Lanczos decomposition of matrix A. Based on this decomposition, it is possible to construct approximations to the solution vector without having to discretize the time variable. Computing these approximations again requires the evaluation of a matrix exponential function, but this time it is the exponent of the tridiagonal matrix that is needed. This function can be calculated by diagonalizing the tridiagonal matrix since the order of this matrix is small and efficient routines are available to diagonalize a real, tridiagonal, and symmetric matrix. A constructed approximation gives an accurate representation of the electric or the magnetic field strength on a certain bounded interval in time. The length of this interval can be extended by performing more steps of the underlying Lanczos algorithm. The order of the tridiagonal matrix then increases but remains much smaller than the order of matrix A. As far as efficiency is concerned, SLDM solves transient diffusive field problems much more efficiently than any other (explicit) time-stepping method. We refer to Druskin and Knizhnerman (1989, 1991) and Druskin et al. (1998) for results on how fast SLDM converges in exact and finite precision arithmetic.

We have proposed a variant of SLDM (see Remis and Van den Berg, 1998). Instead of dealing with a second-order partial differential equation for either the electric or the magnetic field strength, we take the first-order Maxwell system as a point of departure. This approach has several consequences. First, we construct approximations for the electric as well as the magnetic field strength. Second, the matrices that are obtained after the spatial discretization procedure have symmetry properties different from the ones that are based on the second-order approach. Loosely speaking, one can say that symmetric matrices result from the second-order approach, while skew symmetric matrices result from the first-order approach. In our previous work we focussed on two-dimensional configurations. Here we present results for the three-dimensional case.

The basic equations for the diffusive electromagnetic field are given in Section 2, and in Section 3 we construct the approximations for the diffusive electromagnetic field. We call these approximations reduced-order models to emphasize that the order of the tridiagonal matrix generated by the Lanczos algorithm is much smaller than the order of the matrices that result from the spatial discretization of Maxwell's equations. Finally, in Section 4 we present some numerical examples for three-dimensional configurations. The Lanczos algorithm for skew symmetric matrices and a brief description of some of its properties can be found in Appendix A.

The notation used is as follows. To specify position, we employ the vector $\mathbf{x}$ with Cartesian coordinates x_1, x_2, and x_3. Further, ∂_1, ∂_2, and ∂_3 denote differentiation with respect to x_1, x_2, and x_3, respectively, while ∂_t denotes differentiation with respect to the time coordinate t. The vector 2-norm of a real vector x is given by $\|x\| = (x^T x)^{1/2}$, and the 2-norm of a matrix A is written as $\|A\|$. In Appendix A we also make use of the Frobenius norm of a matrix. This matrix norm is given by $\|A\|_F = [\mathrm{tr}(A^T A)]^{1/2}$, where $\mathrm{tr}(A^T A)$ denotes the trace of matrix $A^T A$.

2. BASIC EQUATIONS

The diffusive electromagnetic field, present in an inhomogeneous and isotropic medium, is governed by Maxwell's equations

$$-\nabla\times\mathbf{H}+\sigma\mathbf{E}=-\mathbf{J}^{\text{ext}}, \tag{7.1}$$

$$\nabla\times\mathbf{E}+\mu\,\partial_t\mathbf{H}=-\mathbf{K}^{\text{ext}}. \tag{7.2}$$

In these equations, $\mathbf{E}$ is the electric field strength (V/m), $\mathbf{H}$ is the magnetic field strength (A/m), σ is the conductivity (S/m), μ is the permeability (H/m), $\mathbf{J}^{\text{ext}}$ is the external electric-current density (A/m^2), and $\mathbf{K}^{\text{ext}}$ is the external magnetic-current density (V/m^2). Equations (7.1) and (7.2) can be arranged in the form

$$(\mathcal{D}+\mathcal{M}_\sigma+\mathcal{M}_\mu\partial_t)\mathcal{F}=\mathcal{S}, \tag{7.3}$$

where $\mathcal{F}=\mathcal{F}(\mathbf{x},t)$ is the *field vector* composed of the components of the electric and magnetic field strength as

$$\mathcal{F}=[E_1,E_2,E_3,H_1,H_2,H_3]^T, \tag{7.4}$$

and

$$\mathcal{S}=-[J_1^{\text{ext}},J_2^{\text{ext}},J_3^{\text{ext}},K_1^{\text{ext}},K_2^{\text{ext}},K_3^{\text{ext}}]^T \tag{7.5}$$

is the *source vector* consisting of the components of the external electric- and magnetic-current densities.

The spatial derivatives are contained in the *spatial differentiation operator matrix*

$$\mathcal{D}=\begin{pmatrix} 0 & 0 & 0 & 0 & \partial_3 & -\partial_2 \\ 0 & 0 & 0 & -\partial_3 & 0 & \partial_1 \\ 0 & 0 & 0 & \partial_2 & -\partial_1 & 0 \\ 0 & -\partial_3 & \partial_2 & 0 & 0 & 0 \\ \partial_3 & 0 & -\partial_1 & 0 & 0 & 0 \\ -\partial_2 & \partial_1 & 0 & 0 & 0 & 0 \end{pmatrix}, \tag{7.6}$$

and the time-independent *medium matrices* are given by

$$\mathcal{M}_\sigma=\text{diag}(\sigma,\sigma,\sigma,0,0,0), \quad \text{and} \quad \mathcal{M}_\mu=\text{diag}(0,0,0,\mu,\mu,\mu). \tag{7.7}$$

In addition, we introduce the diagonal matrices δ^{E} and δ^{H} as

$$\delta^{\text{E}}=\text{diag}(1,1,1,0,0,0), \quad \text{and} \quad \delta^{\text{H}}=\text{diag}(0,0,0,1,1,1). \tag{7.8}$$

The structure of the matrices $\mathcal{D}$, $\mathcal{M}_\sigma$, and $\mathcal{M}_\mu$ can be described in terms of the matrices δ^{E} and δ^{H}. In particular, we have

$$\mathcal{D}\delta^{\text{E}}=\delta^{\text{H}}\mathcal{D}, \quad \mathcal{D}\delta^{\text{H}}=\delta^{\text{E}}\mathcal{D}, \tag{7.9}$$

$$\begin{aligned} \mathcal{M}_\sigma\delta^{\text{E}}&=\delta^{\text{E}}\mathcal{M}_\sigma=\mathcal{M}_\sigma, \\ \mathcal{M}_\sigma\delta^{\text{H}}&=\delta^{\text{H}}\mathcal{M}_\sigma=0, \end{aligned} \tag{7.10}$$

and

$$\mathcal{M}_\mu \delta^{\mathrm{E}} = \delta^{\mathrm{E}} \mathcal{M}_\mu = 0,$$
$$\mathcal{M}_\mu \delta^{\mathrm{H}} = \delta^{\mathrm{H}} \mathcal{M}_\mu = \mathcal{M}_\mu. \tag{7.11}$$

When discretizing Maxwell's equations in space we require that Equations (7.9)–(7.11) have a counterpart after discretization, since the construction of our approximate solutions to Maxwell's equations is based on these equations.

In what follows, we consider source vectors of the form

$$\mathcal{S}(\mathbf{x},t) = w(t)\mathcal{Q}(\mathbf{x}), \tag{7.12}$$

where $w(t)$ is a scalar function of the time coordinate t that vanishes for $t < 0$, and $\mathcal{Q} = \mathcal{Q}(\mathbf{x})$ is a time-independent vector. The source vector is said to be of the *electric-current type* if $\mathcal{Q} = \delta^{\mathrm{E}}\mathcal{Q}$, and of the *magnetic-current type* if $\mathcal{Q} = \delta^{\mathrm{H}}\mathcal{Q}$.

Maxwell's equations are now discretized in space on a uniform grid using the standard second-order finite-difference scheme of Yee (see, for example: Yee, 1966; Kunz and Luebbers, 1993; Taflove, 1995; Smith, 1996). In addition, we set the tangential electric field to zero at the boundary of the computational domain. The resulting algebraic matrix equation is given by

$$(D + M_\sigma + M_\mu \partial_t)F(t) = w(t)Q, \tag{7.13}$$

where D, M_σ, M_μ, F, and Q are the discrete counterparts of $\mathcal{D}$, $\mathcal{M}_\sigma$, $\mathcal{M}_\mu$, $\mathcal{F}$, and $\mathcal{Q}$, respectively. Matrix D is real and of order n, and since we have used a uniform grid with the tangential electric field strength set to zero at the boundary of the computational domain, it is skew symmetric as well (see Monk and Süli, 1994, where the extension to a nonuniform grid is also given). The matrices M_σ and M_μ are both real, diagonal, and semi-positive definite.

The reason for choosing the simple Yee scheme is that Equations (7.9)–(7.11) then have a counterpart after discretization. We stress that this is not the only possible choice. Any discretization method will do as long as the structure of Maxwell's equations is preserved. Examples of such methods are higher-order finite difference schemes (see Wang and Tripp, 1996), or finite element methods (see, for example, Lee et al., 1997).

By restricting the source vector to be of the electric- or of the magnetic-current type, we can solve Equation (7.13) for the electromagnetic field quantities. The results are stated below. For a detailed derivation we refer the reader to the paper by Remis and Van den Berg (1998).

2.1. A source vector of the electric-current type

If the source vector satisfies $Q = \delta^{\mathrm{E}}Q$, the electric field strength is given by

$$\delta^{\mathrm{E}} F(t) = \partial_t w(t) * \chi(t) M^{-1/2} \exp(A^2 t) M^{-1/2} Q, \tag{7.14}$$

and for the magnetic field strength we obtain

$$\delta^{\mathrm{H}} F(t) = -w(t) * \chi(t) M^{-1/2} A \exp(A^2 t) M^{-1/2} Q, \tag{7.15}$$

where $M = M_\sigma + M_\mu$ is a diagonal and positive definite matrix, $A = M^{-1/2} D M^{-1/2}$, $\chi(t)$ is the Heaviside unit step function, and the asterisk denotes convolution in time.

2.2. A source vector of the magnetic-current type

If the source vector satisfies $Q = \delta^{\mathrm{H}} Q$, the electric field strength is found as

$$\delta^{\mathrm{E}} F(t) = -w(t) * \chi(t) M^{-1/2} A \exp(A^2 t) M^{-1/2} Q, \tag{7.16}$$

and the magnetic field strength is given by

$$\delta^{\mathrm{H}} F(t) = w(t) * \chi(t) M^{-1/2} \exp(A^2 t) M^{-1/2} Q. \tag{7.17}$$

Observe that the time coordinate appears as a continuous parameter in the above expressions for the diffusive electromagnetic field quantities. To evaluate these expressions, without discretizing the time variable, we could use the eigendecomposition of matrix A. However, this approach is not feasible since the order of matrix A is too large. We therefore construct approximate solutions with the help of a Lanczos algorithm for skew symmetric matrices. We call these approximations reduced-order models and their construction is described in the next section.

3. REDUCED-ORDER MODELS FOR THE DIFFUSIVE ELECTROMAGNETIC FIELD

In the previous section we have seen that all expressions for the electromagnetic field quantities contain the matrix exponential function $\exp(A^2 t)$. The construction of the reduced-order models is based on the Chebyshev expansion of this function. The reason for using this expansion is that it provides us with an estimate of the approximation error that adequately describes the convergence properties of the reduced-order models.

As a first step, we rewrite the exponential function as

$$\exp(A^2 t) = \exp(-\zeta) \exp(\zeta B), \tag{7.18}$$

where

$$B = I + 2(\|A\|^{-1} A)^2, \tag{7.19}$$

and

$$\zeta = \frac{\|A\|^2 t}{2}. \tag{7.20}$$

It is easily verified that all eigenvalues of matrix B are located in the interval $[-1, 1]$. We can therefore use the Chebyshev expansion

$$\exp(\zeta B) = 2 \sum_{k=0}^{\infty}{}' I_k(\zeta) T_k(B), \tag{7.21}$$

where the prime indicates that the first term in this series must be halved, where $I_k(\cdot)$ denotes the modified Bessel function of the first kind and order k, and where $T_k(\cdot)$ is the

Chebyshev polynomial of the first kind and order k. The vector $\exp(A^2t)M^{-1/2}Q$ is now written as

$$\exp(A^2t)M^{-1/2}Q$$
$$= 2\exp(-\zeta)\left[\sum_{k=0}^{\frac{m}{2}-1}{}' I_k(\zeta)T_k(B)M^{-1/2}Q + \sum_{k=\frac{m}{2}}^{\infty} I_k(\zeta)T_k(B)M^{-1/2}Q\right], \tag{7.22}$$

where m is taken to be even. By Theorem (A1) of Appendix A, we can rewrite this equation as

$$\exp(A^2t)M^{-1/2}Q = \beta_1 V_m \exp(H_m^2 t)e_1 + \mathcal{E}'_m(t), \tag{7.23}$$

where the error vector $\mathcal{E}'_m(t)$ is given by

$$\mathcal{E}'_m(t) = 2\exp(-\zeta)\sum_{k=\frac{m}{2}}^{\infty} I_k(\zeta)[T_k(B)M^{-1/2}Q - \beta_1 V_m T_k(B_m)e_1], \tag{7.24}$$

and

$$B_m = I_m + 2(\|A\|^{-1}H_m)^2. \tag{7.25}$$

Similarly, we have

$$A\exp(A^2t)M^{-1/2}Q = \beta_1 V_m H_m \exp(H_m^2 t)e_1 + \mathcal{E}''_m(t), \tag{7.26}$$

where

$$\mathcal{E}''_m(t) = 2\exp(-\zeta)\sum_{k=\frac{m}{2}}^{\infty} I_k(\zeta)[AT_k(B)M^{-1/2}Q - \beta_1 V_m H_m T_k(B_m)e_1]. \tag{7.27}$$

We are now in a position to introduce the reduced-order models. Just as in Section 2, we consider source vectors of the electric- and magnetic-current type separately.

3.1. A source vector of the electric-current type

Substitute Equation (7.23) in Equation (7.14) and Equation (7.26) in Equation (7.15). We obtain

$$\delta^{\mathrm{E}}F(t) = \delta^{\mathrm{E}}F_m(t) + \partial_t w(t) * \chi(t)M^{-1/2}\mathcal{E}'_m(t), \tag{7.28}$$

and

$$\delta^{\mathrm{H}}F(t) = \delta^{\mathrm{H}}F_m(t) - w(t) * \chi(t)M^{-1/2}\mathcal{E}''_m(t), \tag{7.29}$$

where the *reduced-order model* for the electric field strength is given by

$$\delta^{\mathrm{E}}F_m(t) = \partial_t w(t) * \chi(t)\beta_1 M^{-1/2}V_m \exp(H_m^2 t)e_1, \tag{7.30}$$

and for the magnetic field strength we have

$$\delta^{\mathrm{H}}F_m(t) = -w(t) * \chi(t)\beta_1 M^{-1/2}V_m H_m \exp(H_m^2 t)e_1. \tag{7.31}$$

3.2. A source vector of the magnetic-current type

If we substitute Equations (7.23) and (7.26) in Equations (7.17) and (7.16), respectively, we obtain

$$\delta^{\mathrm{E}} F(t) = \delta^{\mathrm{E}} F_m(t) - w(t) * \chi(t) M^{-1/2} \mathcal{E}''_m(t), \tag{7.32}$$

and

$$\delta^{\mathrm{H}} F(t) = \delta^{\mathrm{H}} F_m(t) + w(t) * \chi(t) M^{-1/2} \mathcal{E}'_m(t), \tag{7.33}$$

where we have introduced the reduced-order model for the electric field strength as

$$\delta^{\mathrm{E}} F_m(t) = -w(t) * \chi(t) \beta_1 M^{-1/2} V_m H_m \exp(H_m^2 t) e_1, \tag{7.34}$$

and the reduced-order model for the magnetic field strength is given by

$$\delta^{\mathrm{H}} F_m(t) = w(t) * \chi(t) \beta_1 M^{-1/2} V_m \exp(H_m^2 t) e_1. \tag{7.35}$$

For a discussion of the structure of these models, see Remis and Van den Berg (1998).

By using the asymptotics of the modified Bessel function it is possible to show that the 2-norm of the error vectors $\mathcal{E}'_m(t)$ and $\mathcal{E}''_m(t)$ decreases exponentially with m provided that $\zeta^{1/2} < m < \zeta$. (See Tal-Ezer, 1989, Remis and Van den Berg, 1998, and the papers by Druskin and Knizhnerman, 1989, and Hochbruck and Lubich, 1997. The latter two papers discuss the case $m \geq \zeta$ as well.) From the above result and the definition of ζ (Equation (7.20)) it follows that the number of Lanczos steps needed to reach a given accuracy is proportional to $\|A\| t^{1/2}$.

4. NUMERICAL RESULTS

We illustrate the reduced order modeling technique by some numerical examples for three-dimensional configurations. In all examples we have an electric-current switch-on source of the type

$$J_1^{\mathrm{e}}(\mathbf{x},t) = \chi(t)\delta(\mathbf{x} - \mathbf{x}^{\mathrm{src}}), \quad J_2^{\mathrm{e}}(\mathbf{x},t) = 0, \quad J_3^{\mathrm{e}}(\mathbf{x},t) = 0, \tag{7.36}$$

and

$$K_1^{\mathrm{e}}(\mathbf{x},t) = 0, \qquad K_2^{\mathrm{e}}(\mathbf{x},t) = 0, \quad K_3^{\mathrm{e}}(\mathbf{x},t) = 0. \tag{7.37}$$

The reduced-order model for the electric field strength then becomes

$$\delta^{\mathrm{E}} F_m(t) = \chi(t) \beta_1 M^{-1/2} V_m \exp(H_m^2 t) e_1, \tag{7.38}$$

and the reduced-order model for the electromotive force per unit area $-\partial_t \mathbf{B}$ is obtained as ($\mathbf{B}$ is the magnetic flux density)

$$-\partial_t \mathbf{B}_m = \chi(t) \beta_1 M_\mu M^{-1/2} V_m H_m \exp(H_m^2 t) e_1. \tag{7.39}$$

We note that since we are dealing with pure diffusion problems, we cannot handle configurations in which the conductivity vanishes. One possible way to include such cases in the present technique is discussed by Druskin and Knizhnerman (1988).

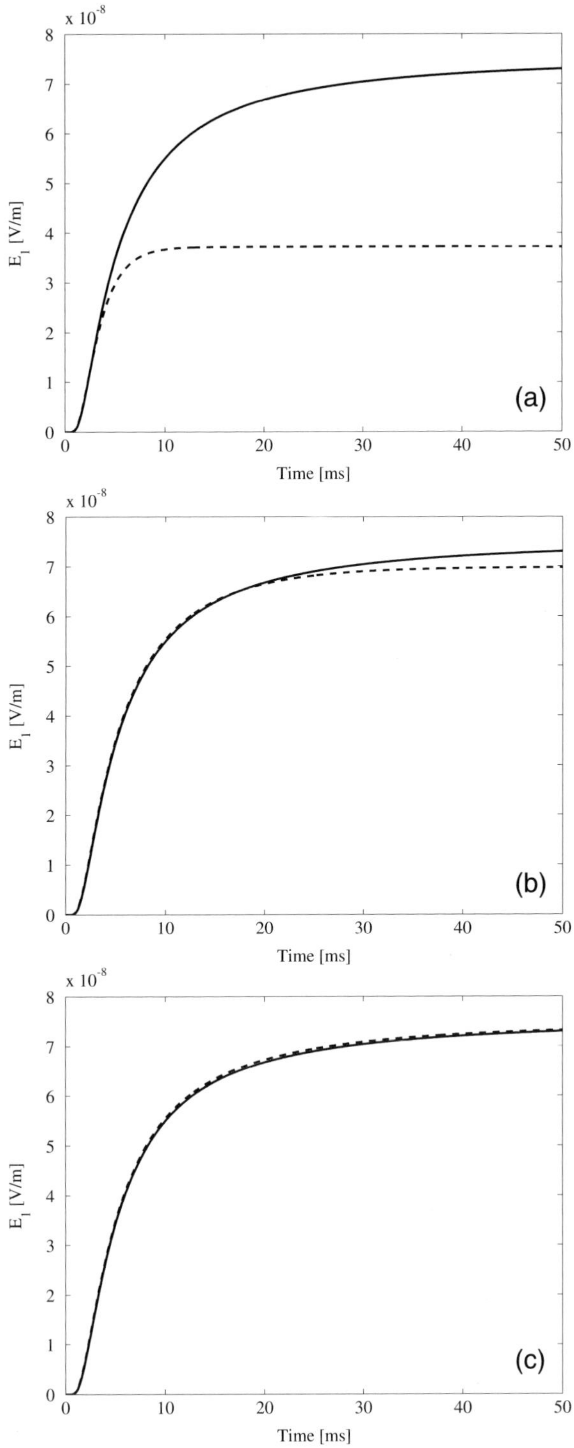
x 10^-8
E1 [V/m]
Time [ms]
(a)
(b)
(c)

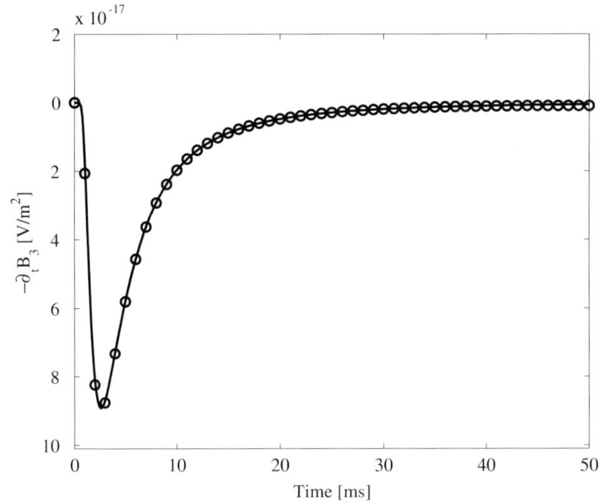

Figure 2. Converged reduced-order model for $-\partial_t B_3$ [V/m^2] as a function of time [ms] (circles). The solid line signifies the exact result.

As a first example, consider a homogeneous medium characterized by a conductivity σ and a permeability μ. For this particular example the electromagnetic field is known in closed form (see Slob, 1994). For example, the E_1 component of the electric field strength is for $t > 0$ given by

$$E_1(\mathbf{x},t) = \frac{1}{4\pi\sigma|\mathbf{x}-\mathbf{x}^{\text{src}}|^3} \cdot \left\{ \left(\frac{x_1 - x_1^{\text{src}}}{|\mathbf{x}-\mathbf{x}^{\text{src}}|} \cdot \frac{x_1 - x_1^{\text{src}}}{|\mathbf{x}-\mathbf{x}^{\text{src}}|} - 1 \right) \frac{4}{\sqrt{\pi}} \left(\frac{\tau}{t} \right)^{3/2} \exp(-\tau/t) \right.$$

$$\left. + \left(3\frac{x_1 - x_1^{\text{src}}}{|\mathbf{x}-\mathbf{x}^{\text{src}}|} \cdot \frac{x_1 - x_1^{\text{src}}}{|\mathbf{x}-\mathbf{x}^{\text{src}}|} - 1 \right) \cdot \left[\text{erfc}(\sqrt{\tau/t}) + \left(\frac{4\tau}{\pi t} \right)^{1/2} \exp(-\tau/t) \right] \right\}, \tag{7.40}$$

with $\mathbf{x} \neq \mathbf{x}^{\text{src}}$. In this equation, erfc($\cdot$) denotes the complementary error function and

$$\tau = \tfrac{1}{4}\sigma\mu|\mathbf{x}-\mathbf{x}^{\text{src}}|^2. \tag{7.41}$$

Moreover, for $\mathbf{x} \neq \mathbf{x}^{\text{src}}$ and $t > 0$ we have

$$-\partial_t B_3 = -\mu \left(\frac{\sigma\mu}{4\pi} \right)^{3/2} \frac{x_2 - x_2^{\text{src}}}{2t^{5/2}} \exp(-\tau/t). \tag{7.42}$$

The solid line in Figure 1 shows the E_1 component of the electric field strength at $(x_1, x_2, x_3) = (100, 0, 0)$ due to a source located at the origin. The conductivity of the homogeneous medium is 2 S/m and its permeability equals that of vacuum throughout the whole configuration. The dashed line in Figure 1a shows the reduced-order model

---

Figure 1. Electric field strength E_1 [V/m] as a function of time [ms]. Dashed lines show the reduced-order models obtained after (a) 100 iterations, (b) 200 iterations, and (c) 300 iterations. The solid line signifies the exact result.

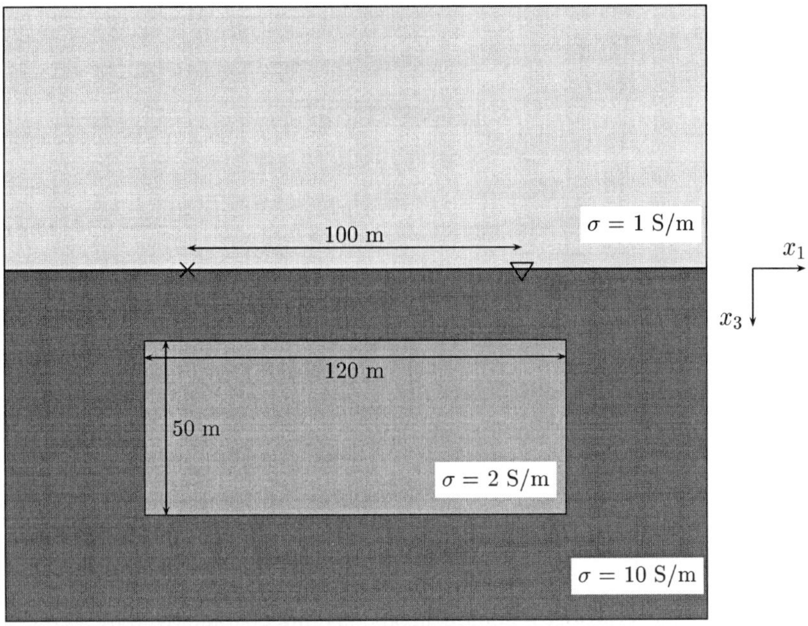

(a) Side view

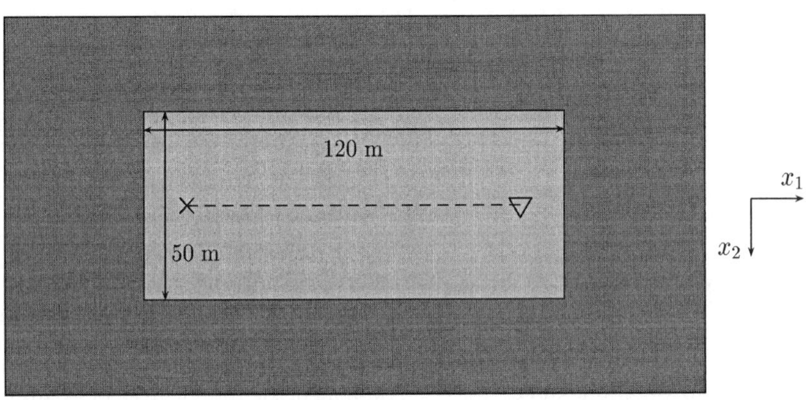

(b) View from above

Figure 3. An electric-current switch-on source and a receiver located symmetrically above an object. The permeability equals that of vacuum throughout the whole configuration.

Figure 4. Electric field strength E_1 [V/m] as a function of time [ms]. Dashed lines show the reduced-order models obtained after (a) 100 iterations, (b) 300 iterations, and (c) 500 iterations. The solid line shows the converged reduced-order model.

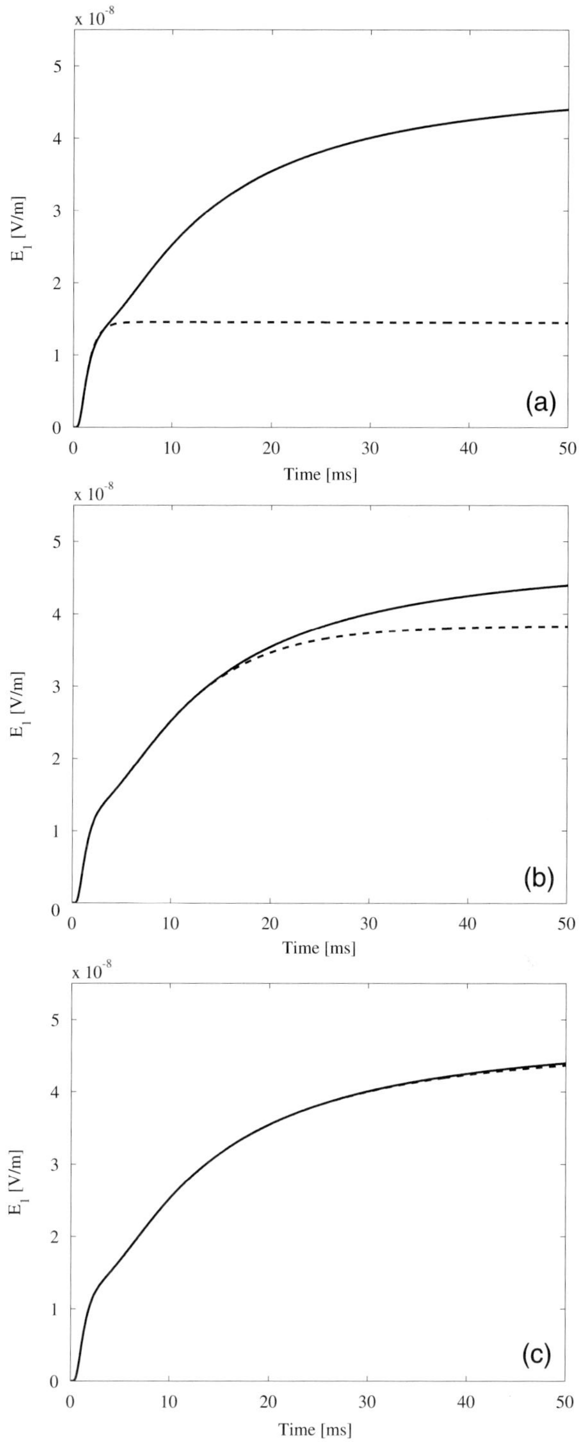
x 10^-8
E_1 [V/m]
Time [ms]
(a)
(b)
(c)

for the E_1 component of the electric field strength after 100 iterations, in Figure 1b after 200 iterations, and in Figure 1c after 300 iterations. The circles in Figure 2 show the converged reduced-order model for $-\partial_t B_3$. The solid line in this figure shows the exact result. For a given spatial discretization of Maxwell's equations, we say that a reduced-order model has converged when its order $m > \zeta^{1/2}$, and improvements of subsequent reduced-order models to the current model are in the order of the machine precision for the latest time instant of interest. A converged model gives an accurate representation of the electromagnetic field provided the spatial discretization of Maxwell's equations was done in a proper way, of course.

As a second example, we consider the configuration of Figure 3. This configuration consists of two lossy half-spaces. The permeability equals that of vacuum throughout the whole configuration. A large object is present in the lower half-space, and the source and the receiver are symmetrically located above this object. Figure 4 shows reduced-order models for the E_1 component of the electric field strength at the receiver location, while Figure 5 shows reduced-order models for $-\partial_t B_2$ at the same receiver location. The dashed lines in Figures 4a and 5a show the reduced-order models obtained after 100 iterations, in Figures 4b and 5b after 300 iterations, and in Figures 4c and 5c after 500 iterations. Just as in the previous example, we observe that the reduced-order models become more accurate on a larger time interval as the number of iterations increases.

5. CONCLUSIONS

We have applied the reduced-order modeling technique to three-dimensional electromagnetic diffusion problems. The technique is based on the first-order Maxwell system and uses a Lanczos algorithm for skew symmetric matrices to construct approximations for the electromagnetic field quantities. These approximations are highly structured and give an accurate representation of the electromagnetic field quantities on a certain bounded interval in time. The length of this interval can be extended by performing more steps of the underlying Lanczos algorithm. The numerical examples illustrated the performance of the technique and showed that reduced-order modeling is an efficient way to solve three-dimensional electromagnetic diffusion problems.

Appendix A. THE LANCZOS ALGORITHM FOR SKEW SYMMETRIC MATRICES

We can apply the following Lanczos algorithm to our diffusion problem since matrix A is skew symmetric. It is based on the recurrence relation $Av_i = \beta_{i+1}v_{i+1} - \beta_i v_{i-1}$ and is

---

Figure 5. Reduced-order models for $-\partial_t B_2$ [V/m^2] as a function of time [ms]. Dashed lines show the reduced-order models obtained after (a) 100 iterations, (b) 300 iterations, and (c) 500 iterations. The solid line shows the converged reduced-order model.

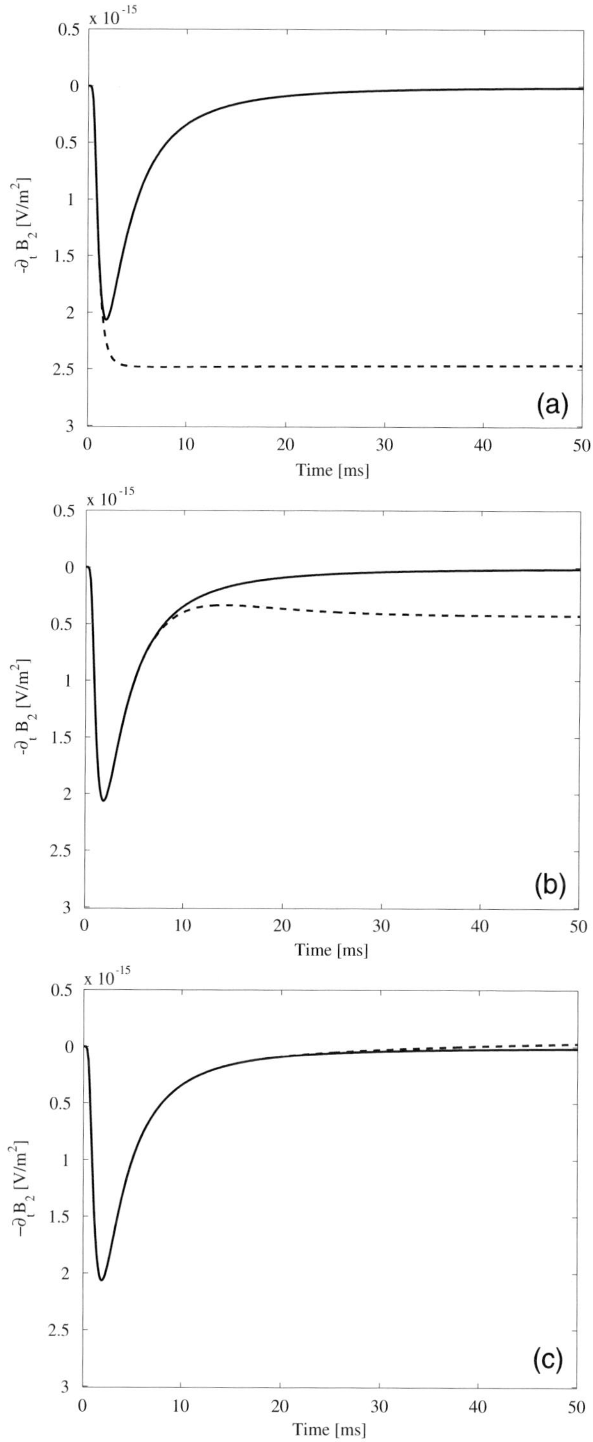
x 10^-15
-∂t B2 [V/m^2]
(a)
Time [ms]
x 10^-15
-∂t B2 [V/m^2]
(b)
Time [ms]
x 10^-15
-∂t B2 [V/m^2]
(c)
Time [ms]

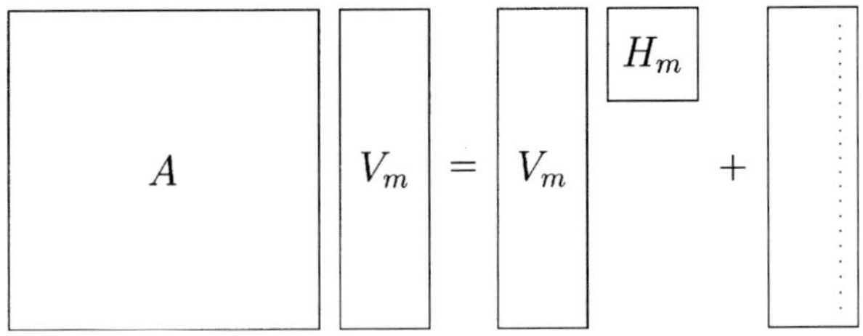

Figure 6. A picture of Equation (7.A2) for $m \ll n$.

given by

$$\beta_1 v_1 = M^{-1/2}Q,$$
$$\beta_{i+1} v_{i+1} = Av_i + \beta_i v_{i-1}, \quad \text{for } i = 1, 2, \ldots, \tag{7.A1}$$

with $v_0 = 0$. This algorithm produces coefficients β_i and vectors v_i known as Lanczos vectors. The coefficients $\beta_i \geq 0$ are determined from the condition $\|v_i\| = 1$ for $i \geq 1$.

After m steps of this algorithm, the first m Lanczos vectors form an orthonormal basis of the Krylov space $\mathcal{K}_m = \{M^{-1/2}Q, AM^{-1/2}Q, \ldots, A^{m-1}M^{-1/2}Q\}$, and the summarizing equation

$$AV_m = V_m H_m + \beta_{m+1} v_{m+1} e_m^T \tag{7.A2}$$

holds. In this equation, matrix H_m is a real, tridiagonal, and skew symmetric m-by-m matrix given by

$$H_m = \begin{pmatrix} 0 & -\beta_2 & & & \\ \beta_2 & 0 & -\beta_3 & & \\ & \beta_3 & \cdot & \cdot & \\ & & \cdot & \cdot & \cdot \\ & & & \cdot & \cdot & -\beta_m \\ & & & & \beta_m & 0 \end{pmatrix}. \tag{7.A3}$$

From Equation (7.A2) and the orthonormality property of the Lanczos vectors it follows that $\|H_m\| \leq \|A\|$. Furthermore, the first m Lanczos vectors form the columns of the n-by-m matrix V_m, and e_m is the mth column of the m-by-m identity matrix I_m. In Section 3, we construct approximations for the electric and magnetic field strength that are based on this Lanczos algorithm. These approximations can be computed efficiently as long as the order of matrix H_m is much smaller than the order of matrix A (see Figure 6).

Some remarks about the behavior of the algorithm are in order. First, the algorithm cannot continue if a coefficient β_{i+1} vanishes. This is called a regular termination of the algorithm. The Krylov space $\mathcal{K}_i$ is then invariant under A, and every eigenvalue of matrix H_i is an eigenvalue of matrix A as well. Unfortunately, a regular termination of the algorithm rarely happens in practice.

Second, the Lanczos vectors are orthonormal by construction from which it follows that $\|V_m\| = 1$. As is well-known, the orthogonality of the Lanczos vectors is completely lost in finite precision arithmetic. Despite this loss of orthogonality, it is still possible to give a bound for the 2-norm of matrix V_m in the finite precision case. To obtain this bound, we make use of the inequality $\|V_m\| \leq \|V_m\|_F$, where $\|\cdot\|_F$ denotes the Frobenius norm, and we assume that the error in normalizing the vectors $Av_i + \beta_i v_{i-1}$ is negligible. In other words, we assume that the Lanczos vectors are of unit length. With this assumption we obtain $\|V_m\|_F = m^{1/2}$, leading to the practical bound $\|V_m\| \leq m^{1/2}$. Usually, this is an overestimate.

Third, rounding errors perturb the computed analogs of Equation (7.A2) and the inequality $\|H_m\| \leq \|A\|$. We assume, however, that these perturbations are so small that the finite precision counterparts of Equation (7.A2) and the estimate $\|H_m\| \leq \|A\|$ are satisfied to working precision. Finally, the Lanczos vectors are highly structured if the source vector is of the electric- or of the magnetic-current type. See Remis and Van den Berg (1998) for a discussion.

To summarize, loss of orthogonality of the Lanczos vectors is severe, and Equation (7.A2) together with the inequality $\|H_m\| \leq \|A\|$ are assumed to hold in finite precision arithmetic.

Now that we have described the main properties of the Lanczos algorithm, we prove a theorem on which the reduced-order modeling technique is based. Note that in proving this theorem, we do not use the orthonormality property of the Lanczos vectors.

Theorem A1. *Let $p_{m-1}(\cdot)$ denote any polynomial with a maximum degree of $m-1$, and assume that m steps of the Lanczos algorithm have been carried out successfully. Then*

$$p_{m-1}(A)M^{-1/2}Q = \beta_1 V_m p_{m-1}(H_m)e_1. \tag{7.A4}$$

Proof (see Druskin and Knizhnerman, 1989). It suffices to show that

$$A^j M^{-1/2}Q = \beta_1 V_m H_m^j e_1, \quad \text{for } j = 0, 1, \ldots, m-1. \tag{7.A5}$$

For $j = 0$ we have

$$M^{-1/2}Q = \beta_1 V_m e_1 = \beta_1 v_1, \tag{7.A6}$$

which is true by definition. Further, by the induction hypothesis, we have

$$\begin{aligned} A^{j+1}M^{-1/2}Q &= AA^j M^{-1/2}Q = \beta_1 A V_m H_m^j e_1 \\ &= \beta_1 (V_m H_m + \beta_{m+1} v_{m+1} e_m^T) H_m^j e_1 = \beta_1 V_m H_m^{j+1} e_1, \end{aligned} \tag{7.A7}$$

since $e_m^T H_m^j e_1 = 0$ for $j \leq m-2$. □

REFERENCES

Druskin, V.L. and Knizhnerman, L.A., 1988. Spectral differential-difference method for numeric solution of three-dimensional nonstationary problems of electric prospecting. Izv., Earth Phys., 24, 641–648.

Druskin, V.L. and Knizhnerman, L.A., 1989. Two polynomial methods of calculating functions of symmetric matrices. USSR Comput. Math. Math. Phys., 29, 112–121.

Druskin, V.L. and Knizhnerman, L.A., 1991. Error bounds in the simple Lanczos procedure for computing functions of symmetric matrices and eigenvalues. USSR Comput. Math. Math. Phys., 31, 20–30.

Druskin, V.L. and Knizhnerman, L.A., 1994. Spectral approach to solving three-dimensional Maxwell's diffusion equations in the time and frequency domains. Radio Sci., 29, 937–953.

Druskin, V.L., Greenbaum, A. and Knizhnerman, L.A., 1998. Using nonorthogonal Lanczos vectors in the computation of matrix functions. SIAM J. Sci. Comput., 19, 38–54.

Hochbruck, M. and Lubich, C., 1997. On Krylov subspace approximations to the matrix exponential operator. SIAM J. Num. Anal., 34, 1911–1925.

Kunz, K.S. and Luebbers, R.J., 1993. The Finite Difference Time Domain Method for Electromagnetics. CRC Press, Boca Raton.

Lee, J.F., Lee, R. and Cangellaris, A.C., 1997. Time-domain finite-element methods. IEEE Trans. Antennas Propagation, 45, 430–442.

Monk, P. and Süli, E., 1994. A convergence analysis of Yee's scheme on nonuniform grids. SIAM J. Num. Anal., 31, 393–412.

Remis, R.F. and Van den Berg, P.M., 1998. Efficient computation of transient diffusive electromagnetic fields by a reduced modeling technique. Radio Sci., 2, 191–204.

Slob, E., 1994. Scattering of Transient Diffusive Electromagnetic Fields. Ph.D. Thesis, Delft University of Technology, Delft.

Smith, J.T., 1996. Conservative modeling of 3-D electromagnetic fields, Part 1. Properties and error analysis. Geophysics, 61, 1308–1318.

Taflove, A., 1995. Computational Electromagnetics, The Finite-Difference Time-Domain Method. Artech House, Norwood.

Tal-Ezer, H., 1989. Spectral methods in time for parabolic problems. SIAM J. Num. Anal., 26, 1–11.

Wang, T. and Tripp, A.C., 1996. FDTD simulation of em wave propagation in 3-D media. Geophysics, 61, 110–120.

Yee, K.S., 1966. Numerical solution of initial boundary value problems involving Maxwell's equations in isotropic media. IEEE Trans. Antennas Propagation, 14, 302–307.

Part II

3-D EM Inversion

Chapter 8

THREE-DIMENSIONAL MAGNETOTELLURIC MODELING AND INVERSION: APPLICATION TO SUB-SALT IMAGING

Gregory A. Newman [a], G. Michael Hoversten [b] and David L. Alumbaugh [c]

[a] *Sandia National Laboratories, Albuquerque, NM 87185, USA*
[b] *Lawrence Berkeley National Laboratory, Berkeley, CA 94729, USA*
[c] *University of Wisconsin-Madison, Madison, WI 53706, USA*

Abstract: Three-dimensional (3D) magnetotelluric (MT) forward and inverse solutions are reviewed and applied in a resolution study for sub-salt imaging of an important target in marine magnetotellurics for oil prospecting. In the forward problem, finite-difference methods are used to efficiently compute predicted data and cost functional gradients. A fast preconditioner is introduced at low induction numbers to reduce the time required to solve the forward problem. We demonstrate a reduction of up to two orders of magnitude in the number of Krylov sub-space iterations and an order of magnitude reduction in time needed to solve a series of test problems. For the inverse problem, we employ a nonlinear conjugate gradient solution developed on massively parallel computing platforms. Solution stabilization is achieved with Tikhonov regularization. To further improve the image resolution of sub-salt structures, we have also incorporated two additional constraints within the inversion process. The first constraint allows for the preservation of known structural boundaries within the inverted depth sections. This type of constraint is justified for the sub-salt imaging problem because the top of salt is constrained by seismic data. The other constraint employed places variable lower bounds on the electrical conductivity above and below the top of salt. Cross-sections of the inversion results over the center of the salt structures indicate that the 3D analysis provides somewhat more accurate images compared to faster 2D analysis, but is computationally much more demanding. On the flanks of the structures, however, 3D analysis is necessary as 2D inversion shows image artifacts arising from the 3D nature of the data. We conclude, however, that 3D inversion may not be cost effective for the sub-salt imaging problem. Very fine data sampling along multiple profiles employed in the 3D analysis yielded only a marginal improvement in image resolution compared to 2D analysis along carefully selected data profiles. The study also indicates that in order to provide resolution that is required to accurately define the base of the salt, additional constraints beyond that employed here, need to be incorporated into the 3D inversion process.

1. INTRODUCTION

With the advent of multiple channel 24 bit data acquisition systems, the ability to acquire large amounts of high-quality magnetotelluric (MT) data is rapidly becoming a reality. This is evidenced by our ability to acquire MT data sets in remote and harsh marine environments (cf. Constable et al., 1998; Hoversten et al., 2000). Accompanying this advance in data acquisition technology has been a significant improvement in the data processing capabilities, including the introduction of a remote reference to reduce bias associated with noise in the magnetic field measurements (Gamble et al., 1979) and robust transfer function response estimation (cf. Egbert and Booker, 1986). Because of these improvements in instrument accuracy and data processing, MT surveys are starting to be designed where data are acquired along several parallel profiles rather than just along one or two which has been traditionally the case. The use of multiple lines allows for three-dimensional (3D) geological structures to be better delineated.

Paralleling the improvement in data acquisition systems has been the increasing speed and memory capability of computers. This has allowed for the development of algorithms that more accurately take into account some of the multi-dimensionality of the MT interpretation problem. For example 2D MT inversion schemes that 10 years ago required a Cray computer for reasonable computation times (cf. DeGroot-Hedlin and Constable, 1990) can now run in a period of a few minutes to an hour or two on standard desktop workstations and PCs. In addition computationally efficient algorithms have been developed that either make subtle approximations to the 2D problem (cf. Smith and Booker, 1991; Siripunvaraporn and Egbert, 2000) or that use efficient iterative gradient algorithms (cf. Rodi and Mackie, 2001) to produce 2D images of the geological structure.

In 3D environments, 2D interpretation of the data is standard practice because of quick processing times and because of the scarcity and time consuming nature of 3D MT modeling and inversion schemes. Thus one may never fully know the advantages, if any, 3D MT inversion can offer without actually applying it to the data. Moreover in some 3D environments it is conceivable that the use of 2D interpretation schemes may result in artifacts appearing in the images that could lead to misinterpretation. Therefore there exists a need for efficient 3D forward modeling and inversion algorithms to be developed. In this paper we outline recent progress that has been made in developing one such set of algorithms (Newman and Alumbaugh, 2000), and employ them on a simulated petroleum exploration problem.

The exploration problem we have selected to demonstrate these algorithms is a resolution study for imaging the base of 3D salt structures. Such structures are encountered in petroleum exploration in the Gulf of Mexico, where seismic imaging beneath these high-velocity formations is a formidable task. The scattering of seismic energy produced by these formations limits the ability of migration methods to delineate the base of salt as well as deeper oil bearing horizons. Additional motivation for pursuing such a study is to compare the 3D results with the 2D MT work already done on the problem (cf. Hoversten et al., 1998, 2000). If 3D data analysis shows significant improvements in our ability to image sub-salt structures, it could justify the extra time and cost needed to acquire and interpret the 3D MT data sets.

Before presenting the resolution study, details on the 3D forward and inversion algorithms employed in the investigation are discussed. This is necessary in order to document the changes we made to the algorithms in the course of this study. The first change we will discuss is the development of a low-frequency preconditioner needed to greatly reduce the computation time required in the forward problem at long periods.

MT simulations at periods greater than 1000 s are used in the resolution study due to the highly conductive seawater and sea sediments, and the new preconditioner has a significant impact in reducing computational run times at the longer periods. The other changes correspond to the incorporation of additional constraints, beyond the standard Tikhonov regularization employed for constructing smooth solutions to the inverse problem. Specific constraints to be discussed include preservation of known structural boundaries, such as the top of salt, within the inverted depth sections, and incorporation of variable lower bounding constraints on the electrical conductivity above and below the top of salt. It will be shown that incorporation of these constraints into the MT inverse problem improves our ability to image the base of salt.

2. THE 3D MT FORWARD PROBLEM

2.1. Governing equations and solution

In solving the 3D MT inverse problem it is critical that the forward-modeling solution simulates the responses arising from realistic 3D geology. Parameterizations of hundreds of thousands of cells are typically required for these types of numerical simulations. Hence highly efficient solution techniques are required. Here we employ finite-difference modeling techniques for the task as outlined in Newman and Alumbaugh (2000).

Assuming a harmonic time dependence of $e^{i\omega t}$, where $i = \sqrt{-1}$ and ω is the angular frequency, the electric field, **E**, satisfies the vector equation

$$\nabla \times \nabla \times \mathbf{E} + i\omega\mu_o\sigma\mathbf{E} = \mathbf{0}. \tag{8.1}$$

In this expression the electrical conductivity is denoted by σ and μ_o represents the magnetic permeability of free space. Note, the equation can be arranged such that the magnetic permeability is also variable (cf. Alumbaugh et al., 1996), but for simplicity we have assumed it to be constant. Dirichlet boundary conditions are applied to Equation (8.1), where the tangential electric-field boundary values are specified on the boundaries of a large prism that includes the investigation domain (the earth) as well as the air. These boundary values arise from a plane wave, with a given source-field polarization, propagating in layered or 2D geologic media assigned at the boundaries of the 3D problem.

When Equation (8.1) is approximated with finite differences using a Yee (1966) staggered grid and symmetrically scaled (Newman and Alumbaugh, 1995), a linear system results:

$$\mathbf{KE} = \mathbf{S}. \tag{8.2}$$

The matrix **K** is complex-symmetric and sparse with 13 non-zero entries per row and **S** is the source vector that depends on the boundary conditions and source-field polarization. This system can be efficiently solved at moderate to high induction frequencies using the quasi-minimum residual (qmr) method with Jacobi preconditioning; solution treatment at low frequencies will be given below. The qmr algorithm belongs to the class of

Krylov sub-space techniques that are highly efficient in iteratively solving sparse linear systems. The reader is referred to Newman and Alumbaugh (1995) for the details on how the solver is implemented. Once the electric fields are determined on the mesh, the magnetic fields, **H**, can be determined from Faraday's law,

$$\mathbf{H} = \nabla \times \mathbf{E} / -i\omega\mu_o, \tag{8.3}$$

by numerically approximating the curl of the electric field at the nodal points. One can then interpolate either the electric or magnetic field nodal values to the detector location. At this point the magnetotelluric impedance tensor, including apparent resistivity and phase are readily computed from the electric and magnetic fields.

2.2. Preconditioning

It is well known that difficulties will be encountered when attempts are made to solve Equation (8.2) as frequencies approach the static limit (cf. Smith, 1996; Newman and Alumbaugh, 2000). In this section we will show how these difficulties can be overcome, with preconditioning. The preconditioner that we will introduce parallels the work of Druskin et al. (1999), who developed a new spectral Lanczos decomposition method (SLDM) with Krylov sub-spaces generated from the inverse of the Maxwell operator. We have successfully implemented it for reducing solution times in induction logging problems (Newman and Alumbaugh, 2002). Here we derive it for MT applications.

Following Druskin et al. (1999) we assume that the electric field can be decomposed into curl-free and divergence-free projections via the Helmholtz theorem, where

$$\mathbf{E} = \boldsymbol{\Psi} + \nabla\varphi \tag{8.4}$$

and

$$\nabla \cdot \boldsymbol{\Psi} = 0 \tag{8.5}$$

Substituting Equations (8.4) and (8.5) into Equation (8.1), and using the vector identity

$$\nabla \times \nabla \times \boldsymbol{\Psi} = -\nabla^2 \boldsymbol{\Psi}, \tag{8.6}$$

along with the fact that $\nabla \cdot \boldsymbol{\Psi} = 0$ and $\nabla \times \nabla \times \nabla\varphi = 0$, we arrive at

$$-\nabla^2 \boldsymbol{\Psi} + i\omega\mu_o\sigma(\boldsymbol{\Psi} + \nabla\varphi) = \mathbf{0}. \tag{8.7}$$

Splitting the electric field into curl-free and divergence-free projections removes the null space of the curl–curl operator in the solution process. When Krylov methods are applied directly to Equation (8.2), this null space is responsible for the poor convergence properties of the solution process as frequency approaches the static limit.

To develop an approximate finite-difference solution to Equation (8.1) at low frequencies, we first estimate the relative sizes of the curl–curl and attenuation operators in Equation (8.1) assuming a finite-difference approximation. Let Δ be the characteristic grid size employed in the finite-difference mesh, then the size of the discrete curl–curl operator is roughly, $1/\Delta^2$, whereas the size of the attenuation operator is approximated as $\omega\mu_o\sigma$. Thus as frequency falls and the grid size is reduced we observe the condition that

$$1/\Delta^2 \gg \omega\mu_o\sigma_{\max} \tag{8.8}$$

or

$$1 \gg \Delta^2 \omega \mu_o \sigma_{\max}, \tag{8.9}$$

where $\sigma_{\max}$ is the maximum conductivity in the mesh. When the finite-difference grid is non-uniform, Δ should be replaced by $\Delta_{\max}$, the maximum cell size used to approximate Equation (8.1). Assuming Equation (8.9) is satisfied, Equation (8.7) can then be decoupled such that

$$-\nabla^2 \mathbf{\Psi} = \mathbf{0}. \tag{8.10}$$

Note that the right hand side of Equation (8.9) is a dimensionless *induction number* (cf. Frischknecht, 1987), which has important implications on the validity of Equation (8.10) and thus the effectiveness of the preconditioner introduced below. More importantly, notice that Equation (8.9) also indicates the importance of the cell size employed within the mesh, in addition to the frequency and conductivity. Thus Equation (8.10) may apply even at moderate to high frequencies (up to 100s of kHz) as long as the grid size employed for the problem is sufficiently small. This property has allowed for the beneficial use of the preconditioner in solving induction logging problems (Avdeev et al., 2002), where very small cell sizes are employed to model a borehole.

The boundary conditions required to solve Equation (8.10) are a mixture of Dirichlet and Neumann types. Dirichlet conditions are applied to the tangential components of $\mathbf{\Psi}$ on the mesh boundaries ($\mathbf{\Psi}_t$ has to be specified). For the normal components, Neumann conditions are applied, where $\partial \mathbf{\Psi}_n / \partial n$ is specified with the constraint that $\nabla \cdot \mathbf{\Psi} = 0$ is discretely satisfied on the mesh boundaries and in turn within the solution domain because the divergence-free field is required to satisfy the constraint equation

$$\nabla^2 (\nabla \cdot \mathbf{\Psi}) = 0. \tag{8.11}$$

Equation (8.11) follows by applying the divergence operator to Equation (8.6). It is well known that when the scalar function, $u = \nabla \cdot \mathbf{\Psi}$, satisfies Laplace's equation, $\nabla^2 u = 0$, on some domain Ω with homogeneous boundary conditions of $u = 0$ prescribed along the boundary Γ, it is identically zero on that domain. Note when applying the Neumann boundary condition, n would specify the direction of the outward normal at the boundary. Thus Equation (8.5) is implicitly enforced with the solution of Equation (8.10).

The vector field $\mathbf{\Psi}$ is not a complete solution to Maxwell's equations since it does not satisfy the auxiliary divergence condition on the current density within the earth. To derive this condition we take the divergence of Equation (8.7) and arrive at

$$\nabla \cdot \sigma \nabla \varphi = -\nabla \cdot \sigma \mathbf{\Psi}. \tag{8.12}$$

Dirichlet boundary conditions will be applied to the discrete version of Equation (8.12), where φ has to be specified on the mesh boundaries; here we assume that $\varphi = 0$. When the air–earth interface is present, however, we employ the Neumann condition, $\partial \varphi / \partial n = 0$, where n again specifies the direction of the outward normal at that interface. This later boundary condition enforces the constraint that current cannot leak from the earth into the air at induction frequencies; frequencies typically less than 1 MHz.

An approximate solution to Equation (8.1) can be obtained at very small induction numbers by first solving Equation (8.10) followed by Equation (8.12) using a staggered finite-difference grid along with conjugate gradient methods. However the real benefit of using this approximation is as a preconditioner for Equation (8.2) in order to obtain a faster solution. This preconditioning involves solving the modified problem

$$\mathbf{M}^{-1}\mathbf{K}\mathbf{E} = \mathbf{M}^{-1}\mathbf{S}, \tag{8.13}$$

where $\mathbf{M}$ is chosen such that $\mathbf{M}^{-1}\mathbf{K}$ approximates the identity matrix (Greenbaum, 1997). This condition will be implicitly satisfied at low induction numbers with the solutions of Equations (8.10) and (8.12). It is important to note that the preconditioning matrix, $\mathbf{M}$, is never actually computed during the preconditioning step, only its effect upon a vector (a matrix–vector multiplication) is needed. We now describe how this is done.

During each iteration of the preconditioned qmr algorithm we substitute the residual, defined by $\mathbf{r} = \mathbf{K}\mathbf{E} - \mathbf{S}$ into the right hand side of the discrete version of Equation (8.10), where $\boldsymbol{\Psi}_\mathrm{t}$ and $\partial\boldsymbol{\Psi}_\mathrm{n}/\partial n$ are set to zero on the mesh boundary. Equations (8.10) and (8.12) are then progressively solved using conjugate gradient techniques along with an incomplete Cholesky factorization. Furthermore, the solutions to Equations (8.10) and (8.12) need not be precise, as test examples have indicated that a crude solution to these equations will still provide a significant impact on reducing the time required for solving Equation (8.2). In Appendix A, we re-derive this low induction number (LIN) preconditioner using a Neumann series expansion. From this series, we are able to provide a rigorous bound on its effectiveness as well as develop higher-order versions.

2.3. Forward model example

To demonstrate the accuracy of the solution, and show the benefits of the preconditioner, we have employed a 0.1 S/m block within a 0.01 S/m half-space. The block's dimensions are 200 m on a side and its depth of burial is 100 m below the surface on which the fields are to be measured. The size of the grid employed in this simulation was 28 cells in x and y by 36 cells in z. The minimum cell size at the center of the grid was 25 m by 25 m by 6.25 m, while at the edges of the grid the largest cell employed was 200 m by 200 m by 200 m. In Figure 1 the apparent resistivities (ρ_{xy}) at 4 Hz and 400 Hz are presented both for the finite-difference solution as well as the integral equation solution of Xiong (1992). In both cases the responses are determined at 25 m increments on the surface over the block, with agreement to within one percent.

In Figure 2, we demonstrate the effectiveness of the preconditioner at 4 Hz. A speed up approaching a factor of 7 is demonstrated compared to a solution employing simple Jacobi scaling for preconditioning; the machine employed in these comparisons is an IBM RS 590 workstation. When the LIN preconditioner is compared with the static-divergence correction procedure of Smith (1996), it is still a factor of 3 faster. At a frequency of 400 Hz, the preconditioner is still effective, but its benefits are not as great, now the speed up is only a factor of 2.

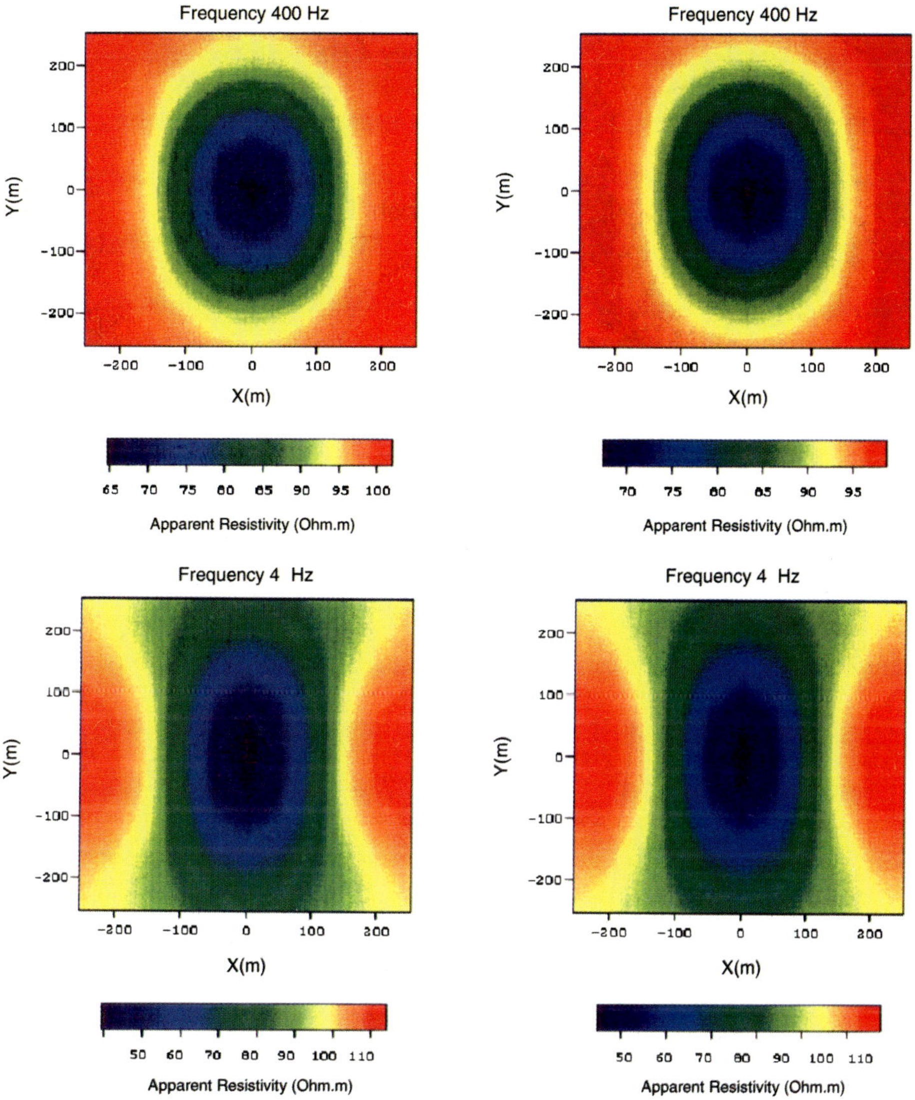

Figure 1. Comparison of finite-difference solution (apparent resistivities at 400 and 4 Hz) with the integral equation solution of Xiong (1992). The finite-difference solution is shown to the right and the integral equation solution to the left.

3. THE 3D MT INVERSE PROBLEM

3.1. Regularized least squares

Following Newman and Alumbaugh (2000), we divide the 3D earth into M cells and assign to each cell an unknown conductivity value. Let $\mathbf{m}$ be a vector of length M that

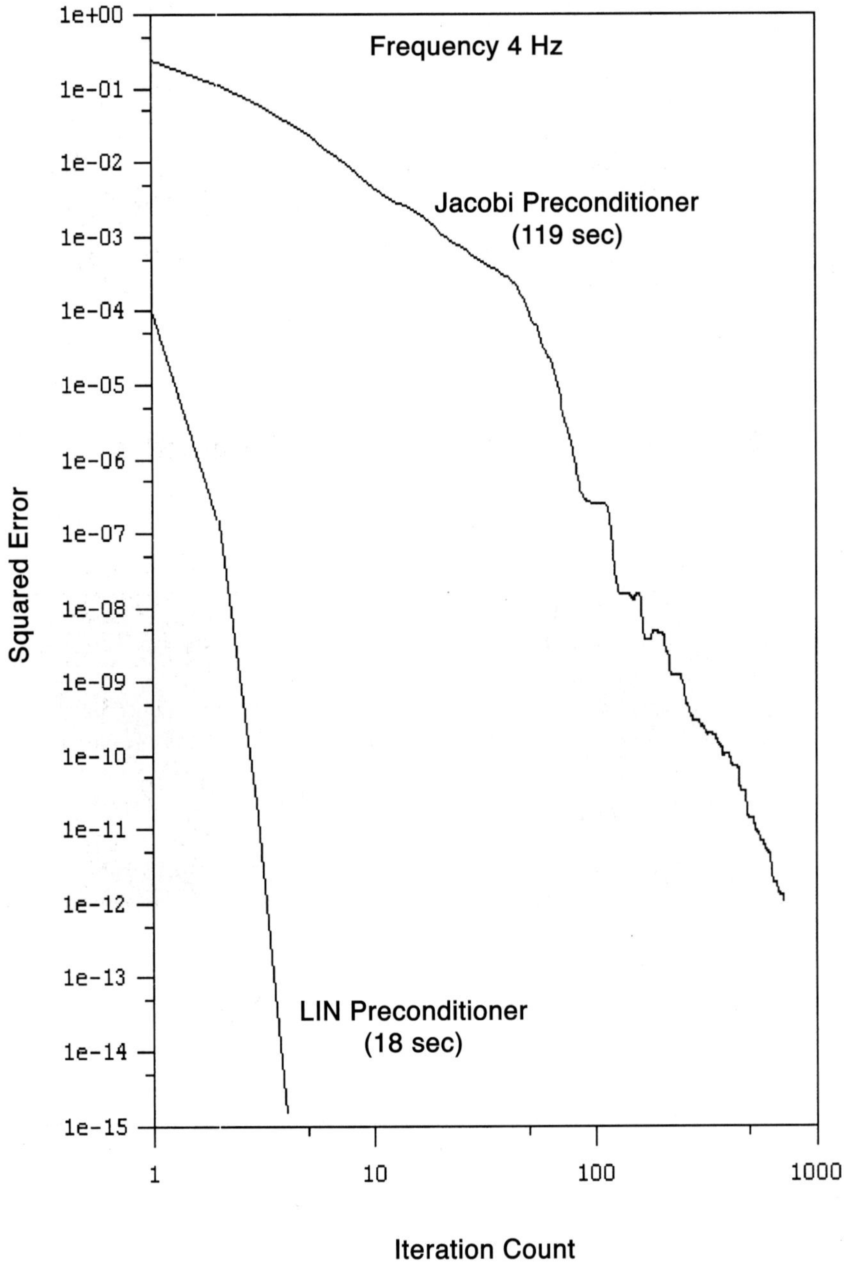

Figure 2. **Convergence rates for different preconditioners, where the squared error is defined as** $||\mathbf{K}E - S||^2/||S||^2$.

describes these values. The cost functional to be minimized in the inversion process, which combines the data error and model smoothness constraint, will be given by

$$\varphi = \{\boldsymbol{\varepsilon}^{-1}(\mathbf{Z}^{\text{obs}} - \mathbf{Z}^{\text{pre}})\}^{\mathsf{H}}\{\boldsymbol{\varepsilon}^{-1}(\mathbf{Z}^{\text{obs}} - \mathbf{Z}^{\text{pre}})\} + \lambda \mathbf{m}^{\mathrm{T}}\mathbf{W}^{\mathrm{T}}\mathbf{W}\mathbf{m}, \tag{8.14}$$

where H denotes the Hermitian operator. In Equation (8.14), $\mathbf{Z}^{\text{obs}}$ and $\mathbf{Z}^{\text{pre}}$ are data vectors that represent the predicted and observed magnetotelluric impedances at different frequencies and locations. These are complex values, and a given entry in the data vector can represent any component of the impedance tensor,

$$\mathbf{Z} = \begin{pmatrix} Z_{xx} & Z_{xy} \\ Z_{yx} & Z_{yy} \end{pmatrix}. \tag{8.15}$$

For a description of the tensor properties as well as how they are derived, see Vozoff (1991). Finally $\boldsymbol{\varepsilon}^{-1}$ in Equation (8.14) is a diagonal matrix that contains the inverse of the data error estimates. Thus noisier data are given smaller weight, or less importance, when forming φ, than good quality data.

The regularization parameters that stabilize the inverse problem (Tikhonov and Arsenin, 1977) enforce a model smoothness constraint. Other constraints are available, some of which will be discussed below, but we employ this particular constraint here because it produces properties in the solution that we desire. In Equation (8.14) the regularization parameters are given by the matrix $\mathbf{W}$, which consists of a finite-difference approximation to the Laplacian (∇^2) operator, and the tradeoff parameter λ. This later parameter is used to control the amount of smoothness to be incorporated into the model. In its selection, we note that a large parameter will produce a highly smooth model, but this model will show poor dependence on the data. A small parameter, on the other hand, will give a superior data fit, but the resulting model may be too rough and non-physical. Following Newman and Alumbaugh (2000), Equation (8.14) is minimized multiple times with different tradeoff parameters that are fixed, and the smoothest model that provides an acceptable match to the data within observational errors is selected as the optimal result.

3.2. Nonlinear conjugate gradients

Because of the size of the inverse problem, gradient methods are the only practical means of solution. The method of steepest descent is the easiest and simplest to implement of the gradient methods. Unfortunately it usually converges very slowly in practice. A better approach is the method of nonlinear conjugate gradients, first proposed by Fletcher and Reeves (1964) for nonlinear optimization, and later improved by Polyak and Ribière (1969) and recently implemented in 2D and 3D MT inversion algorithms (cf. Newman and Alumbaugh, 2000; Rodi and Mackie, 2001). The method is closely related to the linear CG method of Hestenes and Stiefel (1952) and is in fact identical if the objective functional is quadratic. Listed below is a flowchart of the Polyak and Ribière algorithm, that will be used in the analysis to estimate the conductivity model, $\mathbf{m}^*$, which will minimize Equation (8.14).

NLCG Algorithm

(1)	set $i=1$, choose initial model $\mathbf{m}_{(i)}$ and compute $\mathbf{r}_{(i)} = -\nabla\varphi(\mathbf{m}_{(i)})$
(2)	set $\mathbf{u}_{(i)} = \mathbf{M}_{(i)}^{-1}\mathbf{r}_{(i)}$
(3)	find $\alpha(i)$ that minimizes $\varphi(\mathbf{m}_{(i)} + \alpha_{(i)}\mathbf{u}_{(i)})$
(4)	set $\mathbf{m}_{(i+1)} = \mathbf{m}_{(i)} + \alpha_{(i)}\mathbf{u}_{(i)}$ and $\mathbf{r}_{(i+1)} = -\nabla\varphi(m_{(i+1)})$
(5)	Stop when $\|\mathbf{r}_{(i+1)}\|$ is sufficiently small, otherwise go to step (6)
(6)	set $\beta_{(i+1)} = (\mathbf{r}_{(i+1)}^{\mathrm{T}}\mathbf{M}_{(i+1)}^{-1}\mathbf{r}_{(i+1)} - \mathbf{r}_{(i+1)}^{\mathrm{T}}\mathbf{M}_{(i)}^{-1}\mathbf{r}_{(i)})/\mathbf{r}_{(i)}^{\mathrm{T}}\mathbf{M}_{(i)}^{-1}\mathbf{r}_{(i)}$
(7)	set $\mathbf{u}_{(i+1)} = \mathbf{M}_{(i+1)}^{-1}\mathbf{r}_{(i+1)} + \beta_{(i+1)}\mathbf{u}_{(i)}$
(8)	set $i = i+1$ and go to step (3)

We define $\mathbf{M}_{(i)}^{-1}$ and $\mathbf{M}_{(i+1)}^{-1}$ as preconditioning matrices for all i. To use the NLCG method requires that we carefully implement two calculations of the procedure efficiently. These are (1) calculate the gradient of the cost functional, $\nabla\varphi(\mathbf{m})$, and (2) find the value of α that minimizes the expression $\varphi(\mathbf{m}+\alpha\mathbf{u})$ for specified model parameters $\mathbf{m}$ and a given conjugate search direction $\mathbf{u}$. Such procedures along with the specification of the preconditioners can be found in Newman and Alumbaugh (2000), including a massively parallel implementation of the algorithm.

3.3. Solution stabilization via additional constraints

It is commonly known that the MT inverse problem is inherently ill posed. The instability of the problem arises because the problem is underdetermined and the data sets that are used are undersampled and noisy. While regularization discussed above can stabilize the problem by omitting solutions that are not geologically reasonable, incorporation of a priori knowledge can significantly reduce remaining non-uniqueness. In fact, the choice of the Laplacian operator to form $\mathbf{W}$ can be thought of as a form of a priori knowledge as the inversion scheme will only search for smoothly varying models as acceptable solutions. Here, we have allowed three other types of a priori information about the model to be incorporated into the inversion algorithm. As it will be demonstrated later in this paper, addition of a priori information can help improve the resolution of the MT inverse problem.

The first constraint we introduce is to invert against a reference model, $\mathbf{m}_{\mathrm{ref}}$, where we now minimize

$$\varphi = \{\boldsymbol{\varepsilon}^{-1}(\mathbf{Z}^{\mathrm{obs}} - \mathbf{Z}^{\mathrm{pre}})\}^{\mathrm{H}}\{\boldsymbol{\varepsilon}^{-1}(\mathbf{Z}^{\mathrm{obs}} - \mathbf{Z}^{\mathrm{pre}})\} + \lambda(\mathbf{m}-\mathbf{m}_{\mathrm{ref}})^{\mathrm{T}}\mathbf{W}^{\mathrm{T}}\mathbf{W}(\mathbf{m}-\mathbf{m}_{\mathrm{ref}}). \tag{8.16}$$

Justification for this type of constraint is that in many cases we have prior information on the geological units that we expect to encounter in a survey. Inverting against a reference model refines this information with the constraint that an acceptable fit to the data can be achieved. Another useful constraint is to incorporate 'tears' in the smoothing matrix W (cf. Hoversten et al., 1998). This constraint applies when we know the location of a boundary surface between two regions of contrasting conductivity. An example of this situation occurs in marine magnetotellurics when determining the base of sub-salt structures. Here the top of salt is known independently from seismic data and this information can be used to better restrain the base of salt that is to be estimated in the MT inversion process. A third type of a priori knowledge involves lower-bound constraints, which allow the user to designate the lowest conductivity value that the model can attain on a cell by cell basis. This constraint is actually implemented by inverting for the logarithm of the parameters and also insures that the conductivity is positive, which is a physical requirement. Following Newman and Alumbaugh (1995), let us introduce a new parameter n_k that is related to the kth model parameter, m_k using a natural logarithm transformation as follows:

$$n_k = \ln(m_k - \mathrm{lb}_k), \tag{8.17}$$

where lb_k is a lower bounding constraint such that

$$m_k > \mathrm{lb}_k. \tag{8.18}$$

We also can express m_k in terms of n_k, via the expression

$$m_k = \mathrm{e}^{n_k} + \mathrm{lb}_k. \tag{8.19}$$

Hence

$$\partial m_k / \partial n_k = m_k - \mathrm{lb}_k. \tag{8.20}$$

To alter the NLCG algorithm to invert for log parameters, $\mathbf{n}$, we first split the gradient into two components which arise from the data misfit, $\nabla_{\mathrm{d}} \varphi$, and the regularization imposed upon the problem, $\nabla_{\mathrm{m}} \varphi$, where

$$\nabla \varphi = \nabla_{\mathrm{d}} \varphi + \nabla_{\mathrm{m}} \varphi. \tag{8.21}$$

We then simply scale components of the gradient arising from the first term in Equation (8.21), $\nabla_{\mathrm{d}} \varphi$, using Equation (8.20) (that is for the kth component of $\nabla_{\mathrm{d}} \varphi$ we scale it by the factor $m_k - \mathrm{lb}_k$) and replace the model parameter vector $\mathbf{m}$ by its log counterpart $\mathbf{n}$ in the NLCG algorithm.

4. MARINE MT RESOLUTION STUDY

Next we demonstrate the 3D MT inversion code capabilities with a resolution study of imaging 3D sub-salt structures, an important target in marine magnetotellurics for oil exploration. For a detailed description of the marine MT method, including the instrumentation involved and data interpretation practices, we refer readers to the works of Constable et al. (1998) and Hoversten et al. (1998, 2000). The aim of the current study is to determine if the base of salt could be better resolved using 3D data analysis compared with much faster 2D interpretation of the data along selected profiles. An additional point of the investigation is to determine where 2D analysis of the data is inappropriate due to the assumption of 2D geology. The data employed in this study were generated with the finite-difference solution discussed earlier in the paper at 13 frequencies that span the range from 0.125 to 0.0005. A 175-MHz R10000 Octane SGI workstation was employed for the forward calculations. Sections of the model are shown

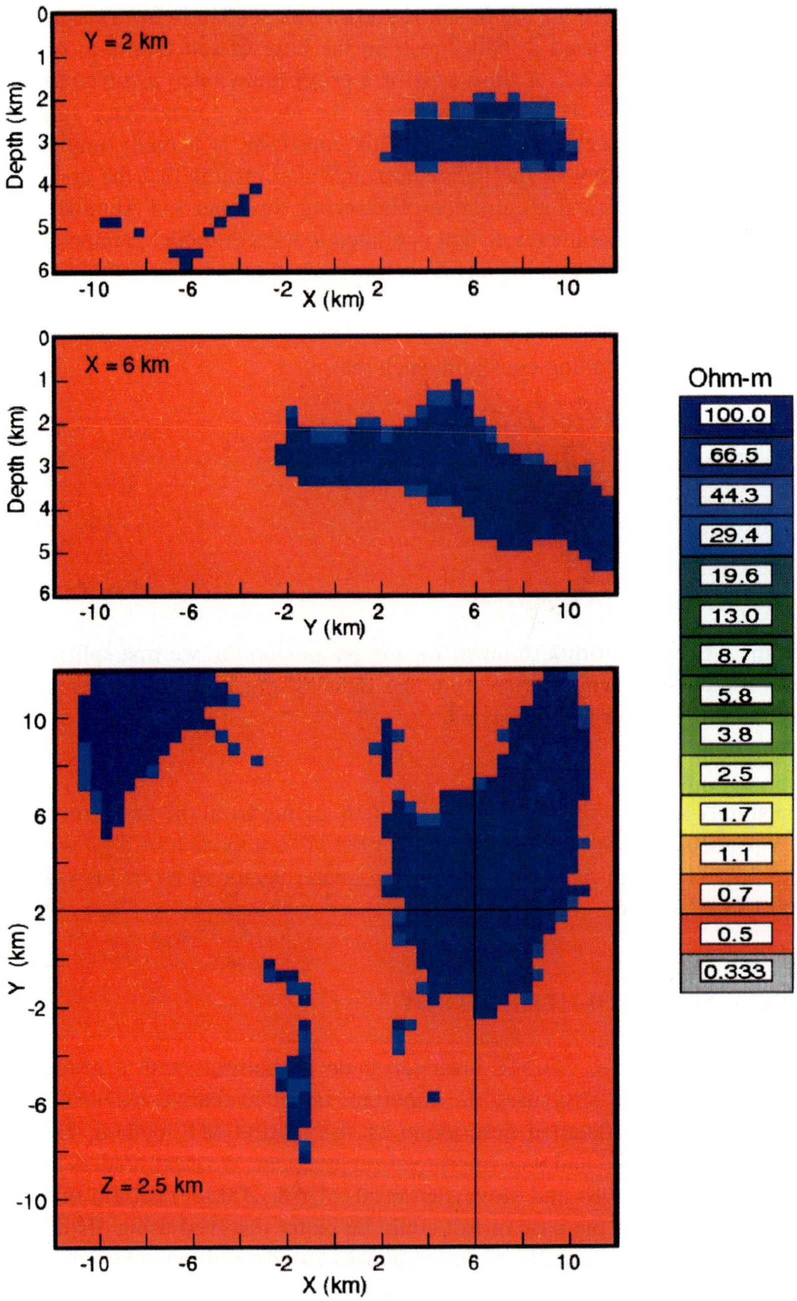

Figure 3. 3D Mahogany salt model used in seafloor MT inversion study. The top two panels show resistivity cross-sections at $Y = 2$ km and $X = 6$ km. The bottom panel illustrates the lateral geometry of the salt bodies at 2.5 km depth; note that the top flank of the salt actually terminates at $Y = 13$ km, off the top of the X–Y map section. The seawater depth is small at 100 m in the model and the seawater is assigned a resistivity of 0.333 Ω m.

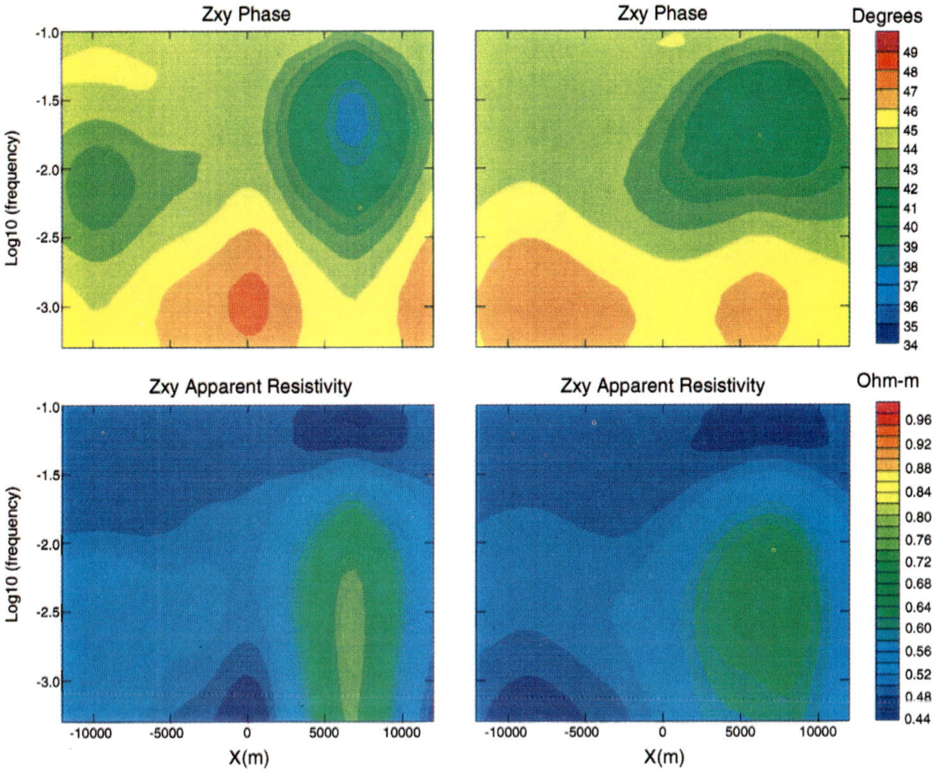

Figure 4. MT apparent resistivity and phase pseudo-sections at $Y = 2$ km. Results are illustrated only for Z_{xy} and Z_{yx} impedances.

in Figure 3, where the sea is quite shallow (100 m) and has been assigned a resistivity of 0.333 Ω m. The 100-Ω m salt structure, embedded within the 0.5-Ω m sediments, is a highly complex 3D feature, and is modeled after the Mahogany prospect, in the Gulf of Mexico. Figures 4 and 5 show selected pseudo-sections of impedance phase and resistivity. The 3D nature of the site is readily apparent in the apparent resistivity responses, which show the 3D salt bodies affecting these responses to arbitrary low frequencies in both the Z_{xy} and Z_{yx} modes. In computing these results the LIN preconditioner was very effective in accelerating the solution at the longer periods. As an example, a reduction of more than a factor of 12 was observed in the computational times at the three lowest frequencies (5×10^{-4}, 7.92×10^{-4} and 1.25×10^{-3} Hz) when compared to the solution that employed a simple Jacobi preconditioner.

Because the 3D finite-difference solution is employed within the 3D NLCG scheme, we used a finer grid in solving the inverse problem. This step insures that the inversion results will not depend upon the grid used in simulating the data, and thereby provides another check on the inversion scheme. In preparing the data for inversion, 5% Gaussian random noise based on the amplitude of the impedance at the seafloor was added to the off diagonal components of the impedance tensor. The diagonal components, Z_{xx}

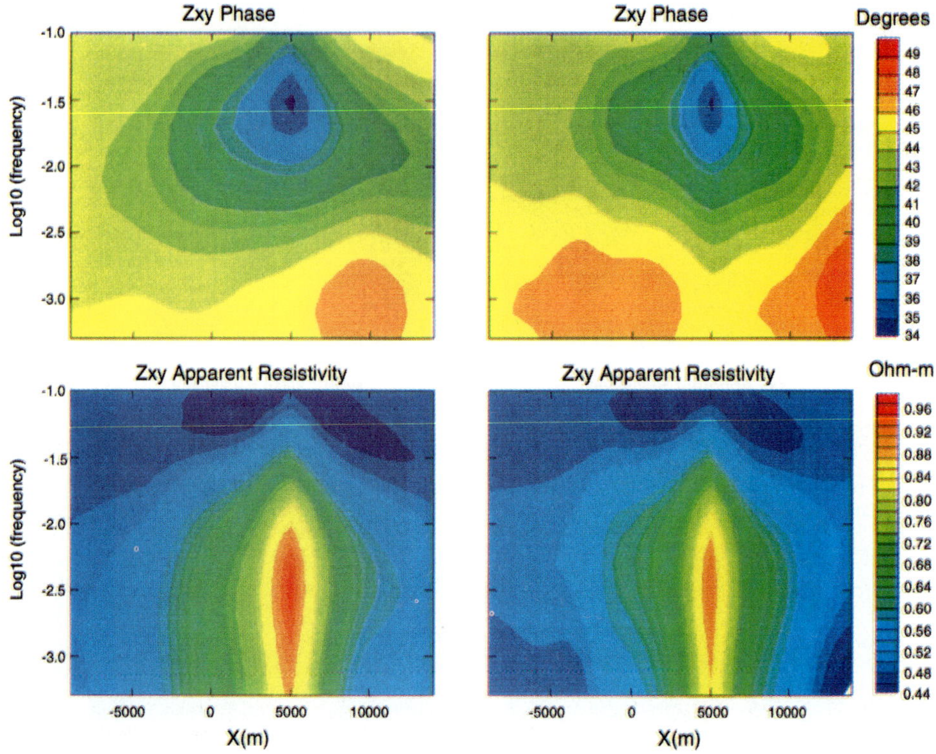

Figure 5. MT apparent resistivity and phase pseudo-sections at $X = 6$ km. Results are illustrated only for Z_{xy} and Z_{yx} impedances.

and Z_{yy}, were omitted in the analysis at all frequencies. Our justification for omitting these data values is based on the observation that they are effectively random noise in the high- to mid-frequency band of the measurement. Common sense then dictates that they be removed before analyzing field data. While it is possible to include these data values at lower-frequency bands in the analysis we have chosen not to at this time. The data set to be inverted is extremely large; 31,200 data points in survey lines that are spaced every kilometer along the X and Y coordinate directions that range from ± 12 km in X and -9 to 14 km in Y. Our justification in using this amount of data in the analysis is to determine in the ideal case the optimal image resolution one can expect from 3D inversion. Thus with the knowledge of the best imaging resolution that can be achieved for the problem, one can then construct data acquisition strategies that tradeoff model resolution against survey costs. To image the 100-Ω m salt structures within the 0.5-Ω m sediments required 129,360 model or conductivity parameters. Even with the efficiencies of the NLCG algorithm it was necessary to run this inversion simulation on Sandia National Laboratories massively parallel terraflop computer using 288 processors. The inversion was launched with a starting model consisting of seawater over 0.5 Ω m sediment. Runs using several different tradeoff parameters, demonstrated that a tradeoff parameter of 10, the smallest employed, yielded a data misfit that

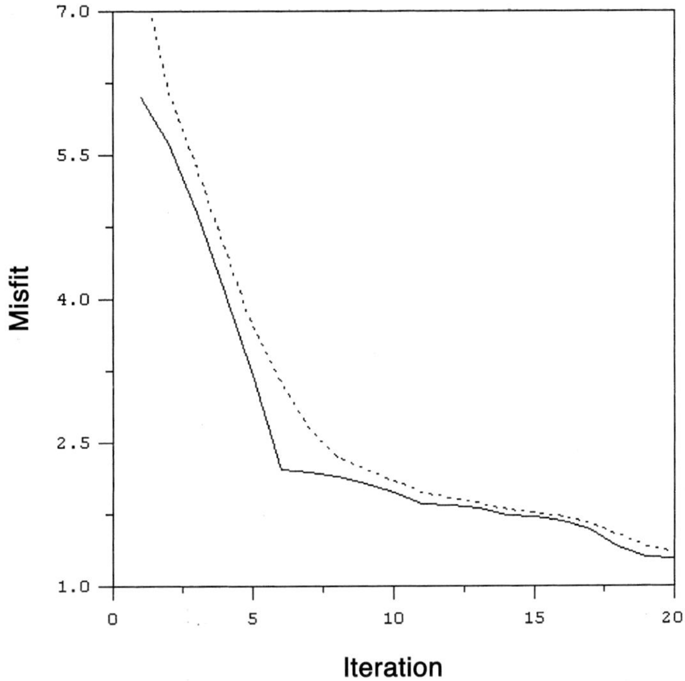

Figure 6. 3D NLCG solution convergence for the marine MT model example. The dashed curve illustrates the reduction in the cost functional (Equation 8.16) with increasing solution iteration. The solid curve represents the squared error, φ_d, where the model smoothness constraint term is omitted in the evaluation of the cost functional. The tradeoff parameter of 10 was specified to obtain these convergence results. Results show that over the first five iterations significant reduction occurs in the misfit φ.

approached the target value of 1, the assumed noise level (Figure 6); larger values yielded larger misfits and the corresponding models were rejected on that basis.

Shown in Figure 7 are cross-sections from the inverse model at 2 km intervals along the North–South direction, coinciding with the y coordinate direction in the model. Each depth section is 20 km in length and 10 km in depth. The top slice is located at $y = 0$ km and the bottom one at 12 km. To improve the resolution of these images we also incorporated a tear in the regularization matrix at the top of salt, and applied lower bounding constraints on the model parameters, but did not invert against a reference model in this example. The lower bounds on the conductivity were set at 2 and 0.01 S/m above and below the top of salt respectively, or equivalently 0.5 and 100 Ω m as upper bounds on the resistivities; because parts of the starting model coincide with the upper bounding constraint of 0.5 Ω m, we perturbed the bound to 0.55 Ω m at those points to insure that Equation (8.17) stays bounded. Incorporation of the tear and lower-bound constraints into the inverse problem is justified since seismic data provide independent estimates on depth to the top of salt.

Notice that the inversion recovers the general shape of the structures, especially at the top of the salt where we have incorporated the tear. However also notice that (1) the

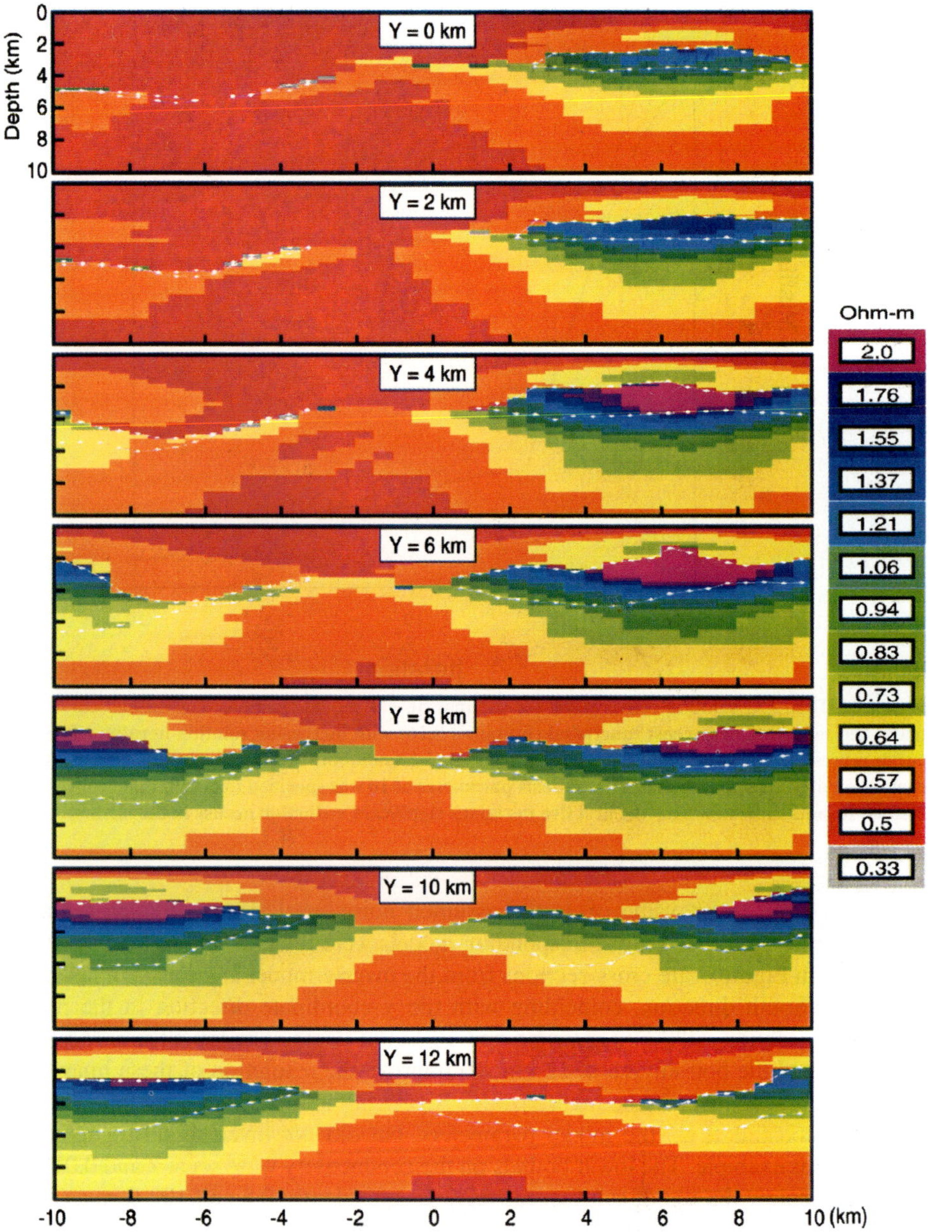

Figure 7. Illustrated here is a three-dimensional reconstruction of the Mahogany salt bodies. Seven cross-sections are illustrated at 2-km intervals along the North–South direction, starting at $Y = 0$ km at the top and ending at 12 km at the bottom. Each depth section is 20 km in length and 10 km in depth. The white lines indicate the true positions of the top and base of salt. Departures from the sediment resistivity of 0.5 Ω m indicate resistive salt features with maximum resistivity estimates of 5 Ω m.

bottom of the salt is imaged to be smoothly varying rather than a sharp interface, and (2) the highest resistivity that is recovered within the salt is well below the known value of 100 Ω m. The first effect is to be expected due to the smoothing nature of the regularization constraints that are imposed, i.e., the problem is not set up to image the location of sharp interfaces. With regard to not being able to recover the actual resistivity of the salt, this too is to be expected as the MT response becomes saturated once the resistivity contrast is large (Hoversten et al., 1998). Fortunately, the resistivity of salt is not of interest in petroleum exploration, only its base location.

In Figure 8, the predicted impedances for the model are compared with the input data at 0.0005 Hz in map view. Figure 9 shows a pseudo-section plot of apparent resistivity and phase at $Y = 9$ km. In both cases the agreement between the predicted and input data are good, with the main difference being the presence of the random noise in the input data which is not present in the predicted results.

We have also done a series of 2D inversions on the Z_{xy} mode data along the $Y = 8$ km section (east–west line) of the 3D data set to illustrate two points. First of all we will compare 2D and 3D inversion over this section to see what if any benefits the 3D inversion offers. Note from Figure 3 and the results in Figure 7 that this profile has been chosen in a region where the structure is somewhat 2D, that is the variation in the y direction is small compared to that in the x direction. Therefore this represents a best-case scenario for 2D inversion to produce reliable results. The second point we wish to consider are the effects of the various constraints that can be imposed in resolving of the base of the salt. In this regard we have not only employed a 2D version of the 3D algorithm described in this paper, but also the Sharp2D algorithm (Smith et al., 1999). This later algorithm is ideally designed to invert for the location of boundaries between regions of different conductivity rather than for a smoothly varying model. Together, these algorithms span a range of possible constraints that can be imposed upon this type of model, and therefore serve to illustrate the considerable impact that different types of a priori knowledge can have on recovering certain features within the image (base of salt in this case).

Five different cases are illustrated in Figure 10. Figure 10a shows the model produced by 2D nonlinear conjugate gradient inversion with no constraints of any kind, and Figure 10b shows the result when a tear in the regularization is applied at the top of the salt. Figure 10c shows the inverse model generated when a tear is employed at the top of the salt along with the same upper bounds on resistivity above and below the top of salt as was used in the 3D case. Figure 10d shows the inverse model produced by the Sharp2D code (Smith et al., 1999) where not only the top of the salt location was fixed but also the sediment resistivity above the top of salt and the resistivity of the salt were fixed. The salt resistivity was fixed at 10 Ω m, which is not equal to the true 100 Ω m but is high enough that the MT response is already saturated. As previously discussed, increasing the salt resistivity has no meaningful effect on the observed MT responses. Finally, Figure 10e shows the 3D inverted depth section at 8 km taken from Figure 7 and re-plotted at an identical scale used for the 2D images.

It is clear from the 2D results in Figure 10 that as more information is added to the inversion in the form of top-of-salt location and bounds on the possible resistivities of different regions, the inverse model improves. When no constraints of any form are

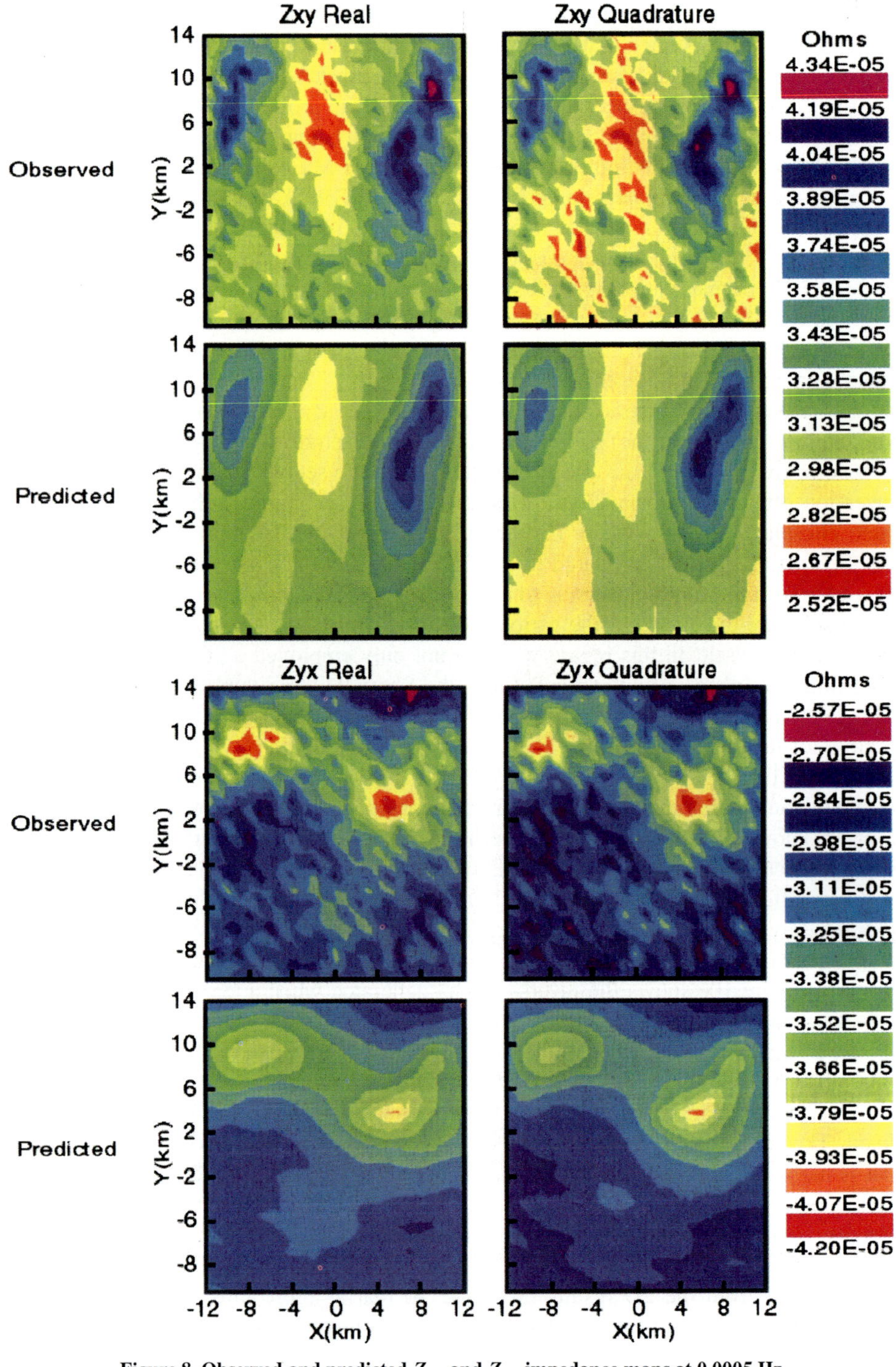

Figure 8. Observed and predicted Z_{xy} and Z_{yx} impedance maps at 0.0005 Hz.

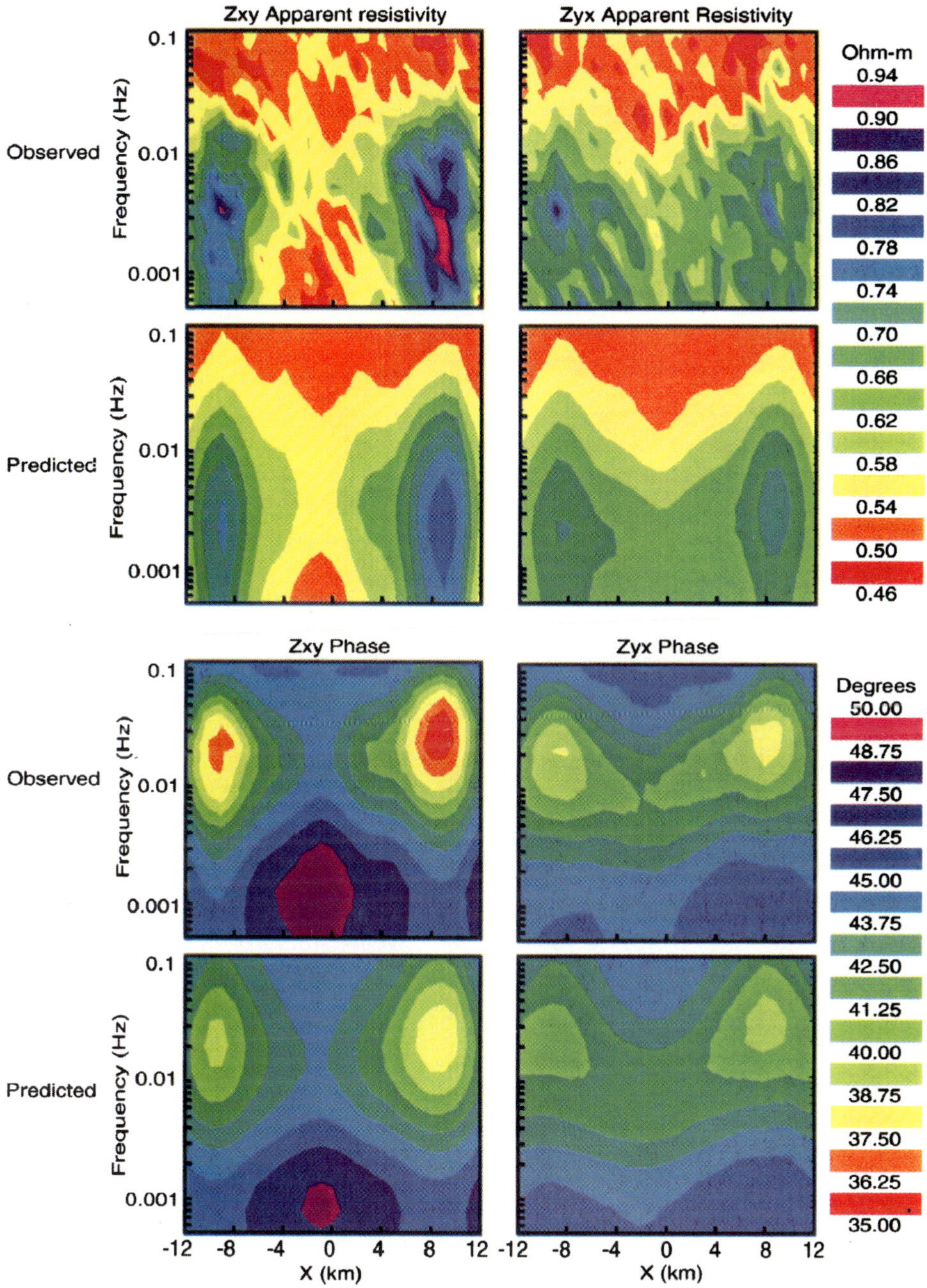

Figure 9. Observed and predicted apparent resistivity and phase soundings, based on Z_{xy} and Z_{yx} impedance data taken at $Y = 9$ km.

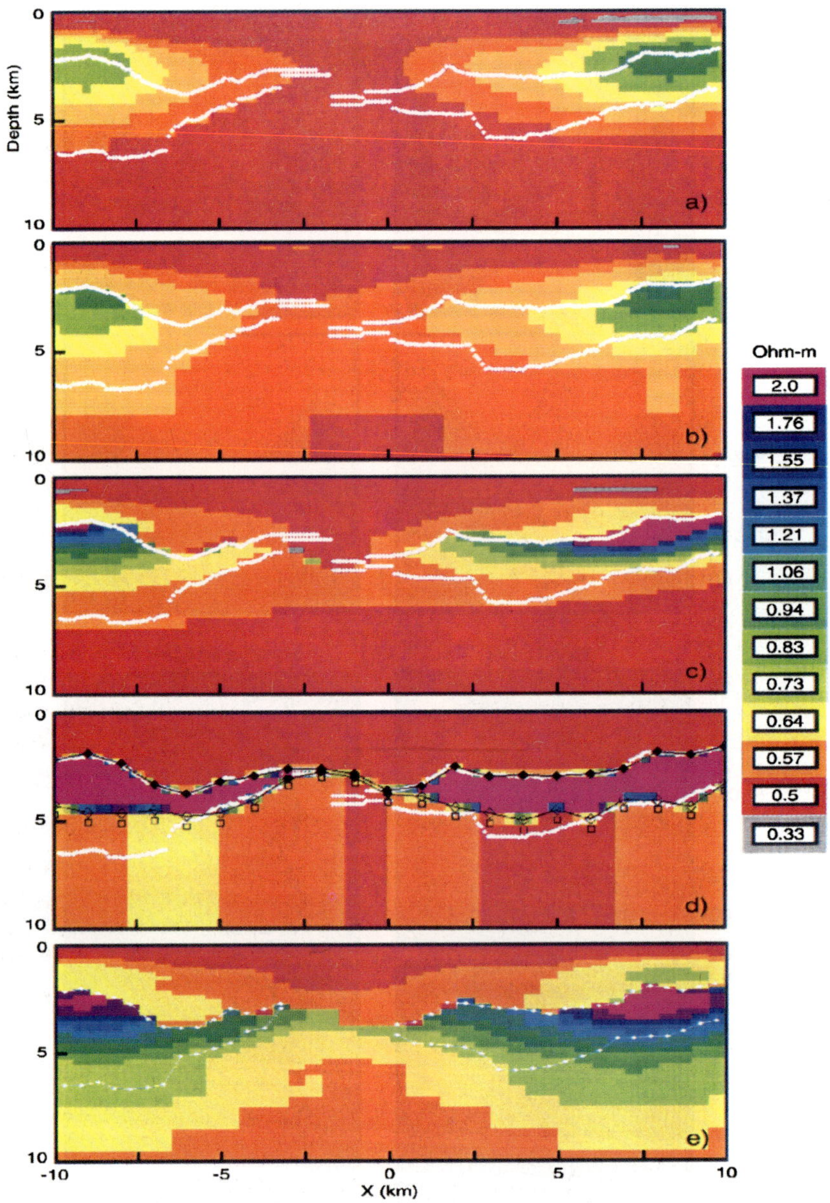

Figure 10. First four panels illustrate 2D inverse models from Z_{xy} mode data at $Y = 8$ km. The starting model was a 0.5-Ω m half-space except where noted otherwise. Panel (a) inversion with no constraints applied. Panel (b) inversion with cell resistivity smoothing eliminated across the top of salt. Panel (c) inversion with cell resistivity smoothing eliminated across the top of salt and upper bounds of 0.5 and 100 Ω m above and below the top of salt respectively. Panel (d) inversion using the Sharp2D algorithm where the resistivity above the top of salt was fixed to the true value, the top salt boundary was fixed at its true location, and the salt resistivity was fixed at 10 Ω m. Only the base of salt and the resistivity below the base of salt were sought in the inversion. Black diamonds represent boundary node locations (filled = fixed, open = variable), black open squares indicate the resistivity below the interface was free to vary. Panel (e) $Y = 8$ km section taken from 3D inversion shown in Figure 7.

used (Figure 10a) the inverse model has high-resistivity zones where the two major salt units are nearest the surface. In this case the distribution of the smoothed high-resistivity zones could lead to interpreting both the top and base of the salt as being too shallow in the section. In addition almost no indication is given of the deeper sills of salt extending from the main salt bodies on either sides into the center of the section. When the location of the top of salt is used to eliminate cell smoothing at this location the high-resistivity zones of the inverse model (Figure 10b) have moved to greater depth, better representing the true distribution. In particular, the transition in the color contours from 0.73 to 0.83 Ω m is now beginning to correctly locate the deeper and thinner salt sills in the center of the section. When the upper resistivity bounds of 0.5 and 100 Ω m, above and below the top of salt respectively, are added to the smoothing tears (Figure 10c) the image again improves. Both salt sills are now better defined when compared to Figure 10a and b.

The final case, Figure 10d, not only increases the constraints placed on the model (resistivity above salt is fixed, top salt is fixed, and salt resistivity is fixed) but also changes the parameterization of the model. Rather than cell resistivity being the parameter of the inversion, the depths of interface–nodes and the resistivity of the layer (defined at the node positions) below an interface are estimated by the inversion process. Both the boundary location and resistivity of the layer below the interface are linearly interpolated between nodes. This parameterization allows sharp jumps in resistivity across boundaries, which more closely matches the geologic model. In this particular example, only the location of the base-of-salt nodes and the resistivities below the salt were determined in the inversion. This represents a realistic case where the top salt is known from seismic data, the sediments above salt are known from well logs, and the salt itself is known to have a resistivity substantially greater than the surrounding sediments. With this amount of a priori information the inversion is able to more accurately locate the base of salt of both bodies as well as clearly define the thinner salt sills in the center of the section. However, notice that the base of the salt body on the left side of the model (the deepest salt) is not well defined by any of the four 2D examples shown here. Thus this either represents a portion of the model which the data is not sensitive to, or a location where 3D effects distort the 2D inversion results.

To compare 2D and 3D inversion results, consider the $Y = 8$ km panel of Figure 7 redrawn as Figure 10e, which should be compared directly with Figure 10c. The same basic inversion algorithm and constraints were used in both cases with the only difference being the dimensionality that is assumed. Remember from Figure 3 that although this profile crosses a region where the variation in the y direction is minimal compared to other areas within the model, there is still significant 3D structure present and thus there will be 3D effects in the data.

In general the two inversion results appear to yield much of the same information. However the 3D inverse model (Figure 10e, $Y = 8$ km) appears to do a better job of indicating the presence and base location of both salt sills in the center of the section than does the 2D inverse model (Figure 10c). In particular, the thin salt sill extending to the center from the left side body is well imaged in the 3D model compared to the 2D model. An argument could also be made that the 3D inverse model better represents the base of the left side salt body better than any of the 2D inversions shown in Figure 10.

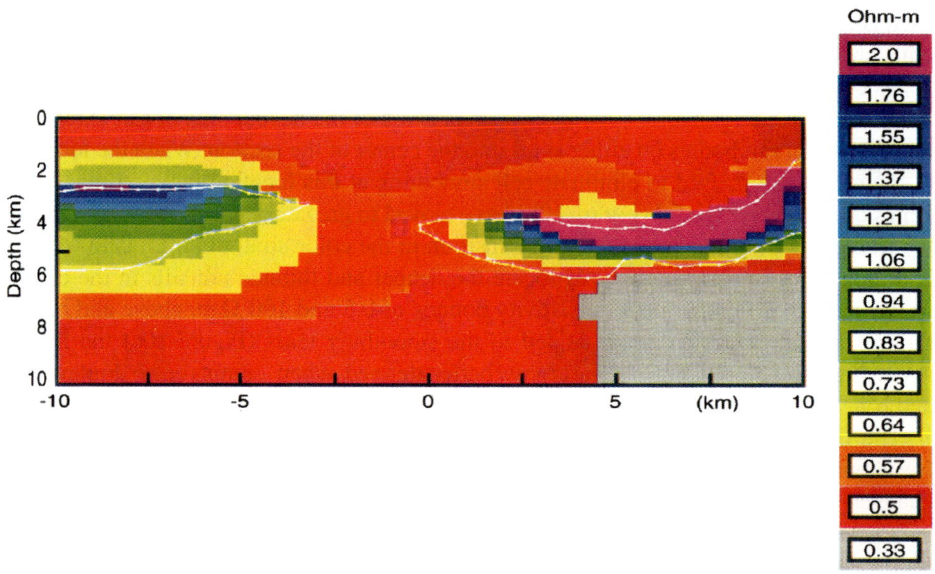

Figure 11. Two-dimensional inverse model from Z_{xy} and Z_{yx} mode data taken over the edge of the salt flank at $Y = 12.5$ km.

However also note that the 2D inversion appears to better represent the right side body toward the $X = 10$ km edge of the inversion mesh than the 3D inversion results.

Nevertheless 2D inversion requires careful selection of data profiles to insure that 3D lateral effects are minimized. A poor selection can produce artifacts as Figure 11 demonstrates. Here the profile selected to be inverted is situated at $Y = 12.5$ km, over the lateral flanks of the salt structures; note that the salt flanks actually extend out to $Y = 13$ km in the lower section of Figure 3. Following the recommendation of Berdichevsky et al. (1998), both Z_{xy} and Z_{yx} modes were used in the inversion to insure the most reliable determination of the subsurface geology, especially given the resistive nature of the 3D salt bodies. While the base of salt is imaged exceedingly well, an artifact appears below the salt in the lower right side of the image. This artifact represents a 34% deviation from the sea sediments and could be erroneously interpreted as an important geological structure. However, one could argue, given the subtle nature of the artifact, this would not lead to a serious misinterpretation of the geology. Nevertheless, this result clearly shows the benefit of 3D inversion, since the corresponding 3D image at 12 km in Figure 7 shows no such artifact and does a reasonable job in imaging the salt.

While 3D inversion results and carefully selected 2D inversion profiles are encouraging, images of the base of salt are not as sharp as would be desired. This finding is somewhat discouraging for 3D data analysis given the amount and coverage of the data employed in the 3D analysis. On the other hand, it suggests that the 3D inversion could be made with fewer sounding locations, given the good results obtained using 2D data analysis. Nevertheless, we are currently investigating additional constraints that can be incorporated into the problem to improve the base salt resolution. The ability to place arbitrary upper and lower bounds on different regions of the model as well as fixing resistivities over portions of the model (comparable to the Sharp2D example) would clearly be beneficial in cases like these.

5. CONCLUSIONS

In this paper we reviewed techniques to model and invert MT data in three dimensions with application to modeling and imaging salt structures, an important target in marine magnetotellurics. To accelerate 3D finite-difference modeling at long periods, we have also introduced a fast preconditioner that is based upon an approximate solution of Maxwell's equations at low induction numbers. Test examples clearly show the benefit of the preconditioner, where an order of magnitude speed up in solution time has been demonstrated on a series of test problems. The unique feature of the preconditioner is its ability to deflate out the effect of the null space of the discrete curl–curl operator in the Krylov solution process. When this null space is active, it is responsible for the poor convergence properties of the forward-modeling solution at long periods.

The resolution analysis has shown that 3D inversion, with carefully crafted constraints that incorporate a priori information, will offer somewhat improved resolution compared to corresponding 2D analysis. Whether the marginal improvements in resolution are cost effective is debatable given the time and amount of high-quality data needed for 3D data inversion. On the other hand, care must be exercised, when interpreting 3D data sets with 2D inversion schemes. As has been demonstrated improperly chosen profiles over the salt flanks, can produce 3D artifacts within the 2D model. While such artifacts are subtle in nature, they have only been observed over the salt flanks and not over the center of the resistive salt bodies. Thus it appears that these artifacts are caused by the assumptions in the 2D modeling and inversion algorithms not being fully able to incorporate the 3D nature of the data.

In spite of our findings on the marginal improvements demonstrated here with 3D data analysis, we believe that with increasing computer power and new algorithms 3D MT modeling and inversion will see increasing use within the induction community. One can well imagine a situation where 3D effects are so severe that proper analysis of the data requires full 3D treatment. Thus the solution techniques presented in this paper are a step in this direction. More importantly, with the incorporation of constraints in the inversion process, 3D MT modeling and inversion offers the potential to realistically image complex geological systems of economic and academic interest.

ACKNOWLEDGEMENTS

We wish to acknowledge the numerous discussions with David Day on solution preconditioning and the critique of Andreas Hördt, which improved the paper. This work was performed at Sandia National Laboratories, Lawrence Berkeley National Laboratory and the University of Cologne in the Federal Republic of Germany. Funding provided by the United States Department of Energy's Office of Basic Energy Sciences,

Division of Engineering and Geoscience and a Mercator grant awarded to G.A. Newman from the Deutsche Forschungsgemeinschaft. Sandia is a multi-program laboratory operated by the Sandia Corporation, a Lockheed Martin Company, for the United States Department of Energy under Contract DE-AC04-94AL85000.

Appendix A

It is possible to generalize the LIN preconditioner and provide a rigorous bound of its effectiveness by recasting the solution of Equation (8.2) as

$$\mathbf{E} = (\mathbf{I} + i\omega\mathbf{A}^{+}\mathbf{D})^{-1}\mathbf{A}^{+}\mathbf{S}. \tag{8.A1}$$

The matrices $\mathbf{I}$, $\mathbf{A}$ and $\mathbf{D}$ are matrices that represent the identity, the discrete curl–curl operator and attenuation terms in Equation (8.A1), where

$$\mathbf{K} = \mathbf{A} + i\omega\mathbf{D}. \tag{8.A2}$$

The matrix $\mathbf{D}$ is also diagonal with real elements and $\mathbf{A}^{+}$ represents the pseudo-inverse of $\mathbf{A}$, such that $\mathbf{A}^{+}\mathbf{A} = \mathbf{I}$ only over those vectors not included in the null space of $\mathbf{A}$; that is for some vector, $\mathbf{v}$, orthogonal to this null space, we have $\mathbf{A}^{+}\mathbf{A}\mathbf{v} = \mathbf{v}$. Matrix–vector products involving the pseudo-inverse, $\mathbf{A}^{+}\mathbf{S}$, are determined by solving discrete versions of Equations (8.10) and (8.12) with appropriate boundary conditions, included explicitly in $\mathbf{S}$. A Neumann series expansion of Equation (8.A1) (cf. Golub and Van Loan, 1989) can formally be written as

$$\mathbf{E} = \sum_{n=0}^{\infty}\{-i\omega\mathbf{A}^{+}\mathbf{D}\}^{n}\mathbf{A}^{+}\mathbf{S}, \tag{8.A3}$$

where the LIN preconditioner would be based on using only the first term in the series. This series will converge provided the spectral radius, denoted by ρ, is less than one, where

$$\rho(\omega\mathbf{A}^{+}\mathbf{D}) < 1. \tag{8.A4}$$

The spectral radius corresponds to the largest eigenvalue of $||\omega\mathbf{A}^{+}\mathbf{D}||$ and its evaluation will provide a measure under the worst possible conditions when the series will converge and the preconditioner will be effective. Unfortunately evaluation of Equation (8.A4) requires the solution of an eigenvalue problem, which is computationally impractical. Nevertheless an estimate of the spectral radius is readily available if we assumed that the conductivity and grid size is constant in the modeling problem. We can then estimate the spectral radius using the smallest non-zero eigenvalue of $\mathbf{A}$ following Newman and Alumbaugh (2002) as

$$\rho(\omega\mathbf{A}^{+}\mathbf{D}) \approx \omega\mu_{o}\sigma L_{\max}^{2}/2\pi^{2}, \tag{8.A5}$$

where $L_{\max}$ is the largest dimension employed in the 3D finite-difference grid. Equation (8.A5) clearly indicates that as frequency is decreased sufficiently, acceleration in the convergence rate of the Neumann series, and an increase in the effectiveness of the preconditioner will be observed. In a final remark, Equation (8.A3) can also be used to develop a higher-order LIN preconditioner at the expense of evaluating more terms in the series.

REFERENCES

Avdeev, D.B., Kuvshinov, A., Prankratov, O.V. and Newman, G.A., 2002. Three-dimensional induction logging problems, Part I: An integral equation solution and model comparisons. Geophysics, 67, 413–426.

Alumbaugh, D.L., Newman, G.A., Prevost, L. and Shadid J.N., 1996. Three-dimensional wide band electromagnetic modeling on massively parallel computers. Radio Sci., 31, 1–23.

Berdichevsky, M.N., Dmitriev, V.I. and Pozdnjakova, E.E., 1998. On two-dimensional interpretation of magnetotelluric soundings. Geophys. J. Int., 133, 585–606.

Constable, S.C., Orange, A.S., Hoversten, G.M. and Morrison, H.F., 1998. Marine magnetotellurics for petroleum exploration Part I: A sea floor equipment system. Geophysics, 63, 816–825.

DeGroot-Hedlin, C. and Constable, S.C., 1990. Occam's inversion to generate smooth two-dimensional models from magnetotelluric data. Geophysics, 55, 1613–1624.

Druskin, V.L., Knizhnerman and Lee, P., 1999. New spectral Lanczos decomposition method for induction modeling in arbitrary 3D geometry. Geophysics, 64, 701–706.

Egbert, G.D. and Booker, J.R., 1986. Robust estimation of geomagnetic transfer functions. Geophys. J. R. Astron. Soc., 87, 173–194.

Fletcher, R. and Reeves, C.M., 1964. Function minimization by conjugate gradients. Comput. J., 7, 149–154.

Frischknecht, F.C., 1987. Electromagnetic physical scale modeling. In: M.N. Nabighain (Ed.), Electromagnetic Methods in Applied Geophysics – Theory, Vol. 1. Soc. Explor. Geophys., Tulsa, OK, pp. 131–311.

Gamble, T.D., Goubau, W.M. and Clarke, J., 1979. Magnetotellurics with a remote reference. Geophysics, 44, 53–68.

Golub, G. and Van Loan, J., 1989. Matrix Computations. John Hopkins University Press, Baltimore, MD.

Greenbaum, A., 1997. Iterative methods for solving linear systems. Society for Industrial and Applied Mathematics, Philadelphia, PA.

Hestenes, M.R. and Stiefel, E., 1952. Methods of conjugate gradients for solving linear systems. J. Res. Natl. Bur. Stand., 49, 409–436.

Hoversten, G.M., Morrison, H.F. and Constable, S., 1998. Marine magnetotellurics for petroleum exploration Part 2: Numerical analysis of subsalt resolution. Geophysics, 63, 826–840.

Hoversten, G.M., Morrison, H.F. and Constable, S.C., 2000. Marine magnetotellurics for base salt mapping: Gulf of Mexico field test at the Gemini structure. Geophysics, 65, 1476–1488.

Newman, G.A. and Alumbaugh, D.L., 1995. Frequency-domain modeling of airborne electromagnetic responses using staggered finite differences. Geophys. Prospect., 43, 1021–1042.

Newman, G.A. and Alumbaugh, D.L., 2000. Three-dimensional magnetotelluric inversion using nonlinear conjugate gradients. Geophys. J. Int., 140, 410–424.

Newman, G.A. and Alumbaugh, D.L., 2002. Three-dimensional induction logging problems, Part II: A finite difference solution. Geophysics, 67, 484–491.

Polyak, E. and Ribière, G., 1969. Note sur la convergence des méthods conjugées. Rev. Fr. Inr. Rech. Oper., 16, 35–43.

Rodi, W. and Mackie, R.L., 2001. Nonlinear conjugate gradients algorithm for 2D magnetotelluric inversion. Geophysics, 66, 174–187.

Siripunvaraporn, W. and Egbert, G., 2000. An efficient data–space inversion for two-dimensional magnetotelluric data. Geophysics, 65, 791–803.

Smith, J.T., 1996. Conservative modeling of 3-D electromagnetic fields; Part II: Biconjugate gradient solution and an accelerator. Geophysics, 61, 1319–1324.

Smith, J.T. and Booker, J.R., 1991. Rapid inversion of two- and three-dimensional magnetotelluric data. J. Geophys. Res., 96, 3905–3922.

Smith, J.T., Hoversten, G.M., Gasperikova, E. and Morrison, H.F., 1999. Sharp boundary inversion of 2-D magnetotelluric data. Geophys. Prospect., 47, 469–486.

Tikhonov, A.N. and Arsenin, V.Y., 1977. Solutions to Ill-Posed Problems. Wiley, New York, NY.

Vozoff, K., 1991. The magnetotelluric method. In: M.N. Nabighain (Ed.), Electromagnetic Methods in Applied Geophysics – Applications, Vol. 2. Soc. Explor. Geophys., Tulsa, OK, pp. 641–712.

Xiong, Z., 1992. Electromagnetic modeling of 3-D structures by the method of system iteration using integral equations. Geophysics, 57, 1556–1561.

Yee, K.S., 1966. Numerical solution of initial boundary value problems involving Maxwell's equations in isotropic media. IEEE Trans. Antennas Propagation, AP-14, 302–309.

Chapter 9

2-D INVERSION OF FREQUENCY-DOMAIN EM DATA CAUSED BY A 3-D SOURCE

Yuji Mitsuhata and Toshihiro Uchida

National Institute of Advanced Industrial Science and Technology, Central No. 7, 1-1-1 Higashi, Tsukuba 305-8567, Japan

Abstract: We present a 2.5-D inversion algorithm for frequency-domain controlled-source electromagnetic data. This algorithm employs 2.5-D forward modeling with a finite-element method, calculation of sensitivities matrix based on the adjoint-equation approach and a linearized least-square method with a smoothness constraint. To find a reasonable balance between the data fitting and the smoothness constraint, we introduce the quasi-linearized Akaike's Bayesian information criterion. The tests for synthetic data show our inversion algorithm improves model resolution as iteration proceeds and can obtain final convergence automatically. Moreover, the algorithm provides a smoother and lower-contrast image of structure from noisy data, and shows dependence of resolution on the location of the source.

1. INTRODUCTION

In interpretation of electromagnetic (EM) data, three-dimensional (3-D) inversion methods must be ultimate tools. However, for elongated targets regarded as being two-dimensional (2-D) by comparison with the size of a transmitting source and the source–receiver offset, a 2-D interpretation is practically effective. In addition, there are some cases in which area-covered measurements for 3-D interpretation are difficult because of steep topography or entrance permission. The 3-D interpretation may be inapplicable for such insufficient data. Thus, there is a role for 2-D interpretation. For magnetotelluric (MT) data and direct current (DC) data, 2-D inversion codes already have been used and widely distributed and they have provided many successful results in various fields. In contrast, the published literature on the 2-D inversion of a 3-D controlled-source electromagnetic (CSEM) data, so-called the 2.5-D inversion, is scarce. Unsworth and Oldenburg (1995) presented a 2.5-D subspace inversion and demonstrated the effectiveness for sea-floor CSEM survey. Torres-Verdin and Habashy (1994) described a 2.5-D inversion using the extended Born approximation for EM tomography. Recently, Lu et al. (1999) developed a 2.5-D rapid relaxation inversion of CSAMT data including data in the transition and near fields.

In this paper we present a 2.5-D inversion method for CSEM data. We use a finite-element method (FEM) for forward modeling and a linearized least-squares technique, employing a smoothness constraint to stabilize the inverse process. Although the smoothness constraint is becoming a common technique as a stabilizer, the adjustment of the constraint in the inversion process is still one of the important problems under research. To find a reasonable weight of the constraint, we use an information criterion based on the Bayes theory. In the beginning of this paper, we will explain the theoretical basis of the inversion and then we will demonstrate the utility of our method by applying it to synthetic CSEM data. We consider influences of an initial model, noise and location of the source in the synthetic data examples.

2. INVERSION THEORY

2.1. 2.5-D forward modeling and sensitivities

We consider the situation shown in Figure 1, where a horizontal electric dipole source parallel with the strike (y-axis) and vertical magnetic receivers are located on the earth surface. The receivers are arranged on the x-axis that is perpendicular to the dipole source. By Fourier transforming the Maxwell equations with respect to the y-coordinate, we obtain the following two coupled differential equations for the along-strike fields, E_y and H_y (Hohmann, 1988),

$$\frac{\partial}{\partial x}\left(\frac{\eta}{k_e^2}\frac{\partial \hat{E}_y}{\partial x}\right)+\frac{\partial}{\partial z}\left(\frac{\eta}{k_e^2}\frac{\partial \hat{E}_y}{\partial z}\right)-\eta\hat{E}_y+ik_y\left[\frac{\partial}{\partial x}\left(\frac{1}{k_e^2}\right)\frac{\partial \hat{H}_y}{\partial z}-\frac{\partial}{\partial z}\left(\frac{1}{k_e^2}\right)\frac{\partial \hat{H}_y}{\partial x}\right]$$
$$=\hat{J}_{ey}-ik_y\left[\frac{\partial}{\partial x}\left(\frac{1}{k_e^2}\hat{J}_{ex}\right)+\frac{\partial}{\partial z}\left(\frac{1}{k_e^2}\hat{J}_{ez}\right)\right]+\frac{\partial}{\partial x}\left(\frac{\eta}{k_e^2}\hat{J}_{mz}\right)-\frac{\partial}{\partial z}\left(\frac{\eta}{k_e^2}\hat{J}_{mx}\right), \quad (9.1)$$

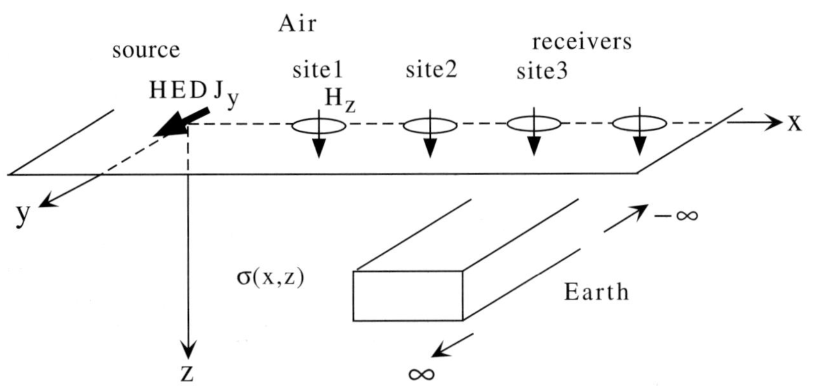

Figure 1. Coordinate system for our 2.5-D situation. The strike direction of a 2-D structure is the y-direction. A horizontal electric dipole (HED) source directed to the y-direction is placed on the y-axis and vertical magnetic receivers are arranged on the x-axis. The conductivity varies only in the x–z plane.

and

$$\frac{\partial}{\partial x}\left(\frac{\varsigma}{k_e^2}\frac{\partial \hat{H}_y}{\partial x}\right)+\frac{\partial}{\partial z}\left(\frac{\varsigma}{k_e^2}\frac{\partial \hat{H}_y}{\partial z}\right)-\varsigma\hat{H}_y+ik_y\left[-\frac{\partial}{\partial x}\left(\frac{1}{k_e^2}\right)\frac{\partial \hat{E}_y}{\partial z}+\frac{\partial}{\partial z}\left(\frac{1}{k_e^2}\right)\frac{\partial \hat{E}_y}{\partial x}\right]$$
$$=-\hat{J}_{my}+ik_y\left[\frac{\partial}{\partial x}\left(\frac{1}{k_e^2}\hat{J}_{mx}\right)+\frac{\partial}{\partial z}\left(\frac{1}{k_e^2}\hat{J}_{mz}\right)\right]+\frac{\partial}{\partial x}\left(\frac{\varsigma}{k_e^2}\hat{J}_{ez}\right)-\frac{\partial}{\partial z}\left(\frac{\varsigma}{k_e^2}\hat{J}_{ex}\right), \tag{9.2}$$

where $\hat{E}_y$ and $\hat{H}_y$ are Fourier transforms of E_y and H_y respectively with respect to the y-coordinate, $\hat{J}_m$ and $\hat{J}_e$ are those of impressed magnetic and electric currents, $\zeta = i\omega\mu_0$, $\eta = \sigma + i\omega\varepsilon$, $k^2 = -\zeta\eta$, and $k_e^2 = k_y^2 - k^2$. Throughout this study, we use the 2.5-D FEM code developed by Mitsuhata (2000) to solve Equations (9.1) and (9.2) numerically. In this modeling code, the EM field and geometry of the mesh are represented by quadratic interpolation in each element, and in order to relieve the singularity of the source term on the right-hand side of Equations (9.1) and (9.2), a distributed source based on Herrmann's pseudo-delta function (Herrmann, 1979; Mitsuhata, 2000) is adopted, which is given by

$$\delta_h(x-x_0)=\frac{1}{2\tau}\begin{cases}0 & (x-x_0)\le -2\tau\\ \dfrac{((x-x_0+2\tau)/\tau)^2}{2} & -2\tau<(x-x_0)\le-\tau\\ \dfrac{-((x-x_0+2\tau)/\tau)^2}{2}+\dfrac{2(x-x_0+2\tau)}{\tau}-1 & -\tau<(x-x_0)\le\tau\\ \dfrac{((x-x_0+2\tau)/\tau)^2}{2}-\dfrac{4(x-x_0+2\tau)}{\tau}+8 & \tau<(x-x_0)\le 2\tau\\ 0 & 2\tau<(x-x_0)\end{cases}, \tag{9.3}$$

where τ is twice as much as a node spacing of the FEM mesh at the location of the source. The pseudo-delta function is described with a quadratic function, and it can be perfectly represented by the quadratic interpolation function of the FEM element when the neighbor of the source is discretized with a regular mesh. In this study, we consider only a vertical magnetic component H_z as a receiving signal. From the along-strike fields, the vertical magnetic component in the k_y-domain is calculated as follows,

$$\hat{H}_z=\frac{1}{k_e^2}\left(-ik_y\frac{\partial \hat{H}_y}{\partial z}-\hat{y}\frac{\partial \hat{E}_y}{\partial x}\right). \tag{9.4}$$

This is transformed to the space domain by the inverse Fourier transformation. Since the receivers are placed on the x-axis, the vertical magnetic component at $y=0$ is evaluated by

$$H_z(x,y=0,z)=\frac{1}{\pi}\int_0^{\infty}\hat{H}_z(x,k_y,z)\,\mathrm{d}k_y. \tag{9.5}$$

This integration is carried out by cubic spline interpolation in the logarithmic k_y-domain.

The adjoint-equation approach is useful for the reduction of a computation load for calculation of sensitivities (McGillivray et al., 1994). Unsworth and Oldenburg (1995) used this approach for the evaluation of the sensitivity of an electrical field. According to the adjoint-equation approach, the sensitivity of the vertical magnetic component at $\mathbf{r}_0$ with respect to conductivity of a volume V_k is given by

$$\frac{\partial H_z(\mathbf{r}_0)}{\partial \sigma_k} = \iiint_{V_k} \mathbf{E}_s \cdot \mathbf{E}_r \, \mathrm{d}v, \tag{9.6}$$

where $\mathbf{E}_s$ is the electric field caused by the original source, and $\mathbf{E}_r$ is the field caused by a unit vertical magnetic current source, $J_{mz} = 1$, located at $\mathbf{r}_0$. In the situation of Figure 1, with respect to the y-coordinate, E_{sx}, E_{rx}, E_{sz} and E_{rz} are odd functions, E_{sy} and E_{ry} are even. Thus, in the k_y-domain Equation (9.6) becomes

$$\begin{aligned}&\frac{\partial H_z(x, y=0, z)}{\partial \sigma_k}\\ &= \frac{1}{\pi}\int_0^\infty \left[\iint_{D_k} \left\{\hat{E}_{sy}(k_y)\hat{E}_{ry}(k_y) - \hat{E}_{sx}(k_y)\hat{E}_{rx}(k_y) - \hat{E}_{sz}(k_y)\hat{E}_{rz}(k_y)\right\} \mathrm{d}A\right] \mathrm{d}k_y, \end{aligned}\tag{9.7}$$

where D_k is a cross section of V_k.

When we place a vertical magnetic source at the receiver site and calculate $\mathbf{E}_r$ with the FEM code, the singularity is caused again there. We require a more flexible pseudo-delta function other than the Herrmann's function since the node spacing in the vicinity of the receiver site is irregular for the convenience of preparation of the FEM mesh depending on the irregular interval of receiver sites. Thus we employ the Gaussian-type pseudo-delta function δ_g:

$$\delta_g(\mathbf{r}, \mathbf{r}_0) = \frac{1}{(4\pi\alpha^2)^{3/2}} \exp\left[-\frac{\rho^2 + (y - y_0)^2}{4\alpha^2}\right], \tag{9.8}$$

where $\rho = \{(x - x_0)^2 + (z - z_0)^2\}^{1/2}$, and the value of α is adjusted in proportion to the smallest node spacing in the vicinity of the receiver site. In the k_y-domain, Equation (9.8) becomes

$$\hat{\delta}_g(\rho, k_y) = \frac{1}{4\pi\alpha^2} \exp(-\tau^2 k_y) \cdot \exp(-\rho^2/4\alpha^2), \tag{9.9}$$

and we distribute the unit vertical magnetic source on the basis of this function.

2.2. Linearized least-squares inversion with a smoothness constraint

As usual, we divide the earth into many blocks and estimate the logarithm of resistivity of each block as an unknown model parameter in the inversion. Our inverse problem is nonlinear because the forward modeling is nonlinear with respect to the resistivity of each block. Therefore, we use a linearized least-squares approach as an inversion algorithm. In addition, our inverse problem tends to be underdetermined since the number of unknown parameters exceeds the number of data or there are some model parameters which are unresolvable by the data. To fix the underdetermined inverse problem and to stabilize the inversion process, we employ a smoothness constraint as an a priori information on the model parameters.

Constrained least-squares inversions have been popular tools in geophysical inverse problems. In most of the application, a target data misfit based on the chi-square distribution for data misfit is considered to adjust the weighing of the constraint properly. This scheme demands us to know an exact variance of the data in advance. In most practical data acquisitions, however, the accurate evaluation of the data variance is not taken into consideration, and hence the estimation of the target misfit is not usually easy. In our study, we assume that N points data contain uncorrelated Gaussian noise and that the true data variance σ_i^2 is written as $\sigma_i^2 = \sigma_c^2 \sigma_i'^2$, $i = 1, \ldots, N$, where $\sigma_i'^2$ is a preliminary predicted value of σ_i^2 and σ_c^2 is a correction factor that is same for all data. In the inversion, only $\sigma_i'^2$ is assumed to be known in advance, and σ_c^2 will be determined. This assumption is more flexible than the use of the target misfit assuming the known true data variance. Under these assumptions, a probability distribution of the data is given by

$$p(\mathbf{d} \mid \mathbf{m}, \sigma_c^2) = \left\{ (2\pi)^N \det(\sigma_c^2 \mathbf{C}_d') \right\}^{-1/2} \exp\left\{ -\frac{1}{2} (\mathbf{d} - \mathbf{f}(\mathbf{m}))^{\mathrm{T}} \frac{\mathbf{C}_d'^{-1}}{\sigma_c^2} (\mathbf{d} - \mathbf{f}(\mathbf{m})) \right\}, \quad (9.10)$$

where $\mathbf{d}$ is an N-dimensional vector composed of N points data $d_1, d_2, \ldots, d_N$, $\mathbf{m}$ is an M-dimensional vector containing M model parameters $m_1, m_2, \ldots, m_M$, $\mathbf{f}(\mathbf{m})$ means the forward modeling, $\mathbf{C}_d'$ is the tentative covariance matrix written as $\mathbf{C}_d' = \mathrm{diag}(\sigma_1'^2, \sigma_2'^2, \ldots, \sigma_N'^2)$, det denotes the determinant of matrix, and T signifies transpose.

Supposing a smoothly varying resistivity structure, we define the a priori distribution of $\mathbf{m}$ as

$$p(\mathbf{m} \mid \lambda, \sigma_c^2) = (\lambda / 2\pi\sigma_c^2)^{(M-4)/2} \exp\left\{ -\frac{\lambda}{2\sigma_c^2} \mathbf{m}^{\mathrm{T}} \mathbf{D}^{\mathrm{T}} \mathbf{D} \mathbf{m} \right\}, \quad (9.11)$$

where λ is called the trade-off parameter or the regularization parameter, and the $M \times M$ matrix $\mathbf{D}$ is a weighted Laplacian operator with the rank of $M - 4$. In electromagnetic explorations from the earth surface, resolution decreases with depth. Model parameters of deeper blocks are insensitive to the data and unresolvable. Therefore, the size of the deeper blocks is often set to be vertically and laterally larger in 2-D inversions. The smoothness a priori information should work for fixing these unresolvable blocks; hence it should be weighted in proportion to the depth. Accordingly, in our study, each element of the matrix $\mathbf{D}$ is multiplied by the following function of the depth:

$$\beta(z) = \begin{cases} 1, & z \le z_0 \\ z/z_0, & z > z_0 \end{cases}, \quad (9.12)$$

where the value of z_0 is properly given by interpreters. deGroot-Hedlin and Constable (1990) and Smith and Booker (1991) used a similar weight increasing with depth. Although the parameter λ adjusts the reliability of the a priori information and is to be specified to stabilize the inversion, we assume that λ and σ_c^2 have already been given here, and their determination will be considered later.

According to Bayes' rule (e.g. Sen and Stoffa, 1995, pp. 13–15), we write the a posteriori distribution:

$$p(\mathbf{m} \mid \mathbf{d}, \lambda, \sigma_c^2) = \frac{p(\mathbf{d} \mid \mathbf{m}, \sigma_c^2) p(\mathbf{m} \mid \lambda, \sigma_c^2)}{p(\mathbf{d} \mid \lambda, \sigma_c^2)}, \tag{9.13}$$

where the numerator of the right-hand side is the joint probability density, and the denominator, which is the marginal probability density and independent of $\mathbf{m}$, is assumed to be a constant (Sen and Stoffa, 1995, pp. 13–15). Consequently, the a posteriori distribution can be written as

$$p(\mathbf{m} \mid \mathbf{d}, \lambda, \sigma_c^2) \propto \left\{(2\pi)^N \det(\sigma_c^2 \mathbf{C}'_d)\right\}^{-1/2} \left(\lambda / 2\pi\sigma_c^2\right)^{(M-4)/2} \exp\left(-U(\mathbf{m})/2\sigma_c^2\right), \tag{9.14}$$

where

$$U(\mathbf{m}) = \{\mathbf{d} - \mathbf{f}(\mathbf{m})\}^{\mathrm{T}} \mathbf{C}'^{-1}_d \{\mathbf{d} - \mathbf{f}(\mathbf{m})\} + \lambda \mathbf{m}^{\mathrm{T}} \mathbf{D}^{\mathrm{T}} \mathbf{D} \mathbf{m}. \tag{9.15}$$

The model existing at the peak of the a posteriori distribution is the most probable model. The search of the peak of the distribution of Equation (9.14) corresponds to the estimation of the model minimizing the objective function $U(\mathbf{m})$ by a least-squares approach.

However, since our forward problem is nonlinear with respect to the model parameters, we use the well-utilized linearization approximation by the Taylor expansion and an iterative process, that is

$$\mathbf{f}(\mathbf{m}_{k+1}) = \mathbf{f}(\mathbf{m}_k) + \mathbf{A}_k \cdot (\mathbf{m}_{k+1} - \mathbf{m}_k) + \mathbf{e}_l, \tag{9.16}$$

where $\mathbf{m}_k$ is the model parameters vector at the k-th iteration, $\mathbf{A}_k$ is the Jacobian matrix for $\mathbf{m}_k$, and $\mathbf{e}_l$ is the error caused by the linearization. Neglecting $\mathbf{e}_l$, we approximate U by

$$U(\mathbf{m}_{k+1}) \cong \left\| \tilde{\mathbf{d}}_k - \mathbf{G}_k \mathbf{m}_{k+1} \right\|^2, \tag{9.17}$$

where

$$\tilde{\mathbf{d}}_k = \begin{bmatrix} \mathbf{W}(\mathbf{d} - \mathbf{f}(\mathbf{m}_k) + \mathbf{A}_k \mathbf{m}_k) \\ \mathbf{0} \end{bmatrix}, \tag{9.18}$$

$$\mathbf{G}_k = \begin{bmatrix} \mathbf{W}\mathbf{A}_k \\ \sqrt{\lambda}\mathbf{D} \end{bmatrix}, \tag{9.19}$$

and the weighting matrix $\mathbf{W}$ is written as $W = \mathrm{diag}(\sigma_1'^{-1}, \sigma_2'^{-1}, \ldots, \sigma_N'^{-1})$. Applying $\partial U(\mathbf{m})/\partial \mathbf{m} = \mathbf{0}$, we obtain the normal equation:

$$\left(\mathbf{G}_k^{\mathrm{T}} \mathbf{G}_k\right) \mathbf{m}_{k+1} = \mathbf{G}_k^{\mathrm{T}} \tilde{\mathbf{d}}_k. \tag{9.20}$$

Factorizing $\mathbf{G}_k = \mathbf{QR}$ by the modified Gram–Schmidt method (Nakagawa and Oyanagi, 1982), in which $\mathbf{R}$ is an $M \times M$ non-singular upper-triangular matrix and $\mathbf{Q}$ is an $(N+M) \times M$ orthogonal matrix, we solve Equation (9.20). Using the solution $\mathbf{m}_{k+1}$, we can easily compute the linearized misfit ϕ_d^{L}:

$$\phi_d^{\mathrm{L}} = \|\mathbf{W}\mathbf{d} - \mathbf{W}\{\mathbf{f}(\mathbf{m}_k) + \mathbf{A}_k(\mathbf{m}_{k+1} - \mathbf{m}_k)\}\|^2. \tag{9.21}$$

However, the linearized misfit is often unacceptable (Oldenburg, 1994) due to the linearization error of Equation (9.16). To avoid such circumstances, although much computing load is demanded, the following misfit should be evaluated through the forward modeling,

$$\phi_d = \|\mathbf{Wd} - \mathbf{Wf}(\mathbf{m}_{k+1})\|^2. \tag{9.22}$$

The root mean square (rms) misfit ϕ_d^{rms} is often used for the evaluation of goodness of fit, which is given by $\phi_d^{\mathrm{rms}} = (\phi_d/N)^{1/2}$. If we know the exact data variance, that is, $\sigma_i'^2 = \sigma_i^2$ and we can estimate true model parameters, the expectation of ϕ_d^{rms} becomes unity. In practice, however, we estimate model parameters from the data with noise, and the true model parameters are not obtainable. Moreover, in application of the least-squares method to overdetermined problems ($N > M$), the chi-squared distribution for ϕ_d indicates that the expectation of ϕ_d is $N - M$, i.e., $\phi_d^{\mathrm{rms}} < 1$ if the exact data variance is known (Pedersen and Rasmussen, 1989). In contrast, in our constrained least-squares inversion, the objective function is not ϕ_d but U. Therefore, we should evaluate the expectation of U for the application of the least-squares method. As shown in Appendix A, when our inversion sufficiently converges and the linearization error $\mathbf{e}_l$ is negligible around the final model parameters $\mathbf{m}^*$, the expectation of $U(\mathbf{m}^*)$ with the known exact data variance is given by

$$\langle U(\mathbf{m}^*)\rangle = N - 4, \tag{9.23}$$

where $\langle\ \rangle$ denotes an expectation. We should note that $\phi_d < U$ by comparison between Equations (9.15) and (9.22), and hence the resultant ϕ_d can be $\phi_d < N - 4$, i.e., $\phi_d^{\mathrm{rms}} < 1$.

There are often some interpretations on overfitting for the case of $\phi_d^{\mathrm{rms}} < 1$ and underfitting for the case of $\phi_d^{\mathrm{rms}} > 1$ in the evaluation of data fitting (Meju, 1994, pp. 71–72). In the overdetermined problems, the condition of $\phi_d = N - M$ that brings $\phi_d^{\mathrm{rms}} < 1$ is used for evaluation of goodness of data misfit (e.g. Li et al., 2000). In our constrained least-squares inversion, Equation (9.23) indicates that the case of $\phi_d^{\mathrm{rms}} < 1$ is reasonably acceptable. However, all these evaluations need the use of correct data variance, and it is difficult in practical situations. In contrast, the minimization of the ABIC explained in the following section can be implemented with the tentative data variance, $\sigma_i'^2$.

2.3. Adjustment of trade-off parameter based on minimum ABIC

The estimation of the optimum $\mathbf{m}$ strongly depends on the value of λ which expresses the reliability of the a priori information. Determining the value of λ is one of the most active research subjects at present. Constable et al. (1987) and deGroot-Hedlin and Constable (1990) set a typical threshold as $\phi_d^{\mathrm{rms}} = 1$, and selected λ so that this condition can be satisfied in 1-D and 2-D magnetotelluric (MT) inversions. Li and Oldenburg (1999) applied the L-curve criterion for the determination of the optimal value of λ in a 3-D DC resistivity inversion. As another approach, we employ the Akaike's Bayesian Information Criterion (ABIC) developed by Akaike (1980). This criterion has been basically applied to linear inverse problems (e.g., Noro, 1989; Murata, 1993; Mitsuhata et al., 2001). For nonlinear problems, Mitsuhata (1994) applied it to a 1-D MT inversion, and Uchida (1993) and Ogawa and Uchida (1996) used it in 2-D MT inversions, and they had successful results.

The basic idea of the ABIC is based on the utility of the marginal probability density obtained by

$$p(\mathbf{d} \mid \sigma_c^2, \lambda) = \int p(\mathbf{d} \mid \sigma_c^2, \mathbf{m})\, p(\mathbf{m} \mid \sigma_c^2, \lambda)\, \mathrm{d}\mathbf{m}. \tag{9.24}$$

This distribution means the probability for the occurrence of **d** with respect to λ and σ_c^2, and it is no longer a function of **m**. We need to choose parameters λ and σ_c^2, such that they maximize $p(\mathbf{d} \mid \lambda, \sigma_c^2)$. On the basis of the entropy maximization principle, Akaike (1980) defined the ABIC as follows,

$$\mathrm{ABIC} = -2\ln p(\mathbf{d} \mid \sigma_0^2, \lambda) + 2 \cdot (\text{number of hyperparameters}), \tag{9.25}$$

where hyperparameters are all unknown parameters in the a priori distribution of **m** that are λ and σ_c^2 in our study. We can determine the optimum values of λ and σ_c^2 by minimizing the ABIC. In the case of linear problems, the integral of Equation (9.24) can be calculated analytically (e.g., Mitsuhata et al., 2001). Unfortunately, for nonlinear problems, the use of numerical integration techniques is computationally demanding and requires many forward model calculation. Consequently, we must use the integration based on the linearized approximation of Equation (9.16) at each iteration. At the k-th iteration, linearized ABIC is given by

$$\mathrm{ABIC}^{\mathrm{L}} = (N-4)\ln(2\pi\sigma_c^2) - 2\ln(\det \mathbf{W}) - (M-4)\ln\lambda + \ln(\det \mathbf{G}_k^{\mathrm{T}}\mathbf{G}_k) + U^{\mathrm{L}}(\lambda)/\sigma_c^2 + 4, \tag{9.26}$$

where U^L is evaluated by substituting $\mathbf{m}_{k+1}$ into Equation (9.17). Considering $\partial\mathrm{ABIC}/\partial\sigma_c^2 = 0$ to estimate the optimum σ_c^2, we obtain

$$\sigma_c^2 = \frac{U^{\mathrm{L}}(\lambda)}{N-4}. \tag{9.27}$$

Moreover, by substituting Equation (9.27) into Equation (9.26), $\mathrm{ABIC}^{\mathrm{L}}$ becomes

$$\mathrm{ABIC}^{\mathrm{L}} = (N-4)\ln\left(2\pi\frac{U^{\mathrm{L}}(\lambda)}{N-4}\right) - (M-4)\ln\lambda + \ln(\det \mathbf{G}_k^{\mathrm{T}}\mathbf{G}_k) - 2\ln(\det \mathbf{W}) + N. \tag{9.28}$$

In our nonlinear problem, the linearization error often makes $\mathrm{ABIC}^{\mathrm{L}}$ unacceptable as well as creates a discrepancy between ϕ_d and ϕ_d^{L}. Therefore, substituting $\mathbf{m}_{k+1}$ into Equation (9.15), we compute $U(\lambda)$ and define quasi-linearized ABIC as follows,

$$\mathrm{ABIC}^{\mathrm{QL}} = (N-4)\ln\left(2\pi\frac{U(\lambda)}{N-4}\right) - (M-4)\ln\lambda + \ln(\det \mathbf{G}_k^{\mathrm{T}}\mathbf{G}_k) - 2\ln(\det \mathbf{W}) + N. \tag{9.29}$$

At each iteration, we search the value of λ minimizing $\mathrm{ABIC}^{\mathrm{QL}}$ by the 1-D golden section method and determine it as the optimum λ, and the model parameters for the optimum λ are selected as the most reliable model parameters. An example of

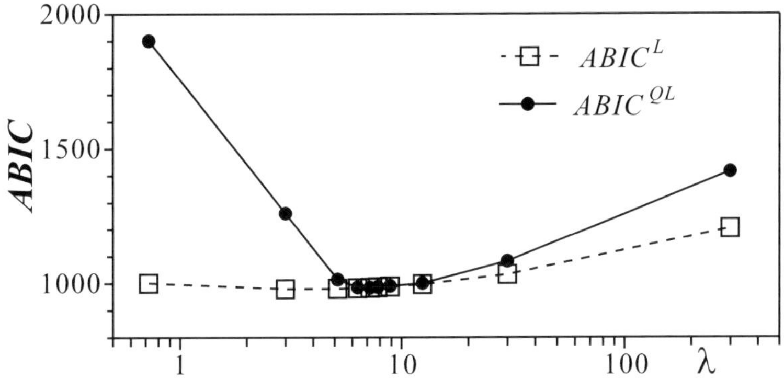

Figure 2. An example of the behaviors of $\mathrm{ABIC^L}$ and $\mathrm{ABIC^{QL}}$ as functions of λ. The significant difference between them occurs for small values of λ due to the linearization error. The minimum point of $\mathrm{ABIC^{QL}}$ is clearer than that of $\mathrm{ABIC^L}$.

discrepancy between $\mathrm{ABIC^L}$ and $\mathrm{ABIC^{QL}}$ is shown in Figure 2. The difference becomes significant for small values of λ, and the minimum point of $\mathrm{ABIC^{QL}}$ is clearer than that of $\mathrm{ABIC^L}$. Therefore, we use $\mathrm{ABIC^{QL}}$ in the following experiments. Our inversion scheme consists of the iterative process. Although there is no guarantee that the value of $\mathrm{ABIC^{QL}}$ decreases with the iteration, overall we have observed the decrease of $\mathrm{ABIC^{QL}}$ with the iteration, and the values of $\mathrm{ABIC^{QL}}$ and ϕ_d^{rms} have converged step by step without any threshold.

3. SYNTHETIC DATA EXAMPLE

3.1. Two-prism model with both-sides sources

We test our inversion method for synthetic data generated from a model shown in Figure 3. The model has two anomalies, one is conductive (10 Ω m) and the other is resistive (1000 Ω m), embedded in a 100 Ω m half-space. The sources are two HED sources directed to the y-direction located at $x = 0$ and $x = 7.5$ km. The vertical magnetic component H_z for the four frequencies of 30, 10, 3 and 1 Hz was calculated for each source at ten receivers located from 1.5 to 6.0 km. We added 1% Gaussian noise to them. The number of data is 160 which consist of real and imaginary parts of H_z. Figure 3 also shows the blocks for the inversion consisting of 140 totally. In each block, the resistivity value is constant and unknown. Throughout all the following numerical experiments, we put $z_0 = 2$ km in Equation (9.12). Hereafter, we call the configuration with both sources the both-sides source.

3.2. Initial model and convergence

Figure 4 shows the model inverted from all data generated by the both-sides sources after ten iterations. The initial model was a 100 Ω m homogeneous half-space. The value

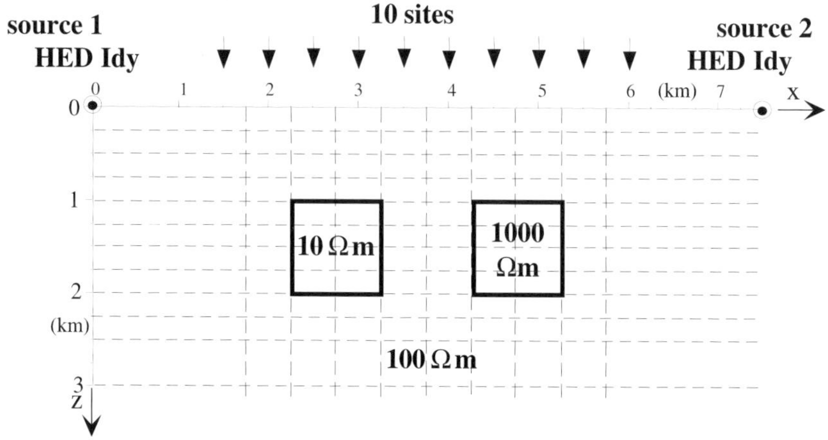

Figure 3. The model used to generate the data for the test of the inversion, consisting of one conductive and one resistive prism embedded in a 100 Ω m half space. The HED sources directed to the y-direction are placed at $x = 0$ and $x = 7.5$ km, and vertical magnetic receivers are arranged at ten sites with the interval of 500 m. The synthetic data are computed for each source for the frequencies of 30, 10, 3 and 1 Hz, and Gaussian noise is added. The underground is divided into many blocks as shown by the dashed lines, and each block has a distinct resistivity value for the inversion.

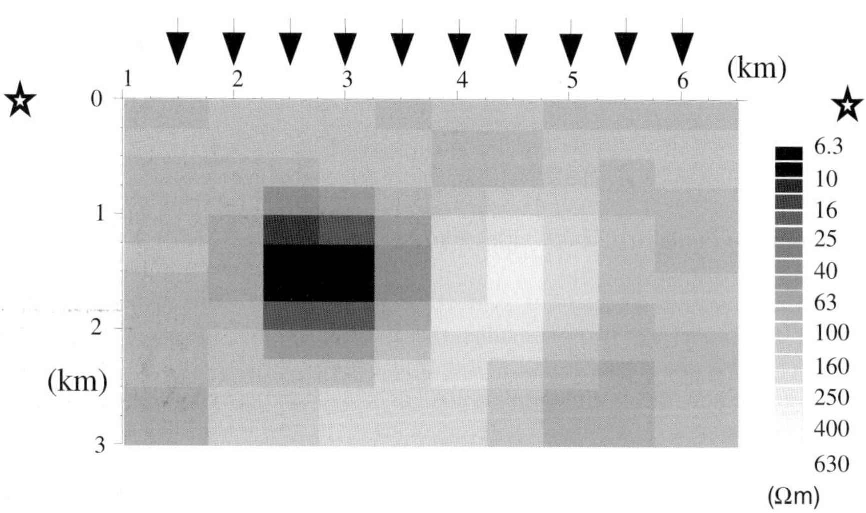

Figure 4. The result obtained by the inversion after ten iterations. It was reconstructed from the data which were generated by the both-sides sources and were added 1% Gaussian noise. The numbers of the data and the blocks are 160 and 140, respectively. The initial model of inversion was a 100 Ω m half space. The value of $\phi_{\rm d}^{\rm rms}$ for the resultant model is 1.0. The star symbols indicate the position of the sources.

of $\phi_d^{\rm rms}$ for the initial model was 79 and it was improved to 1.0 at the 10th iteration. We can recognize two prisms clearly.

To check the dependence of the final model on the initial model, we changed the

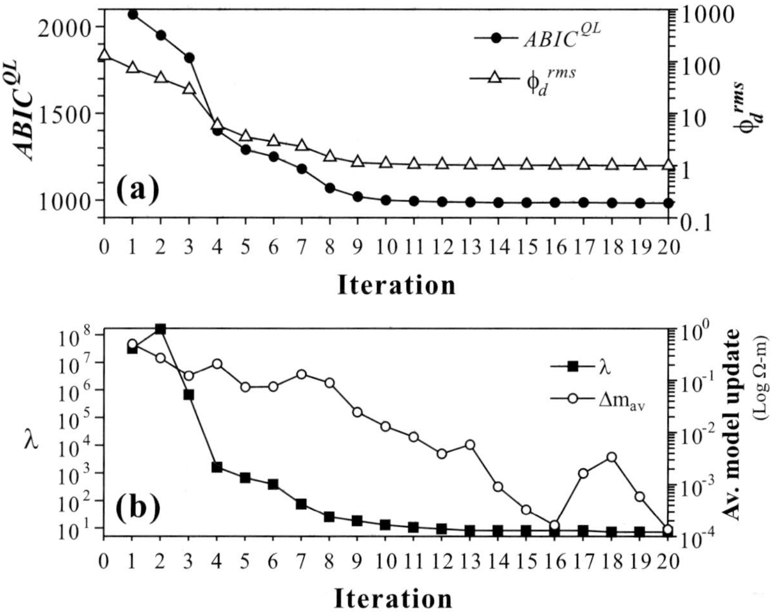

Figure 5. Convergence plots of the inversion for the initial model of a 10 Ω m half space. The data were the same as we used for the result of Fig. 4. (a) ABIC^{QL} and rms misfit, ϕ_d^{rms} versus the iteration number of inversion. (b) The trade-off parameter, λ, and the average model update Δm_{av} versus the iteration number of inversion.

initial model to a 10 Ω m homogeneous half-space and repeated the inversion. The convergence of our inversion process and the model inverted at each iteration are shown in Figures 5 and 6. The values of ABIC^{QL} and ϕ_d^{rms} decrease with iteration and become almost constant after the 10th iteration. The value of ϕ_d^{rms} for the initial model is 128 and converges to 1.0. The value of λ also decreases with iteration and becomes almost flat at the end. The average model update that is defined as $\Delta m_{av} = |\mathbf{m}_k - \mathbf{m}_{k-1}|/M$ decreases logarithmically, but begins to increase at the 17th iteration. However, this increment does not affect the values of ABIC^{QL} and ϕ_d^{rms}. Therefore, a meaningful model update may be more than about 10^{-2}. In Figure 6, the resolution is improved with the iteration. At the first iteration, the model is inverted to almost 30 Ω m homogeneous half-space. A conductive anomaly begins to be visible at the 4th iteration, and it becomes clearer and a resistive prism appears at the 7th iteration. At the 11th iteration, two anomalies appear distinctively and we cannot recognize much difference with the model of the 20th iteration. The final model after 20 iterations was nearly equal to the result of Figure 4.

3.3. Effect of noise

To investigate the effect of noise, we examined two cases when 5% and 10% Gaussian noise were added to the original data. The resultant models of Figure 7 were obtained after 10 iterations for the 5% noise data and 17 iterations for the 10% noise data. The

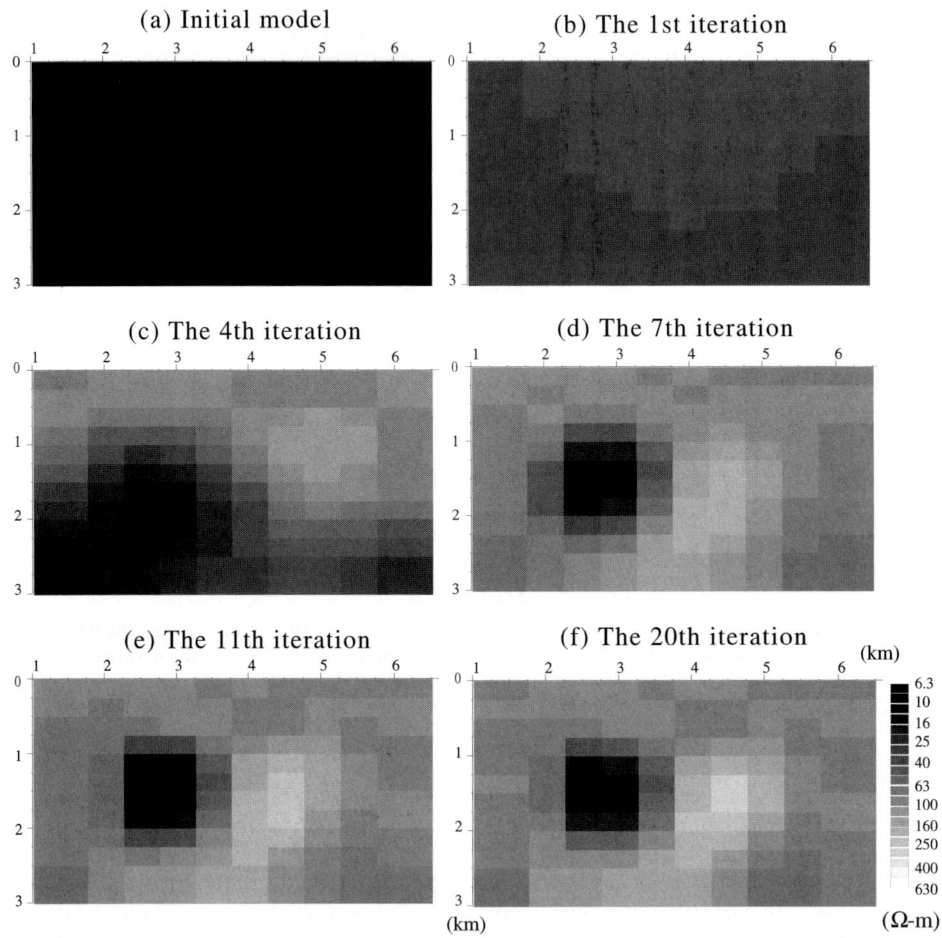

Figure 6. The model reconstructed at each iteration from the same data that were used in Fig. 4. (a) The initial model of a 10 Ω m half space. (b–f) The models reconstructed at the first, the fourth, the seventh, 11th and 20th iteration, respectively.

value of $\phi_d^{\rm rms}$ finally became 1.0 in both cases. The 5% noise makes the conductive anomaly broad and makes the resistive one insignificant. In the case of 10% noise, the resolution becomes poor and the conductive anomaly grows dim with depth and the resistive one almost disappears. In our inversion scheme, the resolution greatly depends on the quality of the data. Poor-quality data make the resolution low and provide a smoother image of resistivity structure.

3.4. Effect of the source location

Unlike the MT method, the CSEM data are strongly affected by placement of the signal source. This effect sometimes appears as the source overprint or the shadow effect in

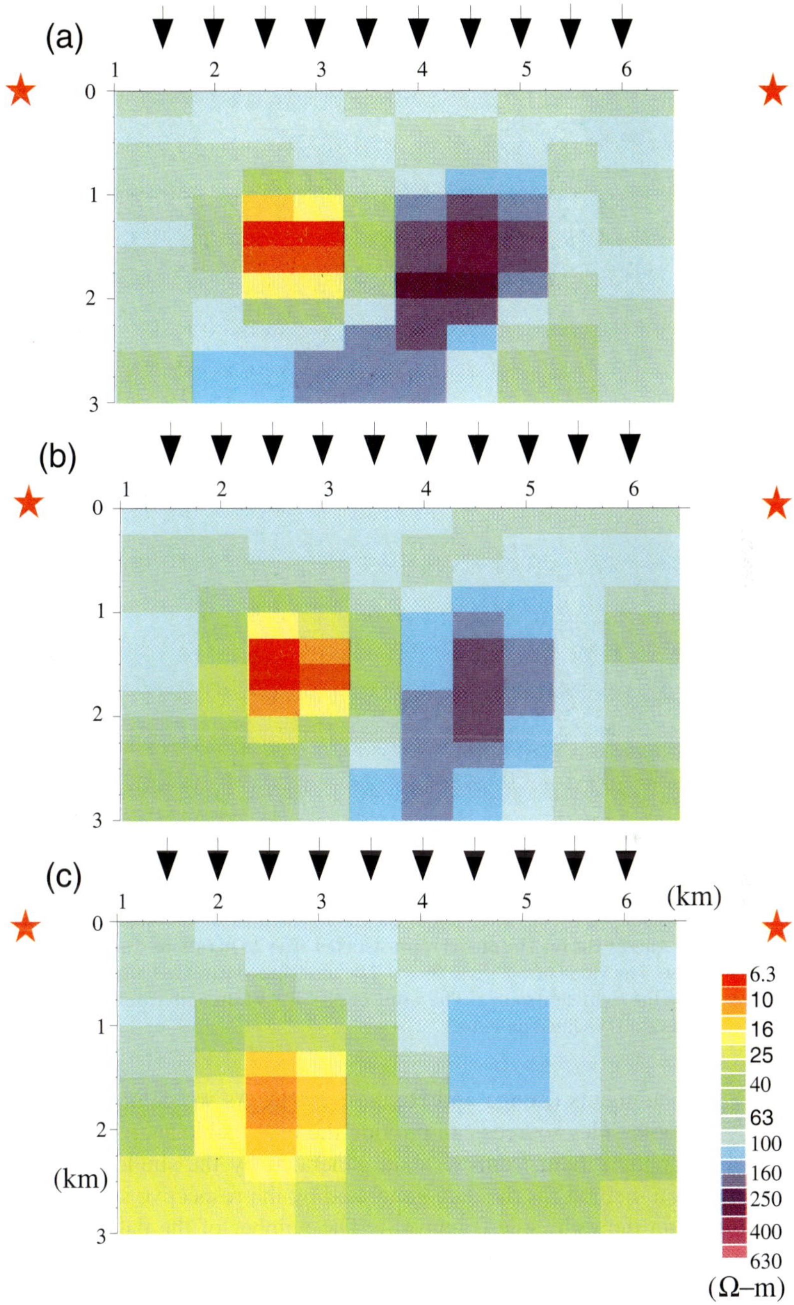

Figure 7. The results obtained by the inversion from the same data that were used in Fig. 4 with different noise levels. (a) The same figure as Fig. 4 (1% Gaussian noise). (b) The model reconstructed after 10 iterations from the data with 5% Gaussian noise. (c) The model reconstructed after 17 iterations from the data with 10% Gaussian noise. In all cases, the value of ϕ_d^{rms} for the model is 1.0.

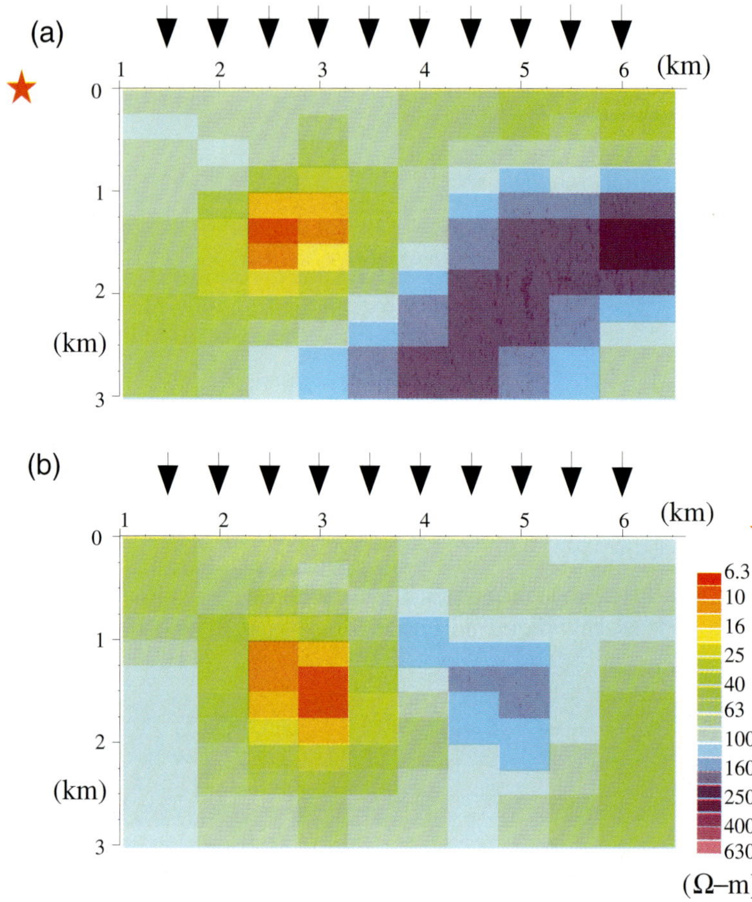

Figure 8. The results obtained by the inversion of the data generated by the single-side source for the frequencies of 30, 10, 3 and 1 Hz. (a) The model reconstructed after 19 iterations from the data generated by the left-side source. The value of ϕ_d^{rms} is 0.86. (b) The model reconstructed after 20 iterations from the data generated by the right-side source. The value of ϕ_d^{rms} is 0.94. In both cases, the number of data is 80 and the data contain 1% Gaussian noise.

the CSAMT measurements (Zonge and Hughes, 1991). As we demonstrated, the data generated by the both-sides sources can provide the two visible anomalies. It is natural to ask if we can obtain them from the data generated by the single-side source. We tested our inversion method for the data generated by the respective sources located on the left side and on the right side (Figure 8). The number of the data was 80 in each case. The result for the left-side source (Figure 8a) was obtained after 19 iterations and the value of ϕ_d^{rms} is 0.86. The convergence was slow. Figure 8(b) is the resultant model at the 20th iteration for the right-side source and the value of ϕ_d^{rms} is 0.94. In Figure 8(a), a conductive anomaly corresponding to the true prism can be reconstructed, but this anomaly extends to depth below the source. We can recognize a significant resistive zone but it is too broad, and it is elongated to depth and to the right. On the other hand, in Figure 8(b), there is a slight anomaly corresponding to the resistive prism. We can see a conductive zone corresponding to the conductive prism, but it is spread to shallow in the left-hand side. In addition, another conductive zone occurs at the lower right end.

These results imply that the conductive body is too sensitive for each source and works as a barrier and decreases resolution behind it, and it makes the image of the resistive prism dim.

4. CONCLUSIONS

We have developed a 2.5-D inversion method for CSEM data and tested its utility for synthetic data. We propose the minimization of the quasi-linearized ABIC to find a reasonable balance between the data fitting and the smoothness constraint for our nonlinear problem. The numerical experiments for the synthetic data show that our inversion algorithm improves resolution with iteration and it converges automatically without any threshold. Moreover, our algorithm reconstructs a smooth, low-contrast image of resistivity structure from noisy data, which is naturally acceptable because high-quality data are generally indispensable in recognizing small structures. The reconstructed images severely depend on the location of the source. Irradiation of the EM field from various directions will be necessary to accurately resolve targets. An inversion algorithm such as ours is recommended for the 2.5-D analysis of real field CSEM data.

ACKNOWLEDGEMENTS

We would like to thank Yasuo Ogawa for his helpful comments on the use of ABIC in nonlinear problems. A part of this work was performed while one of the authors, Y. Mitsuhata, was a visitor in the University of British Columbia. Y.M. would like to thank D.W. Oldenburg for his kind hospitality. The constructive reviews by Philip E. Wannamaker, Ivan M. Varentsov and Zonghou Xiong are greatly appreciated. Computation was carried out on the DEC cluster system at the Tsukuba Advanced Computation Center (TACC), the Agency of Industrial Science and Technology, Tsukuba.

Appendix A. EXPECTATION OF OBJECTIVE FUNCTION IN CONSTRAINED LINEARIZED LEAST-SQUARES METHOD

In this appendix, we derive the expectation of the objective function for a constrained linearized least-squares inversion. Nakagawa and Oyanagi (1982) described a detailed derivation of the expectation of the objective function in overdetermined problems. We extend their derivation into our problem. We assume that observed data and some constraints obtained from a priori information or additional measurements can be written as

$$\mathbf{d}_{\text{obs}} = \mathbf{f}(\mathbf{m}_{\text{true}}) + \mathbf{e}_{\text{d}}, \tag{9.A1}$$

and

$$\mathbf{d}_{\text{con}} = \mathbf{B}\mathbf{m}_{\text{true}} + \mathbf{e}_{\text{c}}, \tag{9.A2}$$

where $\mathbf{d}_{\text{obs}}$ and $\mathbf{d}_{\text{con}}$ are column vectors composed of N points data and K values due to constraints, respectively, $\mathbf{m}_{\text{true}}$ is true model parameters, $\mathbf{f}(\ )$ is a function representing a physical relationship between model parameters and data, $\mathbf{B}$ is a linear operator describing K independent constraint equations, and $\mathbf{e}_{\text{d}}$ and $\mathbf{e}_{\text{c}}$ are data noise and error for the constraints, respectively. For $\mathbf{e}_{\text{d}}$ and $\mathbf{e}_{\text{c}}$, covariance matrixes are defined as

$$\langle \mathbf{e}_{\text{d}}\mathbf{e}_{\text{d}}^{\text{T}} \rangle = \mathbf{C}_{\text{d}}, \tag{9.A3}$$

and

$$\langle \mathbf{e}_{\text{c}}\mathbf{e}_{\text{c}}^{\text{T}} \rangle = \mathbf{C}_{\text{c}}, \tag{9.A4}$$

where $\langle\ \rangle$ denotes an expectation, and $\mathbf{C}_{\text{d}}$ and $\mathbf{C}_{\text{c}}$ are assumed to be positive definite. In the constrained least-squares method, the objective function U to be minimized is

$$U(\mathbf{m}) = \|\mathbf{W}_{\text{d}}\{\mathbf{d}_{\text{obs}} - \mathbf{f}(\mathbf{m})\}\|^2 + \|\mathbf{W}_{\text{c}}(\mathbf{d}_{\text{con}} - \mathbf{B}\mathbf{m})\|^2, \tag{9.A5}$$

where weighting matrixes, $\mathbf{W}_{\text{d}}$ and $\mathbf{W}_{\text{c}}$ are given by the Cholesky factorization: $\mathbf{C}_{\text{d}}^{-1} = \mathbf{W}_{\text{d}}^{\text{T}}\mathbf{W}_{\text{d}}$ and $\mathbf{C}_{\text{c}}^{-1} = \mathbf{W}_{\text{c}}^{\text{T}}\mathbf{W}_{\text{c}}$.

To deal with nonlinear inverse problems, the linearized approximation of Equation (9.16) is commonly used, and estimated model parameters are iteratively updated. After the k-th iteration, the next step model $\mathbf{m}_{k+1}$ is estimated by minimizing

$$U(\mathbf{m}_{k+1}) = \|\mathbf{W}(\mathbf{d}_k - \mathbf{F}_k\mathbf{m}_{k+1})\|^2, \tag{9.A6}$$

where

$$\mathbf{d}_k = \begin{bmatrix} \mathbf{d}_{\text{obs}} - \mathbf{f}(\mathbf{m}_k) + \mathbf{A}_k\mathbf{m}_k \\ \mathbf{d}_{\text{con}} \end{bmatrix}, \tag{9.A7}$$

$$\mathbf{F}_k = \begin{bmatrix} \mathbf{A}_k \\ \mathbf{B} \end{bmatrix}, \tag{9.A8}$$

$$\mathbf{W} = \begin{bmatrix} \mathbf{W}_{\text{d}} & \mathbf{0} \\ \mathbf{0} & \mathbf{W}_{\text{c}} \end{bmatrix}, \tag{9.A9}$$

and $\mathbf{A}_k$ is the Jacobian matrix for $\mathbf{m}_k$. From $\partial U/\partial \mathbf{m}_{k+1} = \mathbf{0}$, the normal equation is derived as

$$(\mathbf{F}_k^{\text{T}}\mathbf{W}^{\text{T}}\mathbf{W}\mathbf{F}_k)\mathbf{m}_{k+1} = \mathbf{F}_k^{\text{T}}\mathbf{W}^{\text{T}}\mathbf{W}\mathbf{d}_k. \tag{9.A10}$$

Assuming that the model parameters converge to $\mathbf{m}^*$ after a sufficiently large iteration number represented by the symbol '∞', and that $\mathbf{m}^*$ is close to $\mathbf{m}_{\text{true}}$, we can write

$$\mathbf{f}(\mathbf{m}_{\text{true}}) \cong \mathbf{f}(\mathbf{m}^*) + \mathbf{A}_\infty(\mathbf{m}_{\text{true}} - \mathbf{m}^*). \tag{9.A11}$$

Moreover, by considering Equations (9.A11), (9.A1) and (9.A2), $\mathbf{d}_\infty$ is given by

$$\mathbf{d}_\infty = \begin{bmatrix} \mathbf{d}_{\text{obs}} - \mathbf{f}(\mathbf{m}^*) + \mathbf{A}_\infty \mathbf{m}^* \\ \mathbf{d}_{\text{con}} \end{bmatrix} = \mathbf{e} + \mathbf{F}_\infty \mathbf{m}_{\text{true}}, \tag{9.A12}$$

where

$$\mathbf{e} = \begin{bmatrix} \mathbf{e}_{\text{d}} \\ \mathbf{e}_{\text{c}} \end{bmatrix}. \tag{9.A13}$$

By substituting Equation (9.A12) into the normal Equation (9.A10), $\mathbf{m}^*$ can be written as

$$\mathbf{m}^* = \mathbf{m}_{\text{true}} + \mathbf{CWe}, \tag{9.A14}$$

where

$$\mathbf{C} = (\mathbf{F}_\infty^{\text{T}} \mathbf{W}^{\text{T}} \mathbf{W} \mathbf{F}_\infty)^{-1} \mathbf{F}_\infty^{\text{T}} \mathbf{W}^{\text{T}}. \tag{9.A15}$$

We define a residual vector $\mathbf{v}$ for $\mathbf{m}^*$ as

$$\mathbf{v} = \begin{bmatrix} \mathbf{d}_{\text{obs}} - \mathbf{f}(\mathbf{m}^*) \\ \mathbf{d}_{\text{con}} - \mathbf{Bm}* \end{bmatrix}, \tag{9.A16}$$

which can be rewritten by using Equations (9.A12) and (9.A14):

$$\begin{aligned} \mathbf{v} &= \mathbf{d}_\infty - \mathbf{F}_\infty \mathbf{m}^* = \mathbf{e} - \mathbf{F}_\infty(\mathbf{m}^* - \mathbf{m}_{\text{true}}) \\ &= \mathbf{e} - \mathbf{F}_\infty \mathbf{CWe} = (\mathbf{I}_{N+K} - \mathbf{F}_\infty \mathbf{CW})\mathbf{e}, \end{aligned} \tag{9.A17}$$

where $\mathbf{I}_{N+K}$ denotes a $(N+K) \times (N+K)$ unit matrix. From Equation (9.A17), the objective function U for $\mathbf{m}^*$ is described as

$$\begin{aligned} U(\mathbf{m}^*) &= \mathbf{v}^{\text{T}} \mathbf{W}^{\text{T}} \mathbf{W} \mathbf{v} = \mathbf{e}^{\text{T}} (\mathbf{I}_{N+K} - \mathbf{W}^{\text{T}} \mathbf{C}^{\text{T}} \mathbf{F}_\infty^{\text{T}}) \mathbf{W}^{\text{T}} \mathbf{W} \mathbf{v} \\ &= \mathbf{e}^{\text{T}} \mathbf{W}^{\text{T}} \mathbf{W} \mathbf{v} - \mathbf{e}^{\text{T}} \mathbf{W}^{\text{T}} \mathbf{C}^{\text{T}} \mathbf{F}_\infty^{\text{T}} \mathbf{W}^{\text{T}} \mathbf{W} \mathbf{v}. \end{aligned} \tag{9.A18}$$

The second term on the right-hand side of Equation (9.A18) is zero since by using the normal Equation (9.A10),

$$\mathbf{F}_\infty^{\text{T}} \mathbf{W}^{\text{T}} \mathbf{W} \mathbf{v} = \mathbf{F}_\infty^{\text{T}} \mathbf{W}^{\text{T}} \mathbf{W} [\mathbf{d}_\infty - \mathbf{F}_\infty \mathbf{m}^*] = \mathbf{0}. \tag{9.A19}$$

Consequently, $U(\mathbf{m}^*)$ becomes

$$\begin{aligned} U(\mathbf{m}^*) &= \mathbf{e}^{\text{T}} \mathbf{W}^{\text{T}} \mathbf{W} (\mathbf{I}_{N+K} - \mathbf{F}_\infty \mathbf{CW}) \mathbf{e} \\ &= \mathbf{e}^{\text{T}} \mathbf{W}^{\text{T}} \mathbf{W} \mathbf{e} - \mathbf{e}^{\text{T}} \mathbf{W}^{\text{T}} \mathbf{W} \mathbf{F}_\infty \mathbf{CWe}. \end{aligned} \tag{9.A20}$$

In order to estimate the expectation of $U(\mathbf{m}^*)$, we calculate the expectation of each term on the right hand side of Equation (9.A20). Before that, we define a joint covariance matrix $\mathbf{C}_{\text{dc}}$ for convenience:

$$\mathbf{C}_{\text{dc}} = \begin{bmatrix} \mathbf{C}_{\text{d}} & \mathbf{0} \\ \mathbf{0} & \mathbf{C}_{\text{c}} \end{bmatrix}, \tag{9.A21}$$

which has a relationship with the weighting matrix $\mathbf{W}$ as $\mathbf{C}_{\text{dc}}^{-1} = \mathbf{W}^{\text{T}} \mathbf{W}$. The expectation of the first term of Equation (9.A20) becomes

$$\langle \mathbf{e}^{\mathrm{T}}\mathbf{W}^{\mathrm{T}}\mathbf{W}\mathbf{e}\rangle = \sum_{i=1}^{N+K}\sum_{j=1}^{N+K} C^{-1}_{\mathrm{cd}(i,j)}\langle e_i e_j\rangle = \sum_{i=1}^{N+K}\sum_{j=1}^{N+K} C^{-1}_{\mathrm{cd}(i,j)} C_{\mathrm{dc}(i,j)}$$
$$= \mathrm{tr}(\mathbf{C}^{-1}_{\mathrm{dc}}\mathbf{C}_{\mathrm{dc}}) = \mathrm{tr}(\mathbf{I}_{N+K})$$
$$= N + K, \tag{9.A22}$$

where 'tr' is the trace of matrix. By replacing as $\mathbf{H} = \mathbf{W}^{\mathrm{T}}\mathbf{W}\mathbf{F}^{\infty}\mathbf{C}\mathbf{W}$, the expectation of the second term can also be written as

$$\langle \mathbf{e}^{\mathrm{T}}\mathbf{W}^{\mathrm{T}}\mathbf{W}\mathbf{F}_{\infty}\mathbf{C}\mathbf{W}\mathbf{e}\rangle = \langle \mathbf{e}^{\mathrm{T}}\mathbf{H}\mathbf{e}\rangle = \sum_{i=1}^{N+K}\sum_{j=1}^{N+K} H_{(i,j)}\langle e_i e_j\rangle$$
$$= \sum_{i=1}^{N+K}\sum_{j=1}^{N+K} H_{(i,j)} C_{\mathrm{dc}(i,j)} = \mathrm{tr}(\mathbf{H}\mathbf{C}_{\mathrm{dc}}). \tag{9.A23}$$

Using the characteristics of the trace: $\mathrm{tr}(\overset{n\times m}{\mathbf{A}}\ \overset{m\times n}{\mathbf{B}}) = \mathrm{tr}(\overset{m\times n}{\mathbf{B}}\ \overset{n\times m}{\mathbf{A}})$, we obtain

$$\mathrm{tr}(\mathbf{H}\mathbf{C}_{\mathrm{dc}}) = \mathrm{tr}(\mathbf{W}^{\mathrm{T}}\mathbf{W}\mathbf{F}_{\infty}\mathbf{C}\mathbf{W}\mathbf{C}_{\mathrm{dc}}) = \mathrm{tr}(\mathbf{C}_{\mathrm{dc}}\mathbf{W}^{\mathrm{T}}\mathbf{W}\mathbf{F}_{\infty}\mathbf{C}\mathbf{W})$$
$$= \mathrm{tr}(\mathbf{I}_{N+K}\mathbf{F}_{\infty}\mathbf{C}\mathbf{W}) = \mathrm{tr}(\mathbf{F}_{\infty}\mathbf{C}\mathbf{W})$$
$$= \mathrm{tr}(\mathbf{C}\mathbf{W}\mathbf{F}_{\infty}) = \mathrm{tr}(\mathbf{I}_M)$$
$$= M, \tag{9.A24}$$

because from Equation (9.A15),

$$\mathbf{C}\mathbf{W}\mathbf{F}_{\infty} = (\mathbf{F}^{\mathrm{T}}_{\infty}\mathbf{W}^{\mathrm{T}}\mathbf{W}\mathbf{F}_{\infty})^{-1}\mathbf{F}^{\mathrm{T}}_{\infty}\mathbf{W}^{\mathrm{T}}\mathbf{W}\mathbf{F}_{\infty} = \mathbf{I}_M. \tag{9.A25}$$

Finally from Equations (9.A20), (9.A22), (9.A23) and (9.A24), we reach

$$\langle U(\mathbf{m}^*)\rangle = N + K - M. \tag{9.A26}$$

In our study, since the rank of $\mathbf{D}$ of Equation (9.11) is $M-4$, Equation (9.23) is obtained by substituting $K = M-4$ into Equation (9.A26).

REFERENCES

Akaike, H., 1980. Likelihood and Bayes procedure. In: J.M. Bernardo, M.H. DeGroot, D.V. Lindley and A.F.M. Smith (Eds.), Bayesian Statistics. University Press, Valencia, pp. 141–166.

Constable, S.C., Parker, R.L. and Constable, C.G., 1987. Occam's inversion, a practical algorithm for generating smooth models from electromagnetic sounding data. Geophysics, 52, 289–300.

deGroot-Hedlin, C. and Constable, S.C., 1990. Occam's inversion to generate smooth two-dimensional models from magnetotelluric data. Geophysics, 55, 1613–1624.

Herrmann, R.B., 1979. SH-wave generation by dislocation source — a numerical study. Bull. Seismol. Soc. Am., 69, 1–15.

Hohmann, G.W., 1988. Numerical modeling for electromagnetic methods of geophysics. In: M.N. Nabighian (Ed.), Electromagnetic Methods in Applied Geophysics, 1. Theory. Society of Exploration Geophysicists, Tulsa, OK, pp. 314–363.

Li, X., Oskooi, B. and Pedersen, L.B., 2000. Inversion of controlled-source tensor magnetotelluric data for a layered earth with azimuthal anisotropy. Geophysics, 65, 452–464.

Li, Y. and Oldenburg, D.W., 1999. Inversion of DC resistivity data using an L-curve criterion. Expanded Abstracts, 69th Annu. Int. Meet., Soc. Explor. Geophys.

Lu, X., Unsworth, M. and Booker, J., 1999. Rapid relaxation inversion of CSAMT data. Geophys. J. Int., 138, 381–392.

McGillivray, P.R., Oldenburg, D.W. and Ellis, R.G., 1994. Calculation of sensitivities for the frequency domain electromagnetic induction problem. Geophys. J. Int., 116, 1–4.

Meju, M.A., 1994. Geophysical Data Analysis: Understanding Inverse Problem Theory and Practice. Society of Exploration Geophysicists Course Notes Ser. 6, SEG Publishers, Tulsa, OK.

Mitsuhata, Y., 1994. Application of prior information of smooth resistivity structure to 1-D inversion of magnetotelluric data by using the ABIC minimization method. Buturi-Tansa (Geophys. Explor.), 47, 358–374.

Mitsuhata, Y., 2000. 2-D electromagnetic modeling by finite-element method with a dipole source and topography. Geophysics, 65, 465–475.

Mitsuhata, Y., Uchida, T., Murakami, Y. and Amano, H., 2001. The Fourier transform of controlled-source time-domain electromagnetic data by smooth spectrum inversion. Geophys. J. Int., 144, 123–135.

Murata, Y., 1993. Estimation of optimum average surficial density from gravity data: an objective Bayesian approach. J. Geophys. Res., 98, 12,097–12,109.

Nakagawa, T. and Oyanagi, Y., 1982. Analysis of Experimental Data by Least-Squares Method. University of Tokyo Press, Tokyo (in Japanese).

Noro, H., 1989. ABIC based method for the trend estimation problem of one dimensional data. Geoinformatics, 14B, 1–10 (in Japanese with English abstract).

Ogawa, Y. and Uchida, T., 1996. A two-dimensional magnetotelluric inversion assuming Gaussian static shift. Geophys. J. Int., 126, 69–76.

Oldenburg, D.W., 1994. Practical strategies for the solution of large-scale electromagnetic inverse problems. Radio Sci., 29, 1081–1099.

Pedersen, L.B. and Rasmussen, T.M., 1989. Inversion of magnetotelluric data: a non-linear least-squares approach. Geophys. Prospect., 37, 669–695.

Sen, M. and Stoffa, P.L., 1995. Global Optimization Methods in Geophysical Inversion. Elsevier, Amsterdam.

Smith, J.T. and Booker, J.R., 1991. Rapid inversion of two- and three-dimensional magnetotelluric data. J. Geophys. Res., 96, 3905–3922.

Torres-Verdin, C. and Habashy, T.M., 1994. Rapid 2.5-D forward modeling and inversion via a new nonlinear scattering approximation. Radio Sci., 29, 1051–1079.

Uchida, T., 1993. Smooth 2-D inversion for magnetotelluric data based on statistical criterion ABIC. J. Geomagnet. Geoelectr., 45, 841–858.

Unsworth, M.J. and Oldenburg, D.W., 1995. Subspace inversion of electromagnetic data: application to mid-ocean-ridge exploration. Geophys. J. Int., 123, 161–168.

Zonge, K.L. and Hughes, L.J., 1991. Controlled source audio-frequency magnetotellurics. In: M.N. Nabighian (Ed.), Electromagnetic Methods in Applied Geophysics, 2. Application, Part B. Society of Exploration Geophysicists, Tulsa, OK, pp. 713–809.

Chapter 10

3-D FOCUSING INVERSION OF CSAMT DATA

Oleg Portniaguine and Michael S. Zhdanov

University of Utah, Salt Lake City, UT 84112, USA

Abstract: We present a method for the solution of 3-D controlled source magnetotelluric (CSAMT) inverse problems. The inverse problem is formulated as the minimization of a Tikhonov parametric functional with a focusing stabilizer. Observed CSAMT apparent resistivities are converted to log-anomalous apparent resistivities, which are linearly connected to anomalous currents via the integral equation. We apply the Born iterative method to solve this integral equation, using a focusing regularized inversion. The focusing is based on a specially selected stabilizing functional which minimizes the area where strong model parameters variations and discontinuities occur. The method is illustrated using examples of 3-D inversion of model CSAMT data, and with a real data example.

1. INTRODUCTION

This paper deals with three particular aspects of the CSAMT inverse problem. The first is use of apparent resistivity data in the inversion, which is solved by using the concept of *log-anomalous apparent resistivity*. The second aspect is nonlinearity of the inverse problem, for the solution of which we apply the Born iterative method (Chew, 1990). The third is the non-uniqueness of the solution, which is addressed by using Tikhonov's regularization theory.

Traditional geophysical inversion methods are often based on the Tikhonov regularization theory (Tikhonov and Arsenin, 1977; Zhdanov, 1993) which provides a stable solution of the inverse problem. This goal is reached, as a rule, by introducing a maximum smoothness stabilizing functional. The obtained solution provides a smooth image of real structures that sometimes looks geologically unrealistic. Recently a new approach to the reconstruction of noisy images has been developed (Rudin et al., 1992; Vogel and Oman, 1998). It is based on a total variation stabilizing functional which requires that the distribution of the model parameters be of bounded variation. This requirement is weaker than one of maximum smoothness because it can be applied to discontinuous functions. However, it still decreases the bounds of the model parameters' variations and therefore distort the image. In papers by Portniaguine and Zhdanov (1998, 1999) we introduced stabilizing functionals, which generate more 'focused' images than conventional methods. We call this approach *focusing of inversion images*. In the present paper we apply this new method to the 3-D CSAMT inversion.

The CSAMT method measures complex impedances, which are often converted to apparent resistivity and phase. For 3-D inversion, the apparent resistivities have to be connected to fields. We solve this problem by converting data to *log-anomalous apparent resistivities*, which are linearly connected to anomalous fields. Such conversion has certain advantages, as opposed to using anomalous fields directly. Often, only the value of apparent resistivity is measured. This makes calculation of the anomalous fields difficult, because recomputing apparent resistivity into anomalous fields requires knowledge of phases. Log-anomalous apparent resistivity can be easily computed without knowing the phase. Another aspect is that the measurements at different frequencies should be properly weighted to invert them together. Conversion to log-anomalous apparent resistivity makes the measurements dimensionless, which provides natural weighting to the fields at different frequencies. These advantages apply to inverse problems both for layered media (finding the background field) and for the subsequent 3-D inversion.

The electromagnetic (EM) inverse problem is nonlinear. For ease of numerical solution, it is often converted to a series of linear problems. One common technique is the Born iterative method (Chew, 1990). This method is based on the fact that the EM inverse problem is actually a bi-linear problem, a special case of a general nonlinear problem. Since the problem is linear to anomalous currents, and the currents are the product of the total electric field and anomalous conductivity, the inverse problem becomes linear with respect to conductivity if the total fields are fixed. The total electric fields, in turn, are connected to anomalous conductivity via the corresponding integral equation (IE). Using these properties, we construct an iterative process whereby the linear inverse problem is solved first for anomalous currents, assuming the total electric fields are fixed. Second, the values of the total fields are updated using the found values of anomalous currents. These new values of the total fields are used in the next iteration of the inverse problem. Convergence of the process is assured by using a modified Green's tensor operator (Zhdanov and Fang, 1997).

Combined with the focusing inversion method, such a strategy allows us to avoid solving the full IE for the whole inversion domain. On the first iteration, the linear inverse problem produces compact bodies, and the total electric field needs to be found only inside the compact domains. The IE for such domains is much smaller than the IE for the whole domain and can be easily solved.

Our discussion is illustrated by applications of the method to synthetic models and real CSAMT data, collected in Hamlin Valley, Nevada, for oil exploration.

2. THE METHOD OF INTEGRAL EQUATIONS

The IE method is a powerful tool for EM modeling and inversion. The basic principles of constructing integral equations in 3-D cases were outlined by Weidelt (1975) and Hohmann (1975). A comprehensive implementation of the IE methods was realized by Xiong (1992) in the SYSEM code. Also, several IE-based approximate methods have been developed recently. These are localized nonlinear approximation (Habashi et al., 1993), quasi-linear approximation (Zhdanov and Fang, 1996a,b), quasi-linear series (Zhdanov and Fang, 1997), and quasi-analytic approximation (Zhdanov et al., 2000).

The main idea of the IE method is the following. The forward EM problem consists of finding the electric and magnetic fields at receivers for a given source and with known distribution of electrical conductivity (Zhdanov and Keller, 1994). We assume that the magnetic permeability μ is constant everywhere and is equal to that of free space $\mu = 4\pi \times 10^{-7}$ H/m. Let us represent a 3-D distribution of conductivity σ as the sum of background (normal) conductivity σ_{b} and anomalous conductivity $\Delta\sigma$, which is nonzero only within the local domain D. This model is excited by a harmonic source of circular frequency ω.

The vectors of total electric $\mathbf{E}$ and magnetic $\mathbf{H}$ fields in this model can be presented as the sum of background (normal) and anomalous (scattered) fields:

$$\mathbf{E} = \mathbf{E}^b + \mathbf{E}^a, \quad \mathbf{H} = \mathbf{H}^b + \mathbf{H}^a. \tag{10.1}$$

The background field $\mathbf{E}^b$ is a field generated by the given sources in the background model σ_{b}, and the anomalous $\mathbf{E}^a$ field is caused by the presence of anomalous conductivity $\Delta\sigma$.

The anomalous field is presented as an integral over the excess currents in the inhomogeneous domain D:

$$\mathbf{E}^a(\mathbf{r}_j) = \iiint_D \widehat{\mathbf{G}}_E(\mathbf{r}_j \mid \mathbf{r}) \cdot \mathbf{j}^a(\mathbf{r})\, dv, \tag{10.2}$$

$$\mathbf{H}^a(\mathbf{r}_j) = \iiint_D \widehat{\mathbf{G}}_H(\mathbf{r}_j \mid \mathbf{r}) \cdot \mathbf{j}^a(\mathbf{r})\, dv, \tag{10.3}$$

where $\widehat{\mathbf{G}}_{\mathbf{E}}(\mathbf{r}_j \mid \mathbf{r})$ and $\widehat{\mathbf{G}}_{\mathbf{H}}(\mathbf{r}_j \mid \mathbf{r})$ are the electric and magnetic Green's tensors defined for an unbounded conductive medium with a background conductivity σ_{b}. Excess (anomalous) current density $\mathbf{j}^a(\mathbf{r})$ at point $\mathbf{r}$ is determined by the equation

$$\mathbf{j}^a(\mathbf{r}) = \Delta\sigma(\mathbf{r})\left(\mathbf{E}^b(\mathbf{r}) + \mathbf{E}^a(\mathbf{r})\right) = \Delta\sigma(\mathbf{r})\mathbf{E}(\mathbf{r}). \tag{10.4}$$

Expression (10.2) becomes the IE with respect to anomalous electric field $\mathbf{E}^a(\mathbf{r})$, if point $\mathbf{r}_j$ is inside D. Solution of this equation yields an anomalous electric field inside domain D. After that, anomalous magnetic fields everywhere and anomalous electric fields outside domain D can be found using Equations (10.2) and (10.3). The background fields are assumed to be known everywhere.

In turn, the anomalous field is a linear combination of anomalous currents $\Delta\sigma(r)\mathbf{E}(\mathbf{r})$:

$$\mathbf{E}^a(\mathbf{r}_j) = \iiint_D \widehat{\mathbf{G}}_E(\mathbf{r}_j \mid \mathbf{r}) \cdot \Delta\sigma(r)\mathbf{E}(\mathbf{r})\, dv, \tag{10.5}$$

$$\mathbf{H}^a(\mathbf{r}_j) = \iiint_D \widehat{\mathbf{G}}_H(\mathbf{r}_j \mid \mathbf{r}) \cdot \Delta\sigma(\mathbf{r})\mathbf{E}(\mathbf{r})\, dv. \tag{10.6}$$

We use discrete analogs of the continuous Equations (10.5), (10.6):

$$\mathbf{E}_{\mathrm{a}} = \widehat{\mathbf{G}}_e \widehat{\mathbf{S}}\mathbf{e}_{\mathrm{t}}, \tag{10.7}$$

$$\mathbf{H}_{\mathrm{a}} = \widehat{\mathbf{G}}_h \widehat{\mathbf{S}} \mathbf{e}_{\mathrm{t}}, \tag{10.8}$$

where discrete vector $\mathbf{e}_{\mathrm{t}}$ contains three components of the total electric fields inside the anomalous domain, and the vectors $\mathbf{E}_{\mathrm{a}}$ and $\mathbf{H}_{\mathrm{a}}$ represent the anomalous fields at the receivers. The anomalous conductivity is stored in a sparse diagonal matrix $\widehat{\mathbf{S}}$. Matrices $\widehat{\mathbf{G}}_{\mathbf{e}}$, $\widehat{\mathbf{G}}_{\mathbf{h}}$ represent the values of corresponding Green's tensors.

If point $\mathbf{r}_j$ is inside the anomalous domain D, Expression (10.7) becomes an integral equation with respect to the electric field:

$$\mathbf{e}_{\mathrm{a}} = \widehat{\mathbf{G}} \widehat{\mathbf{S}} \mathbf{e}_{\mathrm{t}}, \tag{10.9}$$

where $\mathbf{e}_{\mathrm{a}}$ is a vector of an anomalous field, and $\widehat{\mathbf{G}}$ is a matrix of the corresponding Green's tensor inside the domain D. Equation (10.9) establishes a connection between the electric field inside the anomalous domain and the anomalous conductivity. Using

$$\mathbf{e}_{\mathrm{t}} = \mathbf{e}_{\mathrm{a}} + \mathbf{e}_{\mathrm{n}}, \tag{10.10}$$

where $\mathbf{e}_{\mathrm{n}}$ is a vector of normal electric field inside the domain, Equation (10.9) becomes

$$\mathbf{e}_{\mathrm{a}} = \widehat{\mathbf{G}} \widehat{\mathbf{S}} (\mathbf{e}_{\mathrm{a}} + \mathbf{e}_{\mathrm{n}}). \tag{10.11}$$

or

$$\mathbf{e}_{\mathrm{t}} = \widehat{\mathbf{G}} \widehat{\mathbf{S}} \mathbf{e}_{\mathrm{t}} + \mathbf{e}_{\mathrm{n}}. \tag{10.12}$$

The solution to (10.12) can be written in matrix notation:

$$\mathbf{e}_{\mathrm{t}} = (\widehat{\mathbf{I}} - \widehat{\mathbf{G}} \widehat{\mathbf{S}})^{-1} \mathbf{e}_{\mathrm{n}}. \tag{10.13}$$

3. THE CSAMT FORWARD PROBLEM

The scalar CSAMT method produces values of a complex apparent resistivity ρ (Tikhonov, 1950; Cagniard, 1953), which is computed according to

$$\rho = \frac{1}{cf} \left(\frac{E_y}{H_x} \right)^2, \quad c = 2\pi \mu_0, \tag{10.14}$$

where f is frequency, μ_0 is the free space magnetic permeability, E_y is the y-component of an electric field, and H_x is the x-component of a magnetic field. We assume that measurement profiles are parallel to the y axis.

In general, apparent resistivity is a nonlinear function of electric and magnetic fields. It is possible, however, to derive a convenient simplification of the apparent resistivity formula which makes these dependencies linear. Below, we demonstrate that the logarithm of apparent resistivity is linearly connected to the anomalous fields.

We represent the field components as the sum of normal (denoted with subscript 'n') and anomalous parts (denoted by subscript 'a'):

$$E_y = E_{\mathrm{n}} + E_{\mathrm{a}}, \quad H_x = H_{\mathrm{n}} + H_{\mathrm{a}}. \tag{10.15}$$

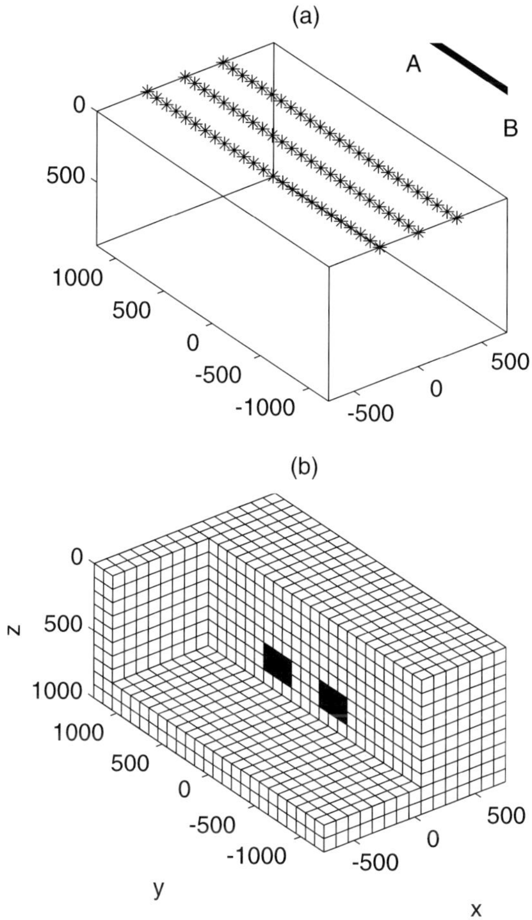

Figure 1. Synthetic EM model: (a) observation system, (b) true model.

Taking the log of Equation (10.14) and using Formula (10.15), we obtain

$$\ln(\rho) = \ln\left(\frac{1}{cf}\left(\frac{E_\mathrm{n}+E_\mathrm{a}}{H_\mathrm{n}+H_\mathrm{a}}\right)^2\right) = 2\ln(E_\mathrm{n}+E_\mathrm{a}) - 2\ln(H_\mathrm{n}+H_\mathrm{a}) - \ln(cf),$$

$$\ln(\rho) = 2\ln(1+E_\mathrm{a}/E_\mathrm{n}) - 2\ln(1+H_\mathrm{a}/H_\mathrm{n}) + 2\ln(E_\mathrm{n}/H_\mathrm{n}) - \ln(cf). \tag{10.16}$$

Provided that the anomalous field is much smaller than the normal field, the log of apparent resistivity can be easily linearized, using

$$\ln(1+\Delta x) \approx \Delta x. \tag{10.17}$$

A *normal apparent resistivity* ρ_n can be introduced to obtain

$$\ln(\rho_\mathrm{n}) = 2\ln(E_\mathrm{n}/H_\mathrm{n}) - \ln(cf) = \ln\left(\frac{1}{cf}\left(\frac{E_\mathrm{n}}{H_\mathrm{n}}\right)^2\right). \tag{10.18}$$

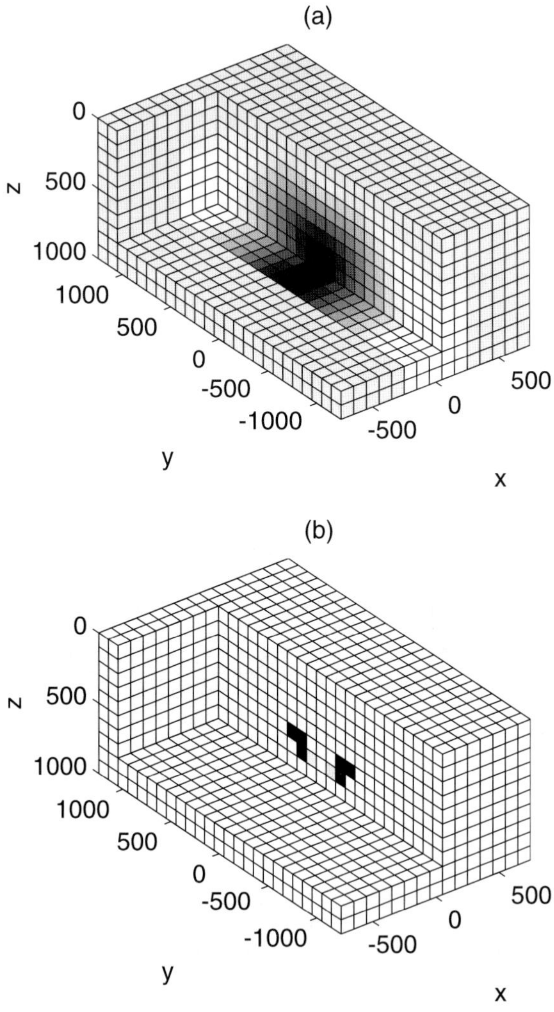

Figure 2. (a) 3-D resistivity model resulting from smooth inversion, 1.5% misfit. (b) Result of focusing inversion with 1.5% misfit.

Combining (10.16), (10.17) and (10.18), we have

$$\ln(\rho) = 2(E_a/E_n - H_a/H_n) + \ln(\rho_n). \tag{10.19}$$

Now we introduce *log-anomalous apparent resistivity* $\ln(\rho_a)$ as the difference between the logs of observed (total) and normal apparent resistivity:

$$\ln(\rho_a) = \ln(\rho) - \ln(\rho_n) = 2(E_a/E_n - H_a/H_n). \tag{10.20}$$

Note that the frequency term $\ln(1/cf)$ in (10.16) cancels. The quantities in (10.20) are dimensionless. Thus, expression (10.20) for different frequencies can be used together in the inversion without additional weighting. If we measure only the absolute value

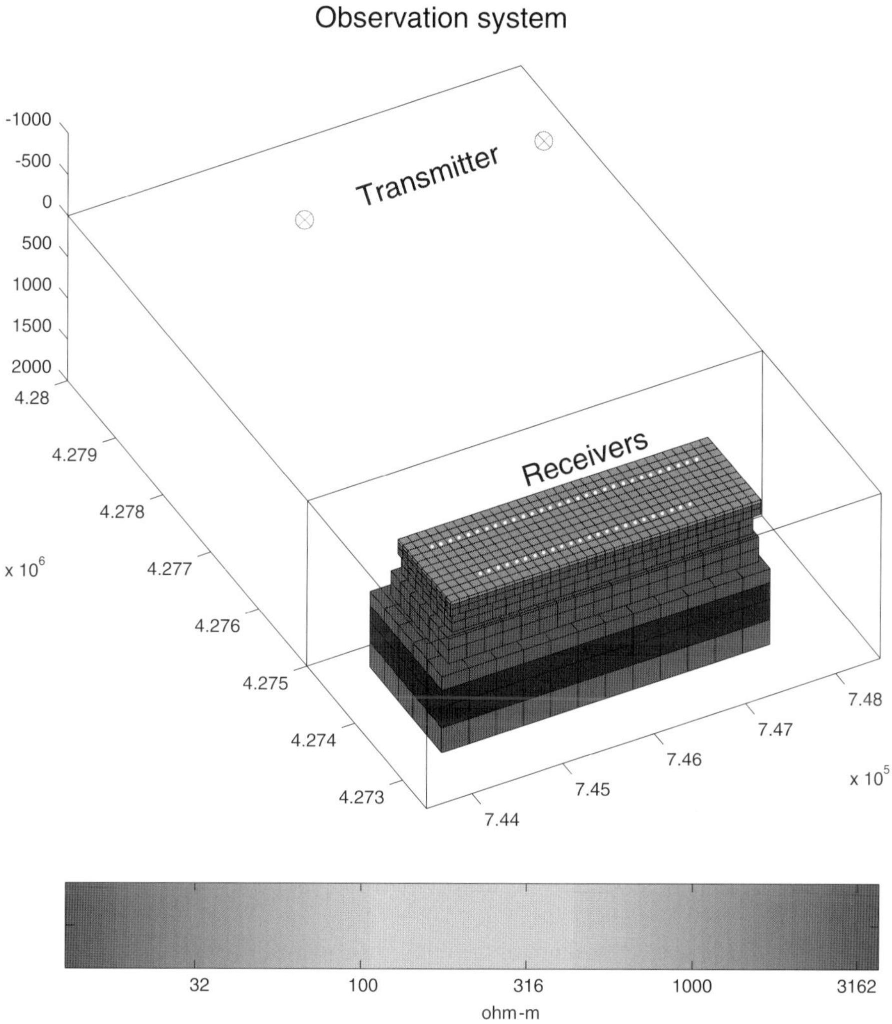

Figure 3. Background model with a resistive layer. The lower half-space is filled with rectangular cells. The size of the cells increases with depth. Stars denote the receiver location. Circles denote the location of the transmitter poles.

of apparent resistivity, but not phase, all we need to use is the real part of logarithmic Equation (10.20):

$$\ln|\rho_{\mathrm{a}}| = 2\,\mathrm{Real}(E_{\mathrm{a}}/E_{\mathrm{n}} - H_{\mathrm{a}}/H_{\mathrm{n}}). \tag{10.21}$$

Note also that log-anomalous apparent resistivity is an approximation. Experiments with synthetic and real data show that the maximum of log-anomalous apparent resistivity is on the order of 0.2. For this value, linearization (10.17) holds with an accuracy of 0.01. The estimated noise in the 3-D inverse problem with respect to

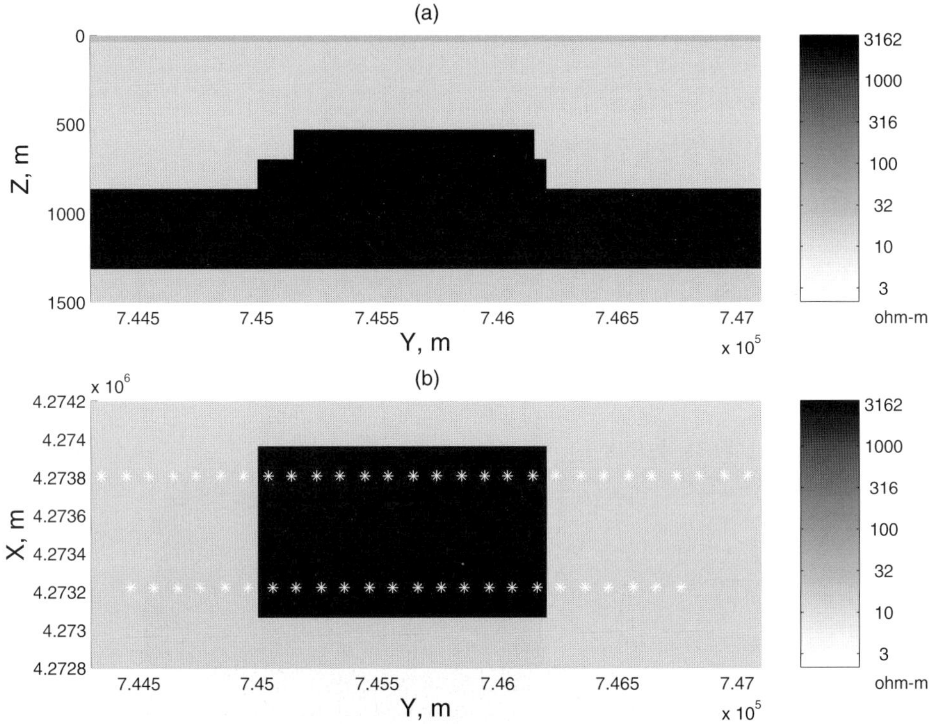

Figure 4. Model with the anticline on top of the resistive layer. (a) Cross-section. (b) Plan view; the slice is taken at 800 m depth. Stars denote the receiver locations.

log-anomalous apparent resistivity is also on the order of 0.01. Thus, we can safely use the notion of log-anomalous apparent resistivity if its maximum value is less than 0.2. For cases with very conductive structures where the value can be more than that, the algorithm shows the correct position of the body but underestimates the conductivity.

4. FORMULATION OF THE INVERSE PROBLEM

The inversion process consists of two major steps. First, we estimate normal apparent resistivity by fitting the log of observed apparent resistivity $\ln(\rho)$ for all stations *and all frequencies* with the log of apparent resistivity derived from a layered model $\ln(\rho_n)$. The residual of fitting is associated with log-anomalous apparent resistivity. As we can see from (10.20), log-anomalous apparent resistivity $\ln(\rho_a)$ is a linear combination of anomalous fields. Further, we represent anomalous field as a linear combination of responses from individual cells and solve the 3-D inverse problem using the method of focusing inversion.

We introduce a vector of the data **d**, which combines the values of log-anomalous apparent resistivity at the receivers, for all frequencies. According to (10.20), it is

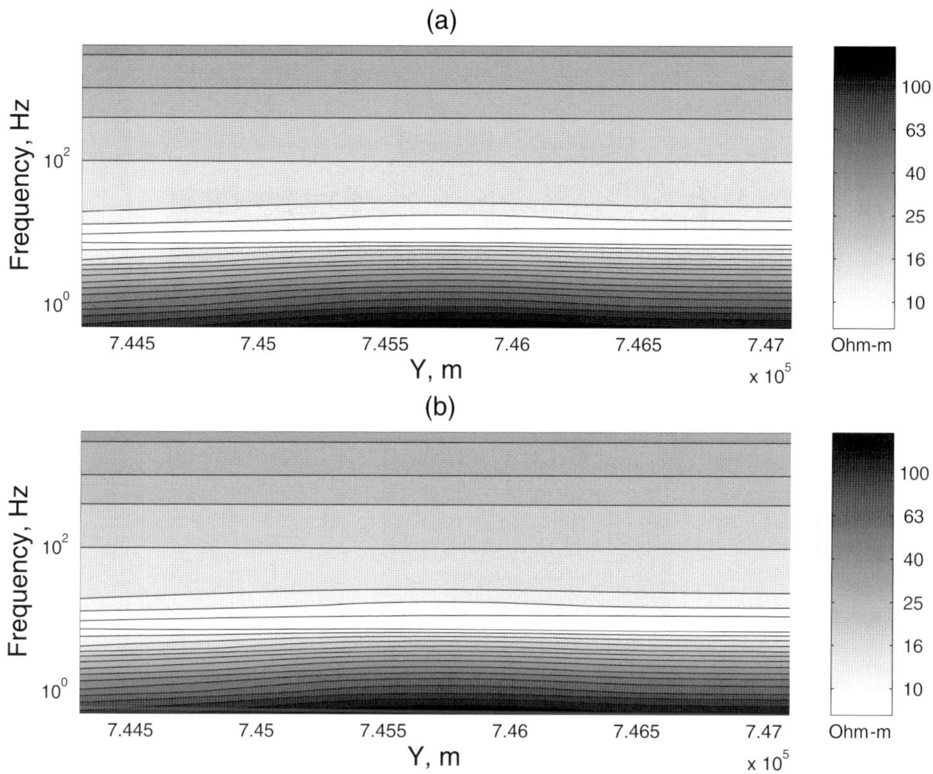

Figure 5. Data for the model shown in Figure 4. (a) Original log-anomalous apparent resistivity for the first profile. (b) Log-anomalous apparent resistivity predicted from the inversion.

a linear combination of anomalous fields, with normal fields as weights. In matrix notations, Formula (10.20) is cast as follows:

$$\mathbf{d} = 2(\mathrm{diag}(\mathbf{E}_\mathrm{n})^{-1}\mathbf{E}_\mathrm{a} - \mathrm{diag}(\mathbf{H}_\mathrm{n})^{-1}\mathbf{H}_\mathrm{a}), \tag{10.22}$$

where $\mathrm{diag}(\mathbf{E}_\mathrm{n})^{-1}$ and $\mathrm{diag}(\mathbf{H}_\mathrm{n})^{-1}$ are sparse matrices containing the inverse normal fields on the main diagonal.

Integral Equations (10.7) and (10.8) connect the anomalous fields at the receivers to anomalous currents $\widehat{\mathbf{S}}\mathbf{e}_\mathrm{t}$ inside the domain. We introduce sensitivity to the currents, matrix $\widehat{\mathbf{G}}_j$, as a linear combination of matrices $\widehat{\mathbf{G}}_e$ and $\widehat{\mathbf{G}}_h$:

$$\widehat{\mathbf{G}}_j = 2\,\mathrm{diag}(\mathbf{E}_\mathrm{n})^{-1}\widehat{\mathbf{G}}_e - 2\,\mathrm{diag}(\mathbf{H}_\mathrm{n})^{-1}\widehat{\mathbf{G}}_h. \tag{10.23}$$

Using (10.22) and (10.23), we convert Equations (10.7) and (10.8) into a single equation:

$$\mathbf{d} = \widehat{\mathbf{G}}_j\widehat{\mathbf{S}}\mathbf{e}_\mathrm{t}. \tag{10.24}$$

We formulate the inverse problem with respect to scaled conductivity $\mathbf{m}$, which is connected to anomalous conductivity via a matrix of expected conductivities $\widehat{\mathbf{S}}_e$ (constraints). The anomalous conductivity is a product of inversion parameter $\mathbf{m}$,

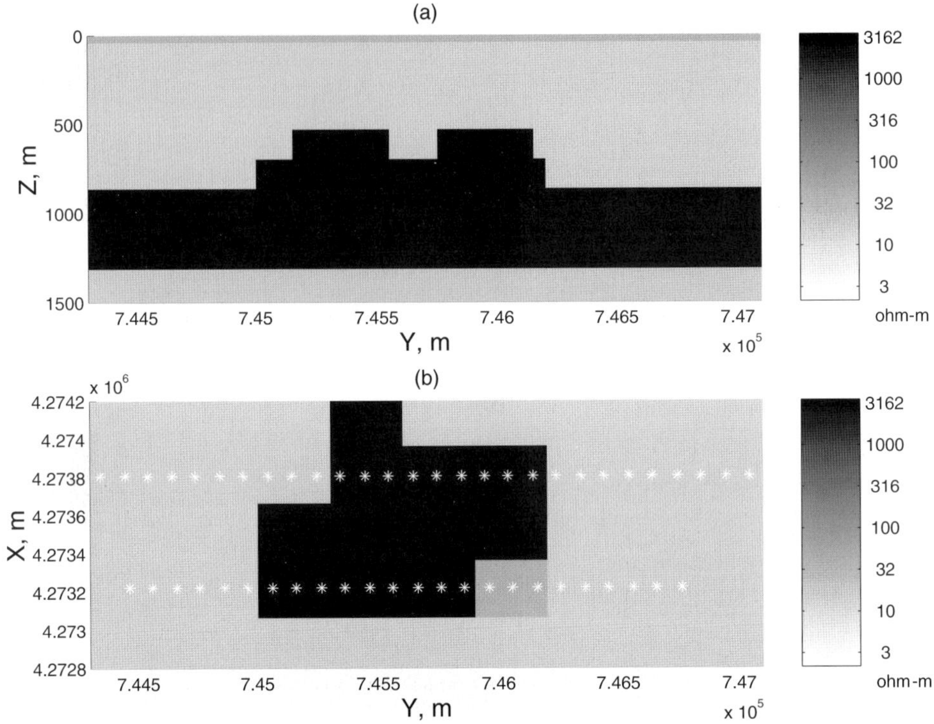

Figure 6. Result of inversion for the body atop the resistive layer. (a) Plan view. (b) Cross-section.

changing from -1 to 1, and the value of constraint, which is stored in the corresponding diagonal of matrix $\widehat{\mathbf{S}}_e$. The following relationships hold:

$$\widehat{\mathbf{S}} = \text{diag}(\widehat{\mathbf{S}}_e\mathbf{m}), \tag{10.25}$$

$$\mathbf{S}\mathbf{e}_t = \text{diag}(\mathbf{e}_t)\widehat{\mathbf{S}}_e\mathbf{m} \tag{10.26}$$

Using (10.26) and (10.24), we establish the equation of the forward problem:

$$\mathbf{d} = \widehat{\mathbf{G}}_j\,\text{diag}(\mathbf{e}_t)\widehat{\mathbf{S}}_e\mathbf{m}. \tag{10.27}$$

Or, expressing $\mathbf{e}_t$ via (10.13), we may rewrite Equation (10.27) as

$$\mathbf{d} = \widehat{\mathbf{G}}_j\,\text{diag}((\widehat{\mathbf{I}} - \widehat{\mathbf{G}}\,\text{diag}(\widehat{\mathbf{S}}_e\mathbf{m}))^{-1}\mathbf{e}_n)\widehat{\mathbf{S}}_e\mathbf{m}. \tag{10.28}$$

Our goal is to find the parameters **m** given the data **d**, that is to solve the inverse problem. Since this inverse problem is ill-posed, we use the Tikhonov regularization method to solve it. We minimize the Tikhonov parametric functional with a minimum support focusing stabilizer (Portniaguine and Zhdanov, 1999):

$$\|\mathbf{d} - \widehat{\mathbf{G}}_\mathbf{j}\,\text{diag}((\widehat{\mathbf{I}} - \widehat{\mathbf{G}}\,\text{diag}(\widehat{\mathbf{S}}_\mathbf{e}\mathbf{m}))^{-1}\mathbf{e}_\mathbf{n})\widehat{\mathbf{S}}_\mathbf{e}\mathbf{m}\|^2 + \alpha\sum_{k-1}^{N_m}\frac{m_k^2}{m_k^2+\beta^2} = \min. \tag{10.29}$$

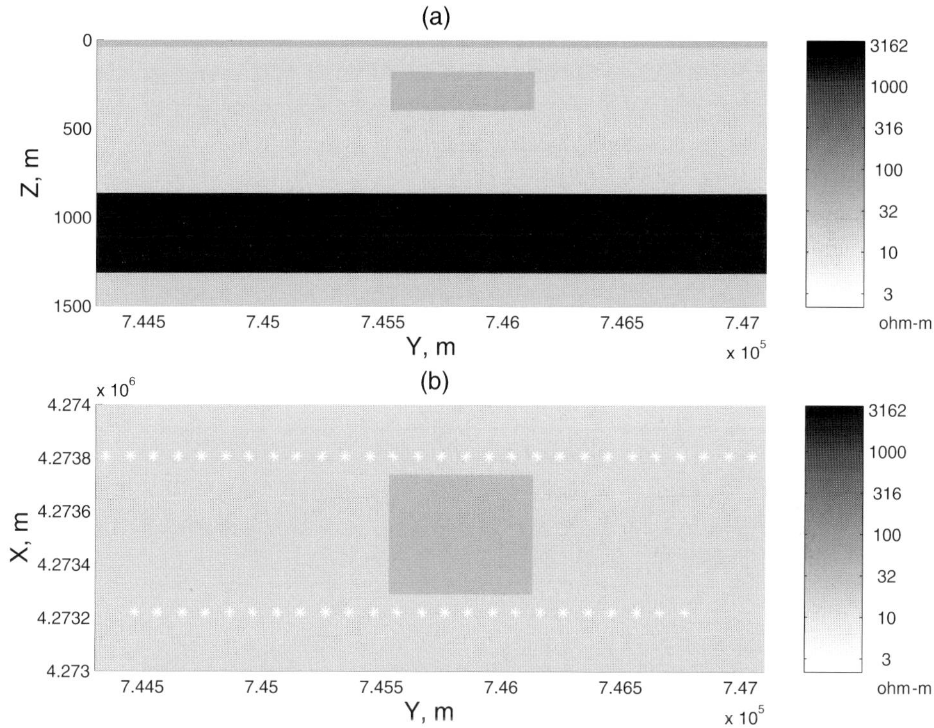

Figure 7. Model with the body inside the conductive layer. (a) Cross-section. (b) Plan view.

Note that problem (10.29) can be expressed as bilinear, using (10.27):

$$\|\mathbf{d} - \widehat{\mathbf{G}}_\mathbf{j}\,\mathrm{diag}(\mathbf{e}_\mathrm{t})\widehat{\mathbf{S}}_\mathrm{e}\mathbf{m}\|^2 + \alpha \sum_{k=1}^{N_m} \frac{m_k^2}{m_k^2 + \beta^2} = \min, \tag{10.30}$$

$$\mathbf{e}_\mathrm{t} = (\widehat{\mathbf{I}} - \widehat{\mathbf{G}}\,\mathrm{diag}(\widehat{\mathbf{S}}_e\mathbf{m}))^{-1}\mathbf{e}_\mathrm{n}. \tag{10.31}$$

The main problem with the IE method is that, for general 3-D cases, matrix $\widehat{\mathbf{G}}$ in Equation (10.31) is full and may be very large. Assume that the anomalous domain D is a rectangular prism. Each side of the prism is divided into N rectangular prismatic cells. Then the number of cells is $N_\mathrm{c} = N^3$. The number of entries in the matrix $\widehat{\mathbf{G}}$ is $(3 \times N_c^3)^2 = 9 \times N^6$. We can see that the number of entries grows as the sixth power of N. This growth is the main limiting factor of the integral equation method. If $N = 5$ the problem is small and readily solvable. Yet, for $N = 10$ the size of the problem becomes very large.

The use of focusing inversion greatly reduces the size of the problem. Numerical solution of (10.30) and (10.31) follows conventional bilinear approaches. First we solve (10.30) for $\mathbf{m}$, using some approximation to multiplier $\mathbf{e}_\mathrm{t}$. Note that if $\mathbf{e}_\mathrm{t}$ is known, problem (10.29) becomes purely linear.

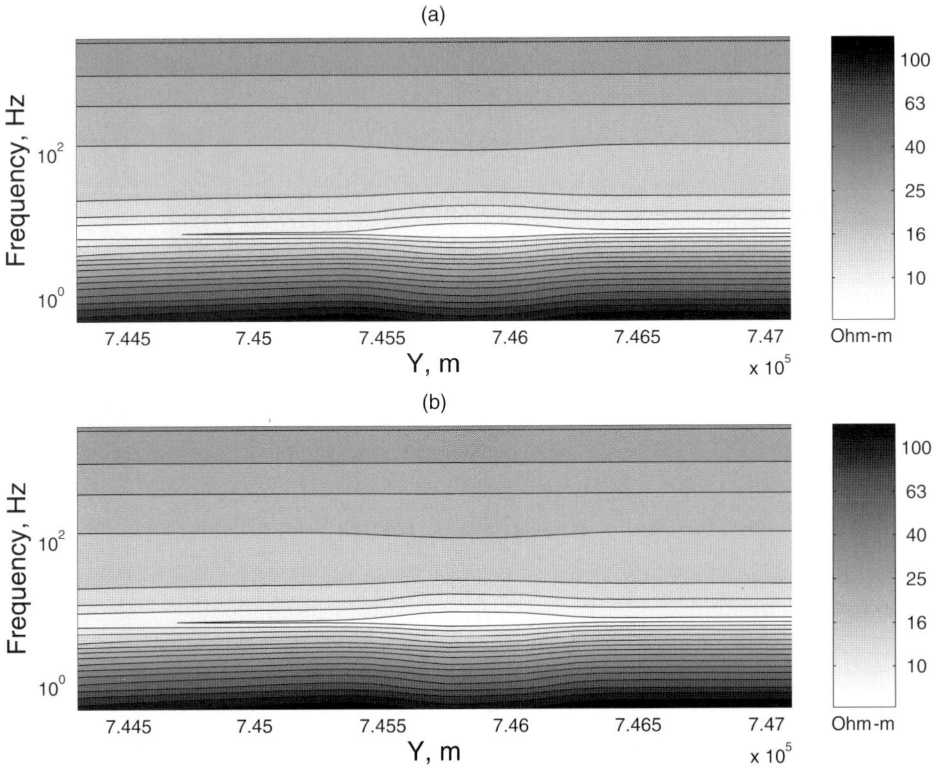

Figure 8. Original apparent resistivity data simulated by SYSEM (case a) and the data predicted from the inversion (case b) for the model with the body inside the conductive layer.

All subsequent iterations deal with a greatly reduced problem. For example, on the second bilinear iteration we use the value of $\mathbf{m}$ to update $\mathbf{e}_t$ using (10.31). But $\mathbf{m}$ has already become sparse, after solving (10.30).

It is very important to find a good initial approximation to the unknown multiplier $\mathbf{e}_t$. The choice of the initial guess determines the course of the inversion. To derive the initial approximation to the multiplier $\mathbf{e}_t$, we use the first iteration of Jacoby's method. According to Jacoby's method, off-diagonal terms (10.13) are neglected:

$$\mathbf{e}_t = \left[\text{diag}(\widehat{\mathbf{I}} - \widehat{\mathbf{G}}\mathbf{S})\right]^{-1} \mathbf{e}_n. \tag{10.32}$$

If we consider an anomalous domain consisting of one cell, (10.32) gives the exact solution. It will also be exact if we consider multiple cells located far away from each other. Physically, this means that the method provides the solution with no electromagnetic interaction between the currents in different cells.

Note also that this approach implicitly assumes the secondary field to be proportional to the primary field. This assumption is similar to the idea of quasi-linear approximation (Zhdanov and Fang, 1996a,b).

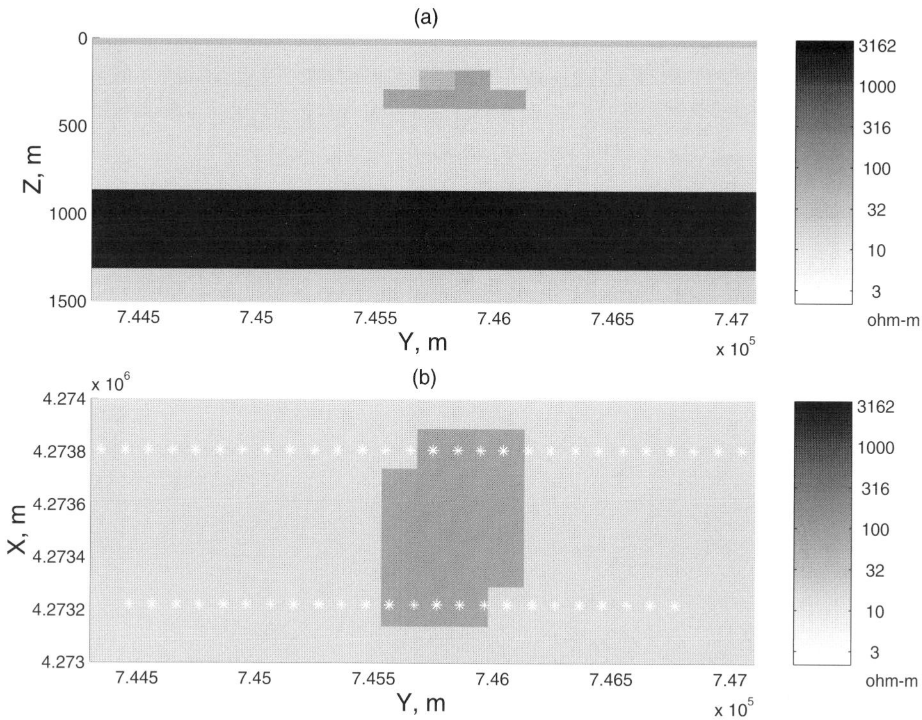

Figure 9. Result of inversion for the model with the body inside the conductive layer. (a) Cross-section. (b) Plan view.

5. BILINEAR CORRECTIONS WITH A MODIFIED GREEN'S OPERATOR

The purpose of subsequent steps in the inversion problem is the following: the nonlinear cross-influence of cells leads to changes in the shape and size of the anomalous bodies but does not change the location of the bodies.

An important question is the convergence of the algorithm. To ensure convergence, we have to modify the Green's matrix $\widehat{\mathbf{G}}$. In other words, we have to condition Equation (10.11) with the conditioners (diagonal matrices) $\widehat{\mathbf{K}}$, $\widehat{\mathbf{A}}$ and $\widehat{\mathbf{B}}$:

$$\mathbf{e}_{\mathrm{a}} = \widehat{\mathbf{G}}\widehat{\mathbf{S}}(\mathbf{e}_{\mathrm{a}} + \mathbf{e}_{\mathrm{n}}); \quad \widehat{\mathbf{K}} = \widehat{\mathbf{A}} - \widehat{\mathbf{B}}; \quad 2\widehat{\mathbf{K}}\widehat{\mathbf{B}} = \widehat{\mathbf{S}}, \tag{10.33}$$

which yields

$$(\widehat{\mathbf{A}} - \widehat{\mathbf{B}})\mathbf{e}_{\mathrm{a}} = \widehat{\mathbf{K}}\widehat{\mathbf{G}}(2\widehat{\mathbf{K}}\widehat{\mathbf{B}})(\mathbf{e}_{\mathrm{a}} + \mathbf{e}_{\mathrm{n}}), \tag{10.34}$$

which we modify further:

$$\widehat{\mathbf{A}}\mathbf{e}_{\mathrm{a}} + \widehat{\mathbf{B}}\mathbf{e}_{\mathrm{n}} = \widehat{\mathbf{K}}\widehat{\mathbf{G}}(2\widehat{\mathbf{K}}\widehat{\mathbf{B}})(\mathbf{e}_{\mathrm{a}} + \mathbf{e}_{\mathrm{n}}) + \widehat{\mathbf{B}}\mathbf{e}_{\mathrm{a}} + \widehat{\mathbf{B}}\mathbf{e}_{\mathrm{n}}. \tag{10.35}$$

Introducing the modified Green's operator $\widehat{\mathbf{G}}_m$ (Zhdanov and Fang, 1997)

$$\widehat{\mathbf{G}}_m(\mathbf{x}) = \widehat{\mathbf{K}}\widehat{\mathbf{G}}2\widehat{\mathbf{K}}(\mathbf{x}) + \mathbf{x}, \tag{10.36}$$

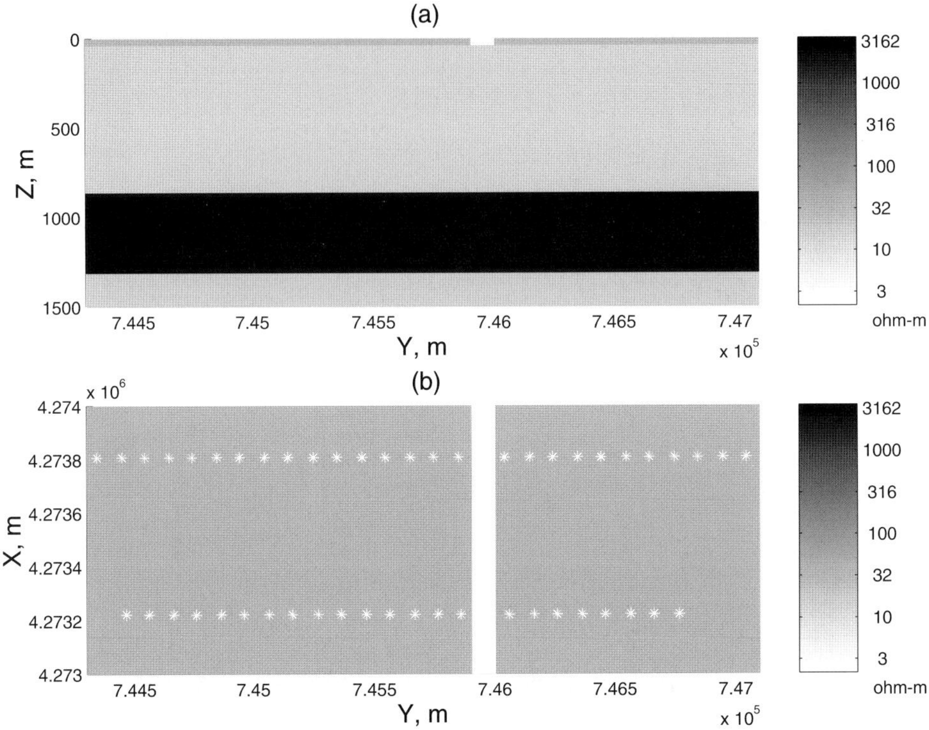

Figure 10. Model with a near-surface local conductive body. (a) Cross-section. (b) Plan view. Stars show the observation points located on the surface. The body is located near coordinate $Y = 746000$ m.

we have

$$\widehat{\mathbf{A}}\mathbf{e}_{\mathrm{a}} + \widehat{\mathbf{B}}\mathbf{e}_{\mathrm{n}} = \widehat{\mathbf{G}}_m(\widehat{\mathbf{B}}\mathbf{e}_{\mathrm{a}} + \widehat{\mathbf{B}}\mathbf{e}_{\mathrm{n}}). \tag{10.37}$$

It has been proven (Zhdanov and Fang, 1997) that the modified operator $\widehat{\mathbf{G}}_m$ has the contraction property

$$\|\widehat{\mathbf{G}}_m(\mathbf{x})\| < \|\mathbf{x}\|, \tag{10.38}$$

provided that the conditioners $\widehat{\mathbf{A}}$ and $\widehat{\mathbf{B}}$ are connected to normal $\widehat{\mathbf{S}}_{\mathrm{n}}$ and anomalous conductivity distribution $\widehat{\mathbf{S}}$ via the relationships

$$\widehat{\mathbf{A}} = \widehat{\mathbf{S}}_{\mathrm{n}}^{-1/2}(2\widehat{\mathbf{S}}_{\mathrm{n}} + \widehat{\mathbf{S}})/2, \quad \widehat{\mathbf{B}} = \widehat{\mathbf{S}}_{\mathrm{n}}^{-1/2}\widehat{\mathbf{S}}/2. \tag{10.39}$$

Property (10.38) ensures convergence of the algorithm. Therefore, instead of (10.31) we iterate Equation (10.37) to solve for $\mathbf{e}_{\mathrm{a}}$, from which we obtain the unknown multiplier $\mathbf{e}_{\mathrm{t}}$, using Equation (10.10).

6. TESTS ON MODELS

We have applied 3-D focusing inversion to synthetic CSAMT data. We consider two sets of models.

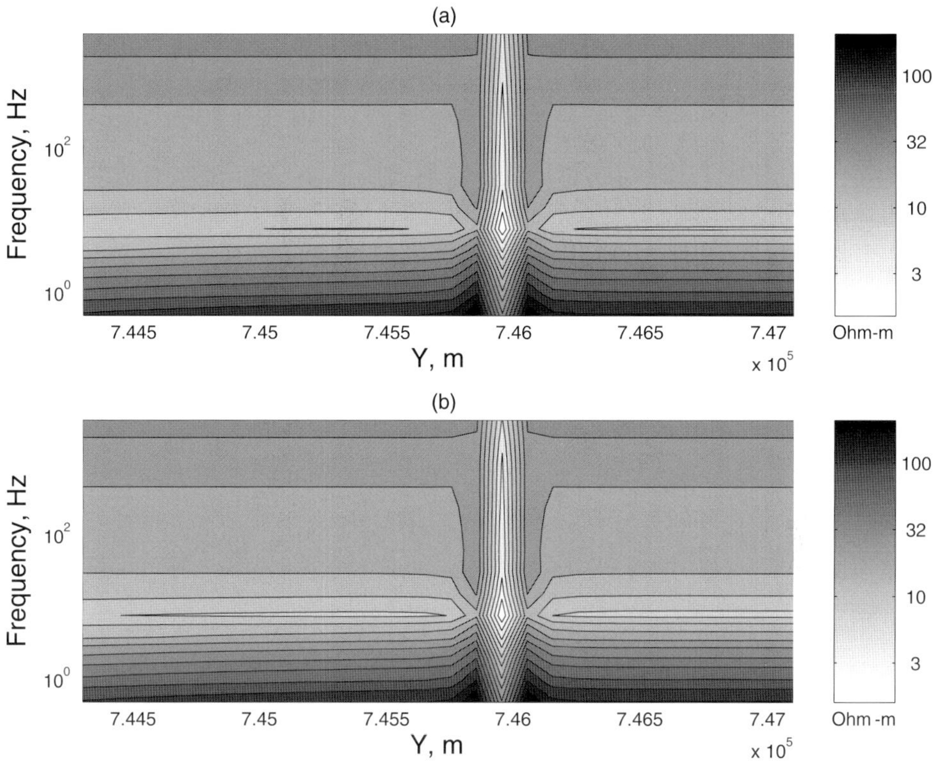

Figure 11. Original data simulated by SYSEM (case a) and the data predicted from the inversion (case b) for the model with the local body near the surface.

The first model contains two conductive (1 Ohm m) bodies embedded in a 20-Ohm m background, with their centers located at a depth of 600 m, 300 m apart. For the full 3-D interpretation we have simulated the data on three parallel profiles (Figure 1, case a). The horizontal electric bipole transmitter AB was located 8 km away from the central profile. Panel b in this figure depicts the true model.

Figure 2 shows the inversion results. Smooth inversion (case a) cannot resolve the bodies. The focusing inversion (case b) resolves the bodies well. Both models in Figure 2 fit the data with the same accuracy of 1.5% (r.m.s. average).

Note that focusing inversion makes strong assumption about compactness of material property distribution. That is why the apparent resolution is so good.

The other model contains a resistive (2600 Ohm m) layer covered by a conductive (10 Ohm m) layer of 900 m in thickness. This model uses the survey geometry which is shown in Figure 3. The same figure shows the discretization for the inverse problem, where the top of the lower half-space is filled with cells increasing in size with depth. We calculate apparent resistivity at 14 frequencies ranging from 0.5 Hz to 4096 Hz, on two profiles located as shown in Figure 3. These parameters are based on the real exploration problem described in the last subsection of this paper.

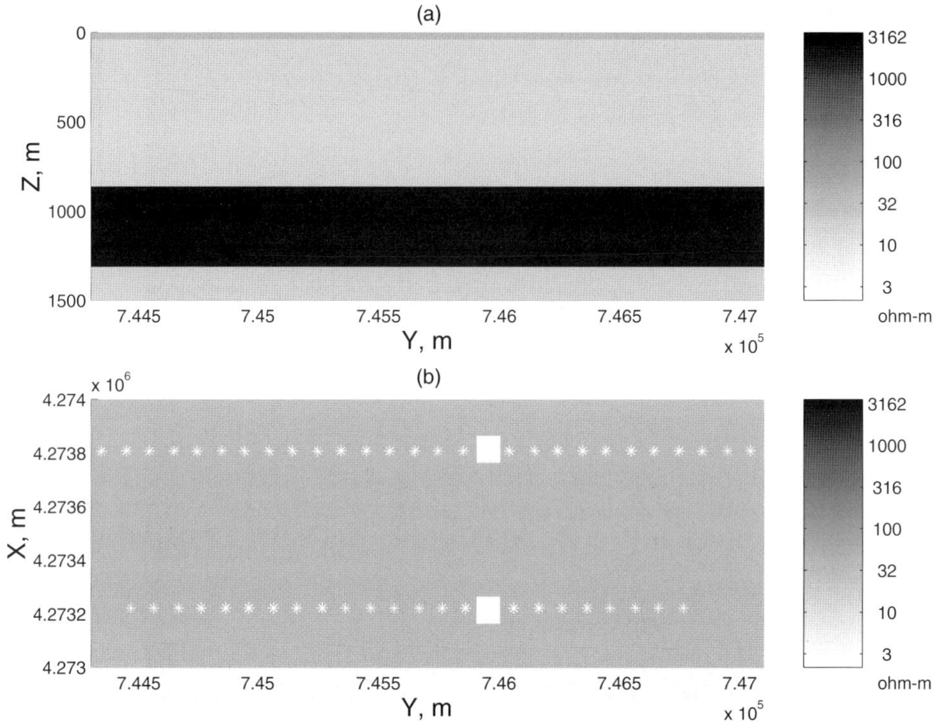

Figure 12. The inversion result for a local, near-surface body. (a) Cross-section. (b) Plan view. Inversion resolves the bodies below the observation profiles but does not resolve the bodies' strike extent away from the profiles.

The problem is to find the topography of the upper boundary of a layer. For example, an anticline structure can be viewed as a departure from the horizontally layered background. It can be modeled as an anomalous resistive body located atop the layer, like the one shown in Figure 4. Figure 5, case a, shows the data (apparent resistivity) computed for this model by the SYSEM full IE forward modeling code (Xiong, 1992). Case b in Figure 5 shows the data predicted from the inversion. The result of inversion for this model is shown in Figure 6.

In a real exploration problem, however, anomalies in the upper layers often overshadow the deeper structures. In the MT and CSAMT method this effect is called 'static shift'. The last two models illustrate this point. Let us consider the model where the observation system and background are the same, but the conductive layer contains a resistive inclusion in it, as shown in Figure 7. Figure 8 shows observed (case a) and predicted apparent resistivity data (case b) for this model. Figure 9 shows the result of inversion for this model. Another model consists of a local near-surface conductive body. This model is depicted in Figure 10. The response from this model is shown in Figure 11. Figure 12 shows the result of inversion for this body. As we can see, inversion can compensate for the static shift effects, but cannot resolve a near-surface structure located between the profile.

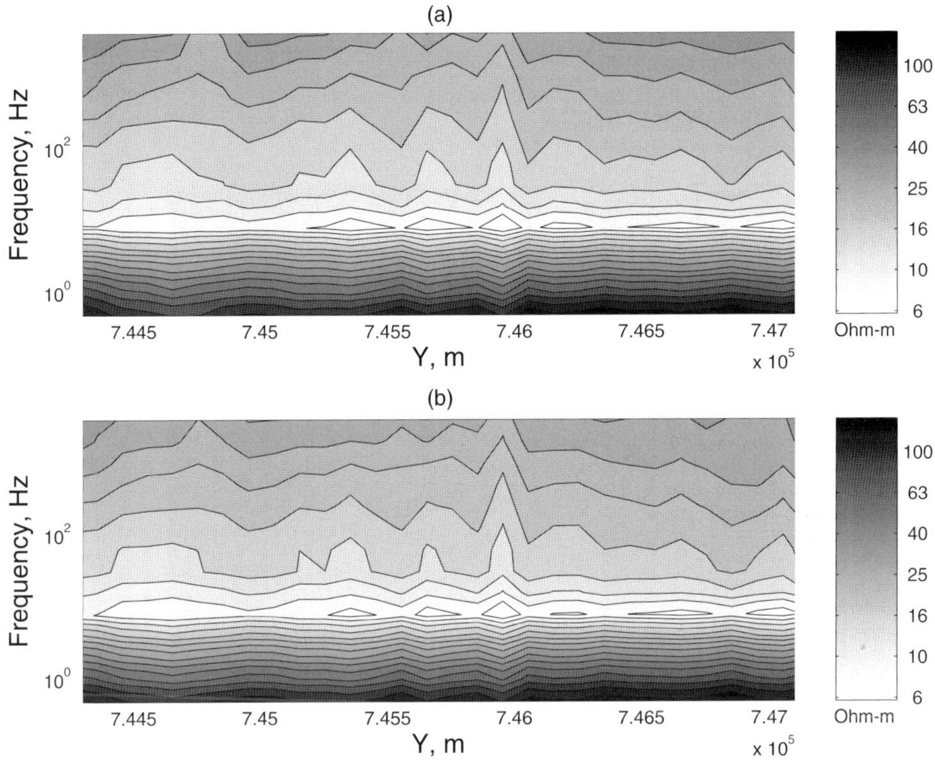

Figure 13. Apparent resistivity on the first profile (a) observed, (b) predicted from 3-D inversion.

7. INVERSION OF REAL DATA

We applied the developed method to the inversion of real 3-D CSAMT data collected in Hamlin Valley for oil exploration. Detailed description of this example is given in Portniaguine (1999).

The observation system is similar to that in Figure 3. We have applied 3-D inversion to the two profiles simultaneously. For the full 3-D interpretation we used the original data uncompensated for static shift (Figure 13). The compensation for static shift is left to the inversion. Panel a in this figure depicts the observed data (apparent resistivity). Panel b depicts the data predicted from inversion. The misfit (global r.m.s. error) was 1%.

Figure 14 shows the inversion results. The stars in panel b are superimposed on the top of the resistive layer. They are picked from the seismic section at the same location and correlated with the Devonian layer. This figure depicts two interesting features. First, we can see that 3-D inversion accurately predicts a down-dropped block in the Devonian. Second, we can see the uplifted part of the fault. Thus, with the 3-D CSAMT inversion, we were able to reach the same conclusions as from the seismic data.

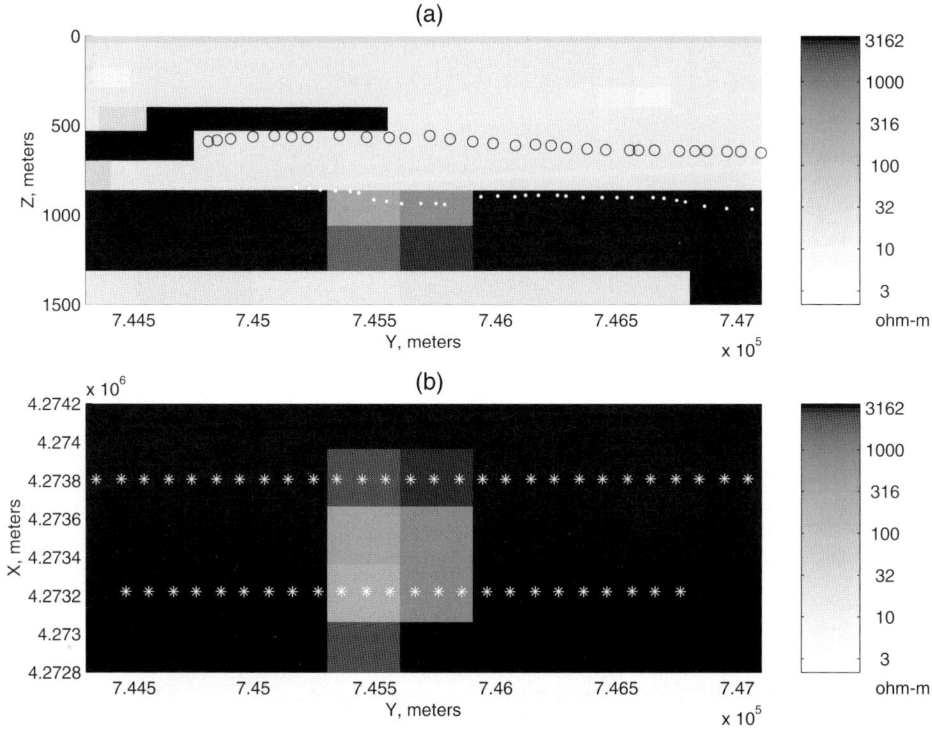

Figure 14. (a) Cross-section of the resulting resistivity model along the first profile. (b) Horizontal slice of the resulting model taken at 800 m depth.

8. CONCLUSION

We have developed a new algorithm of 3-D CSAMT data inversion based on integral equation forward modeling and regularized inversion. This algorithm combines the ideas of the iterative Born method and focusing imaging.

For inversion, we converted the CSAMT data to log-anomalous apparent resistivity, which was linearly connected to anomalous fields. Such a conversion simplifies the inversion algorithm and is convenient to use.

The results of our work demonstrated that the combination of iterative Born method and focusing regularized inversion resulted in a new effective method of CSAMT data interpretation over 3-D geoelectrical structures.

ACKNOWLEDGEMENTS

The authors acknowledge the support of the University of Utah Consortium for Electromagnetic Modeling and Inversion (CEMI), which includes Advanced Power Technologies Inc., AGIP, Baker Atlas Logging Services, BHP Minerals, ExxonMobil Upstream

Research Company, Geological Survey of Japan, INCO Exploration, Japan National Oil Corporation, MINDECO, Naval Research Laboratory, Rio Tinto–Kennecott, 3JTech Corporation, and Zonge Engineering. We thank Kriac Energy, Inc. for providing Hamlin Valley data.

REFERENCES

Cagniard, L., 1953. Basic theory of the Magneto-Telluric method of geophysical prospecting. Geophysics, 18, 605–645.

Chew, W.C., 1990. Waves and Fields in Inhomogeneous Media. Van Nostrand Reinhold, New York.

Habashi, T.M., Groom, R.W. and Spies, B.R., 1993. Beyond the Born and Rytov approximations: a non-linear approach to electromagnetic scattering. J. Geophys. Res., 98, 1759–1775.

Hohmann, G.W., 1975. Three-dimensional induced polarization and EM modeling. Geophysics, 40, 309–324.

Portniaguine, O., 1999. Image Focusing and Data Compression in the Solution of Geophysical Inverse Problems. Ph.D. dissertation, University of Utah.

Portniaguine, O. and Zhdanov, M.S., 1998. Focusing geophysical inversion images. SEG 68 Annual Meeting in New Orleans, Sep. 13–18, Expanded abstracts, pp. 465–468.

Portniaguine, O. and Zhdanov, M.S., 1999. Focusing geophysical inversion images. Geophysics, 64, 874–887.

Rudin, L.I., Osher, S. and Fatemi, E., 1992. Nonlinear total variation based noise removal algorithms. Physica D, 60, 259–268.

Tikhonov, A.N., 1950. On the determination of electrical characteristics of deep layers of the Earth's crust (in Russian). Dokl. Akad. Nauk SSSR, 73, 295–297.

Tikhonov, A.N. and Arsenin, V.Y., 1977. Solution of Ill-Posed Problems. V.II. Winston and Sons, Washington, DC, 223 pp.

Vogel, C.R. and Oman, M.E., 1998. A fast, robust total-variation based reconstruction of noisy, blurred images. IEEE Trans. Image Process., 7, 813–824.

Weidelt, P., 1975. EM induction in three-dimensional structures. Geophysics, 41, 85–109.

Xiong, Z., 1992. EM modeling of three-dimensional structures by the method of system iteration using integral equations. Geophysics, 57, 1556–1561.

Zhdanov, M.S., 1993. Tutorial: regularization in inversion theory. Colorado School of Mines, Golden, CO, 47 pp.

Zhdanov, M.S. and Fang, S., 1996a. Quasi-linear approximation in 3-D EM modeling. Geophysics, 61(N 3), 646–665.

Zhdanov, M.S. and Fang, S., 1996b. 3-D quasi linear electromagnetic inversion. Radio Sci., 31(4), 741–754.

Zhdanov, M.S. and Fang, S., 1997. Quasi-linear series in three-dimensional electromagnetic modeling. Radio Sci., 32, 2167–2188.

Zhdanov, M.S. and Keller, G.V., 1994. The Geoelectrical Methods in Geophysical Exploration. Elsevier, Amsterdam, 873 pp.

Zhdanov, M.S., Dmitriev, V.I., Fang, S. and Hursan, G., 2000. Quasi-analytical approximation and series in electromagnetic modeling. Geophysics, 65, 1746–1757.

Chapter 11

SGILD EM MODELING AND INVERSION IN GEOPHYSICS AND NANO-PHYSICS USING MAGNETIC FIELD EQUATIONS

Ganquan Xie [a,b,c], Jianhua Li [b] and Chien-Chang Lin [c]

[a] *Lawrence Berkeley National Laboratory, Berkeley, USA*
[b] *GL Geophysical Laboratory, Berkeley, USA*
[c] *Nanometer Physical and Technology Laboratory, Huwei Institute of Technology, Taiwan, China*

Abstract: In this paper, we prove the equivalence between the integral equation and Galerkin integro-differential equation solutions for the magnetic field; that the collocation finite element method (FEM) for the integral equation and Garlekin FEM for the differential equation is equivalent, which means the Global Integral and Local Differential algorithm (GILD) is consistent. Based on these properties of the GILD algorithm and the probability moment perturbation theory, we develop a new stochastic GILD electromagnetic (EM) modeling and inversion algorithm (SGILD). SGILD methods, which incorporate statistical properties of measured fields, are useful for interpreting geophysical measurements contaminated by random noise. Tests on synthetic and field data show that the SGILD and GILD algorithms can reliably generate 3-D images of electrical properties. Our SGILD algorithm is useful for geophysics, biochemics, nanometer-material, and nanometer physics, etc., sciences and engineering.

1. INTRODUCTION

Three-dimensional EM modeling and inversion algorithms play an important role in geophysical exploration, oil reservoir management, and environmental site characterization. Many studies of EM modeling and inversion have been presented in the literature. Raiche (1974) proposed an integral equation approach to 3-D modeling. Hohmann (1975) developed a three-dimensional induced polarization and electromagnetic modeling. Xie and Li (1972) developed a 3-D finite element method and first proposed super convergence $o(h^4)$ (Xie, 1975). Xie et al. (1979) developed a mixed finite element method for the electromagnetic field modeling in the large-scale electric motor and transform. Xie (1980a) developed a convex functional finite element analysis to governor the nonlinear underground flow, electromagnetic field, and minimum surface problems. Xie (1980b, 1981) developed an accelerating finite element method for the nonlinear magnetic field induction and its applications. Lee et al. (1981) developed a hybrid three-dimensional electromagnetic modeling scheme by using the traditional electric field integral equation. Wannamaker et al. (1984) developed an electromagnetic modeling of three-dimensional bodies in layered earth using integral equations. Xiong

(1992) developed an electromagnetic modeling of three-dimensional structures by the method of system iterations using integral equations. Lee and Xie (1993) developed a new approach to imaging with low frequency electromagnetic fields using wave transform and tomography. Mackie and Madden (1993) developed an iterative nonlinear inversion of magnetotelluric (MT) data using the conjugate gradient method. Pellerin et al. (1993) developed a 3-D inversion of controlled-source EM data; Zhou et al. (1993) developed a 2-D audio frequency electromagnetic tomography algorithm. Habashy et al. (1994) developed a nonlinear reconstruction of 2-D permittivity and conductivity. Ellis (1995) developed a joint 3-D EM inversion. Li and Xie (1995) developed a new 3-D cubic-hole element using boundary integral equation. Li et al. (1995) developed nonlinear inversion of dc resistivity surveys. Newman (1995) developed a crosswell EM inversion using integral equation. Spichak et al. (1995) developed a three-dimensional inversion of the magnetotelluric field using Bayesian statistics. Oristaglio and Habashy (1996) derived the reciprocity for the electromagnetic field. Zhdanov and Fang (1996) developed 3-D quasi-linear electromagnetic inversion. Hoversten et al. (1998) demonstrated marine magnetotellurics for petroleum exploration. Xie et al. (1995, 1997, 2000) proposed a magnetic integral equation and the global integral and local differential equation (GILD) modeling and inversion.

The GILD algorithm (Xie and Li, 1997, 2000) overcomes the high computational cost and strong ill-posed properties in conventional nonlinear EM inversion. In this paper, we study the relationship of the magnetic field integral equation and differential equation and prove their equivalence in both a continuous system and a numerical discrete system. We describe these theoretical results and several integral equations in Appendices A and B. Based on these properties of the GILD algorithm and probability moment perturbation theory, we develop a new stochastic GILD electromagnetic modeling and inversion (SGILD) algorithm. SGILD methods, which incorporate statistical properties of the measured fields, are useful for interpreting geophysical measurements contaminated by random noise.

The SGILD algorithm works with the differential and integral equations for the magnetic field, where the field and the electromagnetic parameters are stochastic functions (Xie and Li, 1999a). We expand the stochastic field, conductivity, and permittivity into perturbation series and derive integro-differential equations for the statistical moments, ensemble mean, covariance, and cross-covariance. The equations for modeling and inversion are then split into four equations for the field and parameter moments, which are solved by the GILD algorithm (Xie et al., 2000).

After introduction, we describe the SGILD EM modeling. We describe the SGILD EM inversion in the next section. In the following sections, we describe how we split the SGILD into four GILD equations. Then we describe an application and conclusion. Finally, we describe the volume and boundary integral equations in Appendix A and the theoretical equivalence in Appendix B. For expanding the GILD and SGILD algorithms to the time domain, we include the magnetic field integral equations in Appendix A.

2. STOCHASTIC GILD EM MODELING

Suppose that the measured data is a random variable and is contaminated by random noise. We consider the magnetic field H, the electric conductivity σ, and the dielectric permittivity ε to be random quantities. For simplicity, we assume that the background magnetic field H_b, the background conductivity σ_b, background permittivity ε_b, the angular frequency ω, the Green function $E_b^M(r,r)$, and the magnetic permeability μ are deterministic quantities.

2.1. Statistical moment integral equations for magnetic field

We expand the stochastic magnetic field, conductivity, and permittivity into perturbation series. Based on the integral Equation (11.A1) in Appendix A, we derive statistical moment integral equations for the ensemble mean, covariance, and cross-covariance of the magnetic field.

Let

$$\sigma + i\omega\varepsilon = \langle\sigma + i\omega\varepsilon\rangle + (\sigma_s + i\omega\varepsilon_s), \tag{11.1}$$

$$\begin{aligned}\frac{(\sigma-\sigma_b)+i\omega(\varepsilon-\varepsilon_b)}{\sigma+i\omega\varepsilon} \approx \frac{\langle(\sigma-\sigma_b)+i\omega(\varepsilon-\varepsilon_b)\rangle}{\langle\sigma+i\omega\varepsilon\rangle} + \frac{\sigma_s+i\omega\varepsilon_s}{\langle\sigma+i\omega\varepsilon\rangle} \\ - \frac{\langle(\sigma-\sigma_b)+i\omega(\varepsilon-\varepsilon_b)\rangle}{\langle\sigma+i\omega\varepsilon\rangle}\frac{\sigma_s+i\omega\varepsilon_s}{\langle\sigma+i\omega\varepsilon\rangle},\end{aligned} \tag{11.2}$$

and

$$H \approx H_0 + H_1 + H_2. \tag{11.3}$$

Upon substituting the perturbation of the parameters, (11.1), (11.2), and the perturbation of the field, (11.3), into the integral Equation (11.A1), we have the integral equation for zero order mean magnetic field H_0,

$$H_0(r) = H_b(r) - \frac{\aleph}{i\omega\mu}\int_{V_s}\left\{\frac{\langle(\sigma-\sigma_b)+i\omega(\varepsilon-\varepsilon_b)\rangle}{\langle(\sigma+i\omega\varepsilon)\rangle}E_b^M(r',r)\left(\nabla\times H_0(r')\right)\right\}dr', \tag{11.4}$$

and the following three field moment integral equations.

We define FMI to be the field moment integral operator,

$$\begin{aligned}&\mathrm{FMI}\left(C_{H(\sigma+i\omega\varepsilon)}, C_{(\sigma+i\omega\varepsilon)}, H_0, E_b^M, \langle\sigma+i\omega\varepsilon\rangle, \langle\sigma_b+i\omega\varepsilon_b\rangle\right) \\ &\quad = C_{H(\sigma+i\omega\varepsilon)} + \frac{\aleph}{i\omega\mu}\int_{V_s}\left\{\frac{\langle(\sigma-\sigma_b)+i\omega(\varepsilon-\varepsilon_b)\rangle}{\langle\sigma+i\omega\varepsilon\rangle}E_b^M(r',r)\cdot(\nabla\times(C_{H(\sigma+i\omega\varepsilon)}))\right\}dr' \\ &\quad + \frac{\aleph}{i\omega\mu}\int_{V_s}\left\{\frac{\langle\sigma_b+i\omega\varepsilon_b\rangle}{\langle\sigma+i\omega\varepsilon\rangle}\frac{C_{(\sigma+i\omega\varepsilon)}}{\langle\sigma+i\omega\varepsilon\rangle}E_b^M(r',r)\cdot(\nabla\times(H_0))\right\}dr'.\end{aligned} \tag{11.5}$$

The field moment integral equations are

$$\mathrm{FMI}\left(C_{H(\sigma+i\omega\varepsilon)}, C_{(\sigma+i\omega\varepsilon)}, H_0, E_b^M, \langle\sigma+i\omega\varepsilon\rangle, \langle\sigma_b+i\omega\varepsilon_b\rangle\right) = 0, \tag{11.6}$$

$$\mathrm{FMI}\left(C_H, C_{H(\sigma+i\omega\varepsilon)}, H_0, E_{\mathrm{b}}^M, \langle\sigma+i\omega\varepsilon\rangle, \langle\sigma_{\mathrm{b}}+i\omega\varepsilon_{\mathrm{b}}\rangle\right)=0, \tag{11.7}$$

and

$$\mathrm{FMI}\left(\langle H_2\rangle, 1, \hat{C}_{H(\sigma+i\omega\varepsilon)}, E_{\mathrm{b}}^M, \langle\sigma+i\omega\varepsilon\rangle, \langle\sigma_{\mathrm{b}}+i\omega\varepsilon_{\mathrm{b}}\rangle\right)=0, \tag{11.8}$$

where $\langle\cdot\rangle$ is ensemble mean; $H_0=\langle H_0\rangle$ is the zero order mean; $\langle H_1\rangle=0$; $\langle\sigma+i\omega\varepsilon\rangle=\langle\sigma\rangle+i\omega\langle\varepsilon\rangle$; $\langle\sigma\rangle$, the mean conductivity; $\langle\varepsilon\rangle$, the mean permittivity; $\langle\sigma_{\mathrm{b}}+i\omega\varepsilon_{\mathrm{b}}\rangle=\langle\sigma_{\mathrm{b}}\rangle+i\omega\langle\varepsilon_{\mathrm{b}}\rangle$; $\langle\sigma_{\mathrm{b}}\rangle$, the background mean conductivity; $\langle\varepsilon_{\mathrm{b}}\rangle$, the background mean permittivity; $E_{\mathrm{b}}^M(r',r)$, the background electric field exciting by the magnetic dipole source; $C_{(\sigma+i\omega\varepsilon)}=\langle(\sigma+i\omega\varepsilon)_{\mathrm{s}}(r)(\sigma+i\omega\varepsilon)_{\mathrm{s}}(\Re)\rangle$, the covariance of $(\sigma+i\omega\varepsilon)$. In the above equations, the $\langle\sigma+i\omega\varepsilon\rangle$, $\langle\sigma_{\mathrm{b}}+i\omega\varepsilon_{\mathrm{b}}\rangle$, E_{b}^M, and $C_{(\sigma+i\omega\varepsilon)}$ are known. The unknown field of Equation (11.4) is H_0. The unknown field of Equation (11.6) is, $C_{H(\sigma+i\omega\varepsilon)}$, where $C_{H(\sigma+i\omega\varepsilon)}=\langle H_1(r)(\sigma+i\omega\varepsilon)_{\mathrm{s}}(\Re)\rangle$. The unknown field of Equation (11.7) is, C_H, where $C_H=C_{HH}=\langle H_1(r)H_1(\Re)\rangle$. The unknown field of Equation (11.8) is, $\langle H_2\rangle$, the second order correction of the ensemble mean; where, $\hat{C}_{H(\sigma+i\omega\varepsilon)}=\langle H_1(r)(\sigma+i\omega\varepsilon)_{\mathrm{s}}(\Re)\rangle|_{r=\Re}$, r and $\Re$ are space variables.

2.2. Statistical moment Galerkin equations for magnetic field

Similarly, upon substituting the perturbation of the parameters, (11.1), (11.2), and the perturbation of the field, (11.3), into the Galerkin Equation (11.A3) in Appendix A, we derive the field moment Galerkin equation for zero order mean magnetic field H_0,

$$\int\limits_{\Omega_\phi}\frac{1}{\langle\sigma+i\omega\varepsilon\rangle}\nabla\times H_0\cdot\nabla\times\phi\,\mathrm{d}r+\oint\limits_{\partial\Omega_\phi}\frac{1}{\langle\sigma+i\omega\varepsilon\rangle}\nabla\times H_0\times\phi\cdot\mathrm{d}s+i\omega\int\limits_{\Omega_\phi}\mu H_0\phi\,\mathrm{d}r$$
$$=\int\limits_{\Omega_\phi}Q_{\mathrm{s}}\cdot\phi\,\mathrm{d}r, \tag{11.9}$$

and the following three field moment Galerkin equations.

We define FMG to be the field moment Galerkin operator,

$$\mathrm{FMG}\left(C_{H(\sigma+i\omega\varepsilon)}, C_{(\sigma+i\omega\varepsilon)}, H_0, \phi, \langle\sigma+i\omega\varepsilon\rangle, \langle\sigma_{\mathrm{b}}+i\omega\varepsilon_{\mathrm{b}}\rangle\right)$$
$$=\int\limits_{\Omega_\phi}\frac{1}{\langle\sigma+i\omega\varepsilon\rangle}\nabla\times C_{H(\sigma+i\omega\varepsilon)}\cdot\nabla\times\phi\,\mathrm{d}r$$
$$+\oint\limits_{\partial\Omega_\phi}\frac{1}{\langle\sigma+i\omega\varepsilon\rangle}\nabla\times C_{H(\sigma+i\omega\varepsilon)}\times\phi\cdot\mathrm{d}S+i\omega\int\limits_{\Omega_\phi}\mu C_{H(\sigma+i\omega\varepsilon)}\phi\,\mathrm{d}r$$
$$+\int\limits_{\Omega_\phi}\left(\frac{C_{(\sigma+i\omega\varepsilon)}}{\langle\sigma+i\omega\varepsilon\rangle}-\frac{1}{\langle\sigma+i\omega\varepsilon\rangle}\frac{C_{(\sigma+i\omega\varepsilon)}}{\langle\sigma+i\omega\varepsilon\rangle}\right)\nabla\times H_0\cdot\nabla\times\phi\,\mathrm{d}r$$
$$+\oint\limits_{\partial\Omega_\phi}\left(\frac{C_{(\sigma+i\omega\varepsilon)}}{\langle\sigma+i\omega\varepsilon\rangle}-\frac{1}{\langle\sigma+i\omega\varepsilon\rangle}\frac{C_{(\sigma+i\omega\varepsilon)}}{\langle\sigma+i\omega\varepsilon\rangle}\right)\nabla\times H_0\times\phi\cdot\mathrm{d}S. \tag{11.10}$$

The field moment Galerkin equations are

$$\mathrm{FMG}\left(C_{H(\sigma+i\omega\varepsilon)}, C_{(\sigma+i\omega\varepsilon)}, H_0, \phi, \langle\sigma+i\omega\varepsilon\rangle\right)=0, \tag{11.11}$$

$$\mathrm{FMG}\left(C_H, C_{H(\sigma+i\omega\varepsilon)}, H_0, \phi, \langle\sigma+i\omega\varepsilon\rangle\right)=0, \tag{11.12}$$

and

$$\mathrm{FMG}\left(\langle H_2\rangle, 1, \hat{C}_{H(\sigma+i\omega\varepsilon)}, \phi, \langle\sigma+i\omega\varepsilon\rangle\right)=0, \tag{11.13}$$

where ϕ is the base function; Ω_ϕ is the support domain of ϕ; $\partial\Omega_\phi$ is the boundary of Ω_ϕ. In the above equations, the sources, $\langle\sigma+i\omega\varepsilon\rangle$, $C_{(\sigma+i\omega\varepsilon)}$ and ϕ are known. The unknown field of Equation (11.9) is H_0. The unknown field of Equation (11.10) is $C_{H(\sigma+i\omega\varepsilon)}$, where $C_{H(\sigma+i\omega\varepsilon)}=\langle H_1(r)(\sigma+i\omega\varepsilon)_{\mathrm{s}}(\Re)\rangle$. The unknown field of Equation (11.11) is C_H, where $C_H=C_{HH}=\langle H_1(r)H_1(\Re)\rangle$. The unknown field of Equation (11.12), $\langle H_2\rangle$, where $\hat{C}_{H(\sigma+i\omega\varepsilon)}=\langle H_1(r)(\sigma+i\omega\varepsilon)_{\mathrm{s}}(\Re)\rangle|_{r=\Re}$.

2.3. SGILD EM modeling

Similarly, we can prove that the FEM equations of the moment integral and differential equation pair, (11.4) and (11.9), (11.6) and (11.11), (11.7) and (11.12), and (11.8) and (11.13) are equivalent, respectively. Using GILD decomposition, we solve the integral and differential equation pair of the FMI and FMG, in order: (11.4) and (11.9) for H_0; (11.6) and (11.11) for $C_{H(\sigma+i\omega\varepsilon)}$; (11.7) and (11.12) for C_H; and (11.8) and (11.13) for $\langle H_2\rangle$. Then we have

$$\langle H\rangle=\langle H_0\rangle+\langle H_2\rangle. \tag{11.14}$$

3. STOCHASTIC GILD EM INVERSION

Geophysical data are incomplete measurements contaminated by random noise. These data can be interpreted by stochastic inversion which incorporate the statistical properties of the measured fields.

In this section, we describe an SGILD inversion for recovering electric conductivity and permittivity.

3.1. Statistical parameter moment integral equation for inversion

Substituting the perturbation of the parameters,

$$\frac{1}{\sigma+i\omega\varepsilon}=\frac{1}{\langle\sigma+i\omega\varepsilon\rangle}-\frac{1}{\langle\sigma+i\omega\varepsilon\rangle}\frac{\sigma_{\mathrm{s}}+i\omega\varepsilon_{\mathrm{s}}}{\langle\sigma+i\omega\varepsilon\rangle}, \tag{11.15}$$

the perturbation of increment parameters,

$$\delta(\sigma+i\omega\varepsilon)=\delta(\sigma+i\omega\varepsilon)_0+\delta(\sigma+i\omega\varepsilon)_1+\delta(\sigma+i\omega\varepsilon)_2, \tag{11.16}$$

and the perturbation of the field,

$$H=\langle H\rangle+H_{\mathrm{s}}, \tag{11.17}$$

into the variation equation of MIE (11.A1) in Appendix A, by splitting the stochastic variation integral equation, we have the integral equation for zero order increment

$\delta(\sigma+i\omega\varepsilon)_0$, where

$$\delta\langle H_d\rangle(r) = -\frac{\aleph}{i\omega\mu}\int_{V_s}\left\{\frac{\delta(\sigma_0+i\omega\varepsilon_0)(\sigma_b+i\omega\varepsilon_b)}{\langle\sigma+i\omega\varepsilon\rangle^2}E_b^M(r',r)\cdot(\nabla\times\langle H\rangle(r'))\right\}dr'$$

$$-\frac{\aleph}{i\omega\mu}\int_{V_s}\left\{\frac{\langle(\sigma-\sigma_b)+i\omega(\varepsilon-\varepsilon_b)\rangle}{\langle\sigma+i\omega\varepsilon\rangle}E_b^M(r',r)\cdot(\nabla\times\delta\langle H\rangle(r'))\right\}dr', \tag{11.18}$$

and the following integral equations for the increment of parameter covariant moments.

We define PMI to denote the parameter moment integral operator,

$$\mathrm{PMI}\left(\delta C_{(\sigma+iw\varepsilon)H},\delta\sigma_0+i\omega\delta\varepsilon_0,\delta C_H(r_d),\delta C_H,C_H,C_{(\sigma+i\omega\varepsilon)H}\right)$$

$$=\delta C_H(r_d)+\frac{\aleph}{i\omega\mu}\int_{V_s}\left\{\frac{\delta C_{(\sigma+i\omega\varepsilon)H}}{\langle\sigma+i\omega\varepsilon\rangle}\frac{(\sigma_b+i\omega\varepsilon_b)}{\langle\sigma+i\omega\varepsilon\rangle}E_b^M(r',r)\cdot(\nabla\times\langle H\rangle(r'))\right\}dr'$$

$$-2\frac{\aleph}{i\omega\mu}\int_{V_s}\left\{\frac{\delta(\sigma_0+i\omega\varepsilon_0)}{\langle\sigma+i\omega\varepsilon\rangle}\frac{(\sigma_b+i\omega\varepsilon_b)}{\langle\sigma+i\omega\varepsilon\rangle}\frac{C_{(\sigma+i\omega\varepsilon)H}}{\langle\sigma+i\omega\varepsilon\rangle}E_b^M(r',r)\cdot(\nabla\times\langle H\rangle(r'))\right\}dr'$$

$$+\frac{\aleph}{i\omega\mu}\int_{V_s}\left\{\frac{\delta(\sigma_0+i\omega\varepsilon_0)}{\langle\sigma+i\omega\varepsilon\rangle}\frac{(\sigma_b+i\omega\varepsilon_b)}{\langle\sigma+i\omega\varepsilon\rangle}E_b^M(r',r)\cdot(\nabla\times C_H(r'))\right\}dr'$$

$$+\frac{\aleph}{i\omega\mu}\int_{V_s}\left\{\frac{\langle(\sigma-\sigma_b)+i\omega(\varepsilon-\varepsilon_b)\rangle}{\langle\sigma+i\omega\varepsilon\rangle}E_b^M(r',r)\cdot(\nabla\times\delta C_H(r'))\right\}dr'$$

$$+\frac{\aleph}{i\omega\mu}\int_{V_s}\left\{\frac{(\sigma_b+i\omega\varepsilon_b)}{\langle\sigma+i\omega\varepsilon\rangle}\frac{C_{(\sigma+i\omega\varepsilon)H}}{\langle\sigma+i\omega\varepsilon\rangle}E_b^M(r',r)\cdot(\nabla\times\delta\langle H\rangle(r'))\right\}dr'. \tag{11.19}$$

The parameter moment integral equations are

$$\mathrm{PMI}\left(\delta C_{(\sigma+i\omega\varepsilon)H},\delta\sigma_0+i\omega\delta\varepsilon_0,\delta C_{H,}(r_d),\delta C_H,C_H,C_{(\sigma+i\omega\varepsilon)H}\right)=0, \tag{11.20}$$

$$\mathrm{PMI}\left(\delta C_{(\sigma+i\omega\varepsilon)},\delta\sigma_0+i\omega\delta\varepsilon_0,\delta C_{H(\sigma+i\omega\varepsilon)}(r_d),\delta C_{H(\sigma+i\omega\varepsilon)},C_{H(\sigma+i\omega\varepsilon)},C_{(\sigma+i\omega\varepsilon)}\right)=0, \tag{11.21}$$

and

$$\mathrm{PMI}\left(\delta\langle\sigma+i\omega\varepsilon\rangle_2,1,0,\delta\hat{C}_{H(\sigma+i\omega\varepsilon)},\hat{C}_{H(\sigma+i\omega\varepsilon)},\delta\hat{C}_{(\sigma+i\omega\varepsilon)}\right)=0, \tag{11.22}$$

where, $\langle H_d(r)\rangle$ is the measured mean magnetic data, $\langle H(r)\rangle$ is the modeled mean magnetic data; the $\delta(\sigma+i\omega\varepsilon)_0=\delta\sigma_0+i\omega\varepsilon_0$ in Equation (11.18) is the unknown increment. The increment cross covariant $\delta C_{(\sigma+i\omega\varepsilon)H}$ in (11.20), the increment covariant $\delta C_{(\sigma+i\omega\varepsilon)}$ in (11.21), and the increment $\delta\langle\sigma_2+iw\varepsilon_2\rangle$ in Equation (11.22) are calculated. V_s is the space domain here $\langle\sigma+i\omega\varepsilon\rangle\neq 0$.

3.2. Statistical parameter moment Galerkin equation for inversion

Similarly, upon substituting the perturbation of the parameters, (11.15), the perturbation of the increment parameters, (11.16), and the perturbation of the field, (11.17), into the Galerkin Equation (11.A3) in Appendix A, we derive the field moment Galerkin equation for zero order increment of the electric parameters, $\delta(\sigma + i\omega\varepsilon)_0$, where

$$\int_{\Omega_\phi} \frac{\delta(\sigma_0 + i\omega\varepsilon_0)}{\langle\sigma + i\omega\varepsilon\rangle^2} \nabla \times \langle H\rangle \cdot \nabla \times \phi \,\mathrm{d}r + \oint_{\partial\Omega_\phi} \frac{\delta(\sigma_0 + i\omega\varepsilon_0)}{\langle\sigma + i\omega\varepsilon\rangle^2} \nabla \times \langle H\rangle \times \phi \cdot \mathrm{d}S$$
$$= \int_{\Omega_\phi} \frac{1}{\langle\sigma + i\omega\varepsilon\rangle} \nabla \times \delta\langle H\rangle \cdot \nabla \times \phi \,\mathrm{d}r + \oint_{\partial\Omega_\phi} \frac{1}{\langle\sigma + i\omega\varepsilon\rangle} \nabla \times \delta\langle H\rangle \times \phi \cdot \mathrm{d}S$$
$$+ i\omega \int_{\Omega_\phi} \mu\delta\langle H\rangle\phi \,\mathrm{d}r, \tag{11.23}$$

and the following three parameter moment Galerkin equations.

We define PMG to denote the parameter moment Galerkin operator,

$$\mathrm{PMG}\left(\delta C_{(\sigma+i\omega\varepsilon)H}, \delta\sigma_0 + i\omega\delta\varepsilon_0, \delta C_H, C_H, C_{(\sigma+i\omega\varepsilon)H}\right)$$
$$= \int_{\Omega_\phi} \frac{\delta C_{(\sigma+i\omega\varepsilon)H}}{\langle\sigma + i\omega\varepsilon\rangle^2} \nabla \times \langle H\rangle \cdot \nabla \times \phi \,\mathrm{d}r + \oint_{\partial\Omega_\phi} \frac{\delta C_{(\sigma+i\omega\varepsilon)H}}{\langle\sigma + i\omega\varepsilon\rangle^2} \nabla \times \langle H\rangle \times \phi \cdot \mathrm{d}S$$
$$+ \int_{\Omega_\phi} \frac{\delta(\sigma_0 + i\omega\varepsilon_0)}{\langle\sigma + i\omega\varepsilon\rangle^2} \nabla \times C_H \cdot \nabla \times \phi \,\mathrm{d}r + \oint_{\partial\Omega_\phi} \frac{\delta(\sigma_0 + i\omega\varepsilon_0)}{\langle\sigma + i\omega\varepsilon\rangle^2} \nabla \times C_H \times \phi \cdot \mathrm{d}S$$
$$- \int_{\Omega_\phi} \frac{1}{\langle\sigma + i\omega\varepsilon\rangle} \nabla \times \delta C_H \cdot \nabla \times \phi \,\mathrm{d}r - \oint_{\partial\Omega_\phi} \frac{1}{\langle\sigma + i\omega\varepsilon\rangle} \nabla \times \delta C_H \times \phi \cdot \mathrm{d}S$$
$$+ \int_{\Omega_\phi} \frac{1}{\langle\sigma + i\omega\varepsilon\rangle} \frac{C_{(\sigma+i\omega\varepsilon)H}}{\langle\sigma + i\omega\varepsilon\rangle} \nabla \times \delta\langle H\rangle \cdot \nabla \times \phi \,\mathrm{d}r$$
$$+ \oint_{\partial\Omega_\phi} \frac{1}{\langle\sigma + i\omega\varepsilon\rangle} \frac{C_{(\sigma+i\omega\varepsilon)H}}{\langle\sigma + i\omega\varepsilon\rangle} \nabla \times \delta\langle H\rangle \times \phi \cdot \mathrm{d}S - i\omega \int_{\Omega_\phi} \mu\delta C_H\phi \,\mathrm{d}r. \tag{11.24}$$

The parameter moment Galerkin equations are

$$\mathrm{PMG}\left(\delta C_{(\sigma+i\omega\varepsilon)H}, \delta\sigma_0 + i\omega\delta\varepsilon_0, \delta C_H, C_H, C_{(\sigma+i\omega\varepsilon)H}\right) = 0, \tag{11.25}$$

$$\mathrm{PMG}\left(\delta C_{(\sigma+i\omega\varepsilon)}, \delta\sigma_0 + i\omega\delta\varepsilon_0, \delta C_{H(\sigma+i\omega\varepsilon)}, C_{H(\sigma+i\omega\varepsilon)}, C_{(\sigma+i\omega\varepsilon)}\right) = 0, \tag{11.26}$$

and

$$\mathrm{PMG}\left(\delta\langle\sigma_2 + i\omega\varepsilon_2\rangle, 1, 0, \delta\hat{C}_{H(\sigma+i\omega\varepsilon)}, \delta\hat{C}_{(\sigma+i\omega\varepsilon)H}\right) = 0. \tag{11.27}$$

3.3. Posterior probability optimization

We rewrite the stochastic EM inversion to the following posterior probability distribution optimization (Xie and Li, 1999b).

$$P\left((\sigma, \varepsilon) \mid H(r_\mathrm{d})\right) = \max, \tag{11.28}$$

where $P((\sigma,\varepsilon)|H(r_\mathrm{d}))$ is the condition probability distribution of electric conductivity, σ and permittivity ε, for measured magnetic data $H(r_\mathrm{d})$. Given the data $H(r_\mathrm{d})$, we find parameters (σ,ε) such that the condition posterior probability distribution, $P((\sigma,\varepsilon)|H(r_\mathrm{d}))$, is maximum. By Bayes theorem, we have

$$P((\sigma,\varepsilon)|H(r_\mathrm{d})) = \frac{P(H(r_\mathrm{d})|(\sigma,\varepsilon))\,P(\sigma,\varepsilon)}{\int P(H(r_\mathrm{d})|(\sigma,\varepsilon))\,P(\sigma,\varepsilon)\,\mathrm{d}\eta}, \tag{11.29}$$

where η is a random variable, magnetic data $H(r_\mathrm{d})$ and parameters (σ,ε) are random variables, which depend on η. $P(H(r_\mathrm{d})|(\sigma,\varepsilon))$ is the condition probability distribution function, $P(\sigma,\varepsilon)$ is the prior probability distribution function of conductivity and permittivity. Suppose that the probability distributions are Gaussian distributions, we rewrite (11.28) into

$$-\log(P((\sigma,\varepsilon)|H(r_\mathrm{d}))) = \min. \tag{11.30}$$

Substitute (11.29) into (11.30), we have the following regularizing stochastic optimization,

$$\|H(r_\mathrm{d}) - H_D(r_\mathrm{d})\| + \alpha(R(\sigma + i\omega\varepsilon),(\sigma - i\omega\varepsilon)) = \min, \tag{11.31}$$

where $H(r_\mathrm{d})$ is the calculated stochastic magnetic field at the location r_d, $H_D(r_\mathrm{d})$ is the magnetic field data at the location r_d with random noises. $\|\cdot\|$ is the square integral norm of the function space, $(\cdot)$ is the inner product, R and α are two parameters depended on the parameter of the prior probability distribution, $P(\sigma,\varepsilon)$, α is a regularizing parameter. The statistical parameter moment integral and Galerkin Equations (11.18)–(11.22) and (11.23)–(11.27) in the last subsections are used to solve the stochastic optimization (11.31) by the GILD approach (Xie et al., 2000). For non-Gaussian distribution, the Bayesian statistics Formula (11.28), (11.29) and SGILD formulas can be used to develop the stochastic inversion for the electric parameter moments.

3.4. SGILD parameter moments inversion

In the SGILD algorithm, the cells of the domain are decomposed into the two subdomains, *CSI* and *CSII*. The detailed decomposition is described in Xie et al. (2000). Its structure allows a very efficient solution on parallel computers using a nested domain decomposition. The Galerkin integro-differential equation in the GILD and SGILD not only saves computational storage and time, but also improves the ill-posed condition. Based on the stochastic optimization (11.31), the PMI parameter moment integral Equations (11.18), (11.20), (11.21) and (11.22) in *CSI* and the PMG parameter moment Galerkin Equations (11.23), (11.25), (11.26) and (11.27) in *CSII* are combined for the increment parameter moments, $\delta(\sigma + i\omega\varepsilon)_0 = \delta\sigma_0 + i\omega\varepsilon_0$, $\delta C_{(\sigma+i\omega\varepsilon)H}$, $\delta C_{(\sigma+i\omega\varepsilon)}$, and $\delta\langle\sigma_2 + i\omega\varepsilon_2\rangle$, in order, then we have the second order improvement formulas,

$$\delta\langle\sigma + i\omega\varepsilon\rangle = \delta(\sigma_0 + i\omega\varepsilon_0) + \delta\langle\sigma_2 + i\omega\varepsilon_2\rangle, \tag{11.32}$$

and parameter moment updating

$$\langle\sigma + i\omega\varepsilon\rangle_{n+1} = \langle\sigma + i\omega\varepsilon\rangle_n + \delta\langle\sigma + i\omega\varepsilon\rangle, \tag{11.33}$$

$$\left(C_{(\sigma+i\omega\varepsilon)H}\right)_{n+1} = \left(C_{(\sigma+i\omega\varepsilon)H}\right)_n + \delta C_{(\sigma+i\omega\varepsilon)H}, \tag{11.34}$$

and

$$\left(C_{(\sigma+i\omega\varepsilon)}\right)_{n+1} = \left(C_{(\sigma+i\omega\varepsilon)}\right)_n + \delta C_{(\sigma+i\omega\varepsilon)}. \tag{11.35}$$

4. APPLICATION

Our SGILD algorithm has widely applications in geophysics, biochemics, nanometer-material, and nanometer physics, etc., sciences and engineering. In this section, we will use the synthetic model and the real model to describe SGILD applications. We use the synthetic model in Figure 1a to test our two and a half dimension GILD and SGILD modeling and inversion. The length in the x direction of the model is 240 m, the length in the y direction is also 240 m, the background resistivity of the half-space is $\rho_3 = 20\ \Omega\,\text{m}$. The model contains two rectangle blocks with size 40 m $\times$ 40 m, one has resistivity $\rho_1 = 10\ \Omega\,\text{m}$ which is located in the left-top part of the model, the other one has resistivity $\rho_2 = 100\ \Omega\,\text{m}$ and is located in the lower right of the model (Figure 1a). The permittivity $\varepsilon = \varepsilon_0$, the magnetic permeability $\mu = \mu_0$.

We use 18 frequencies, 6 electric line sources on the surface, and 21 receivers in the vertical borehole to make the synthetic data using our 2.5 D GILD modeling; among the 18 frequencies, 15 frequencies are logarithmically spaced frequencies between 1 Hz and 100 kHz and three other frequencies of 50, 5000, and 50,000 Hz. The synthetic data are shown in Figures 2 and 3. The amplitude and phase of the horizontal magnetic field Hx is shown in Figure 2. The amplitudes in sub-figures (2a, …, 2g) and phases in sub-figures (2b, …, 2h) are shown for frequencies 50 Hz, …, 50,000 Hz, respectively. The amplitudes and phases of the vertical magnetic field Hz are shown in Figure 3.

For testing the SGILD inversion, we include 5% Gaussian noise in the data and choose the homogeneous half-space background model as the initial model with resistivity $\rho = 20\ \Omega\,\text{m}$. The 2-D mesh is 256×256. By using 64 *CPUs* $\times$ 30 *minutes* in MPP and 58 iterations, we obtain the ensemble mean resistivity image in Figure 1b. The local standard deviation (LSTD) of the resistivity of the target in the left-top corner is 6%. The LSTD of the resistivity in the right-lower corner is 18%. The upper block is better resolved because it is shallower (high frequency content) and is more surrounded by sources and receivers, producing better data coverage. The optimum mean regularizing parameter α is 0.687×10^{-6}. The resistivity image in subdomain *CSI* is shown in Figure 1c. The image in the *CSII* is shown in Figure 1d.

In Figure 4, we show a resistivity image of the field data using 3-D GILD EM inversion for the Very Early Time EM (VETEM) project. The data configuration is described in the paper by Lee et al. (1995). In that paper, we used the traditional electric integral equation for nonlinear inversion. In this paper, we use the VETEM data to test the GILD inversion. The data were obtained using 3 traverses and 16 frequencies from 5 Hz to 750 kHz. We used 64 ($4 \times 4 \times 4$) CPU and 35 min on a Cray T3D and a Cray SPP at NERSC center to obtain the resistivity image (Figure 4a). The sub-resistivity image on subdomain *CSI* is shown in Figure 4b. The sub-resistivity image on subdomain *CSII*

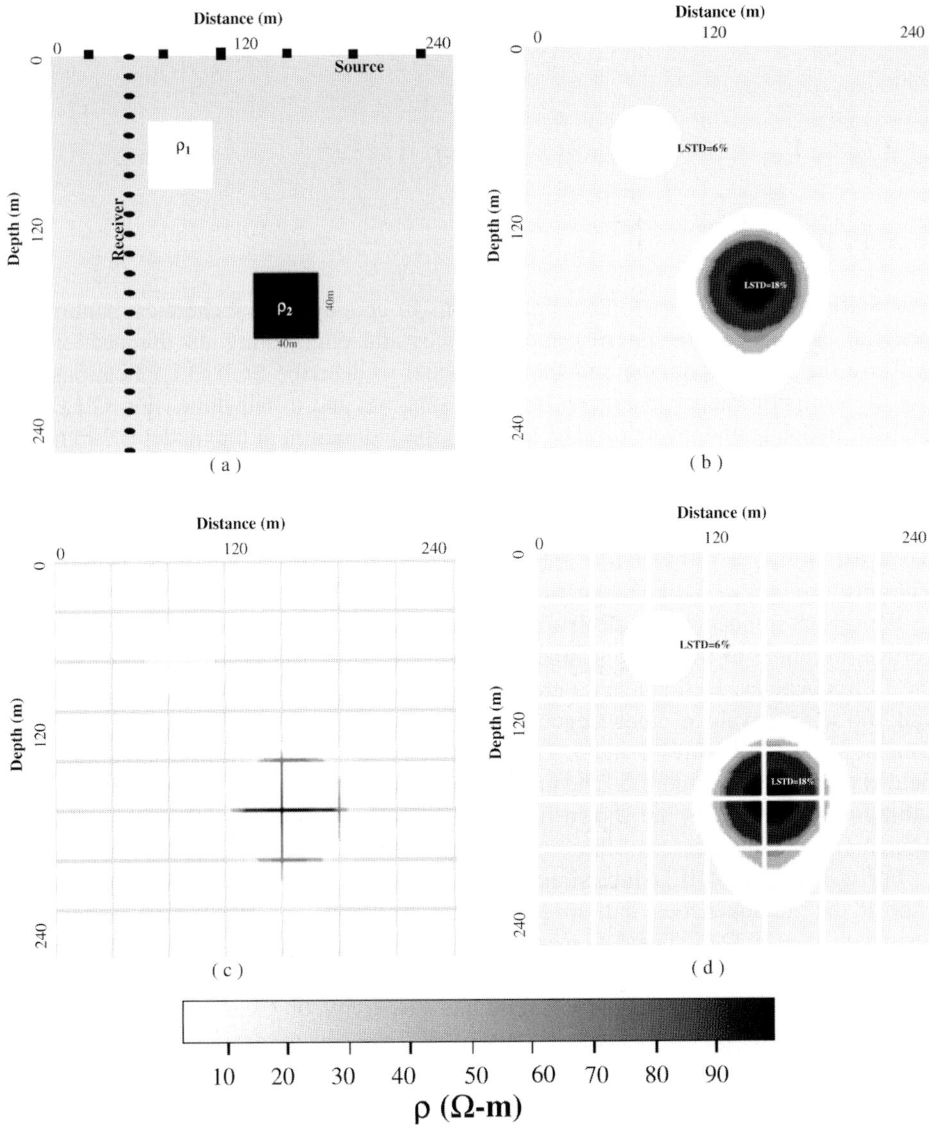

Figure 1. Mean resistivity image of the 2.5-D SGILD EM inversion for synthetic data: (a) model, $\rho_1 = 10$ Ω m, $\rho_2 = 100$ Ω m, 6 sources are located on the surface, 20 receivers are located in the borehole; (b) resistivity image with the LSTD 6% in the left-top part and the LSTD 18% in the right-down part; (c) sub-resistivity image of the model in the subdomain *CSI*; (d) image of the model in the subdomain *CSII*; image in (b) is the combination of the image in (c) and image in (d).

is shown in Figure 4c. The mesh size is $66 \times 21 \times 66$ in x, y, z, respectively. After 36 iterations, we obtained the image. The regularizing parameter is 6.9865×10^{-3}.

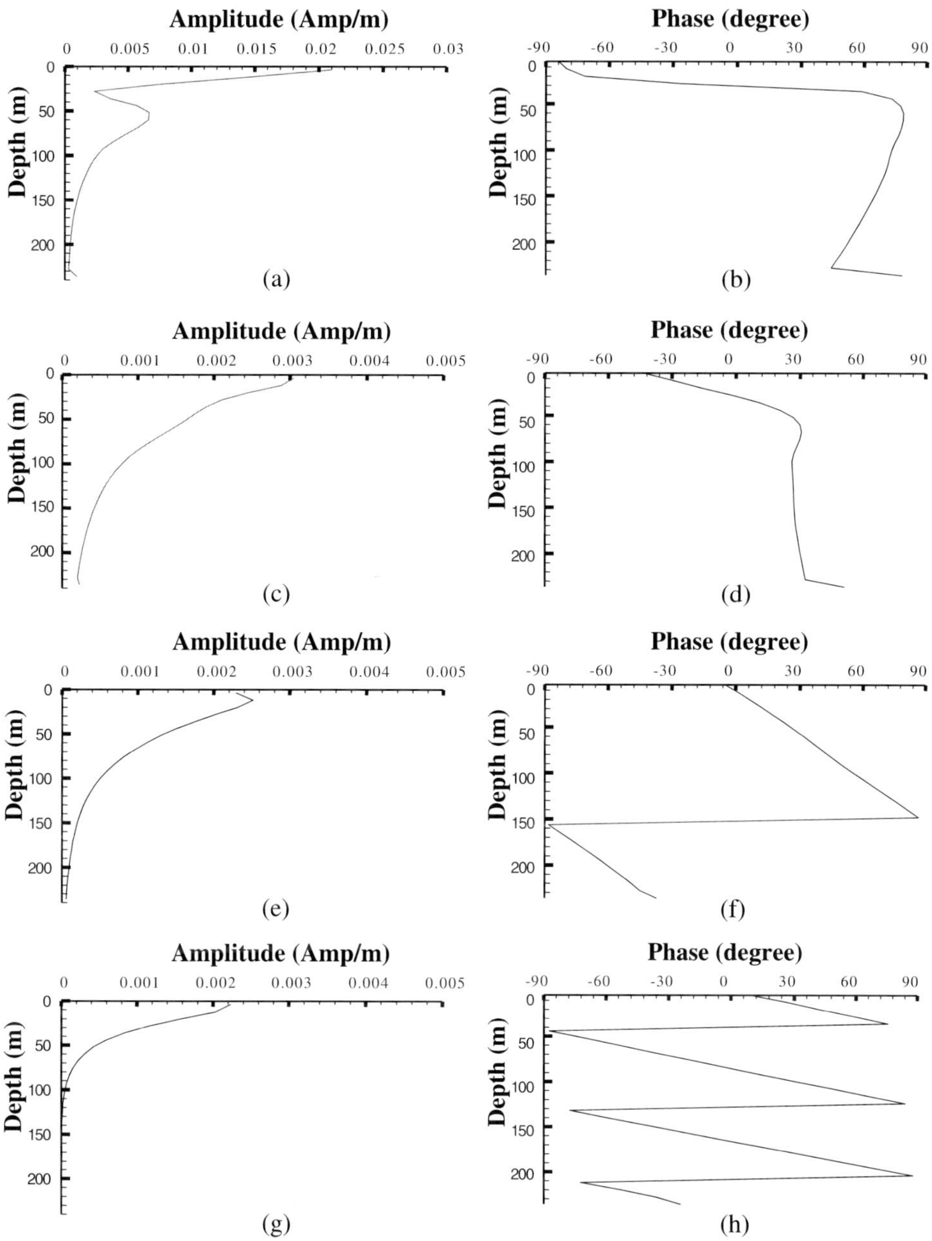

Figure 2. The horizontal magnetic field (Hx): when frequency is 50 Hz, (a) the amplitude, (b) the phase; when frequency is 500 Hz, (c) the amplitude, (d) the phase; when frequency is 5000 Hz, (e) the amplitude, (f) the phase; when frequency is 50,000 Hz, (g) the amplitude, (h) the phase.

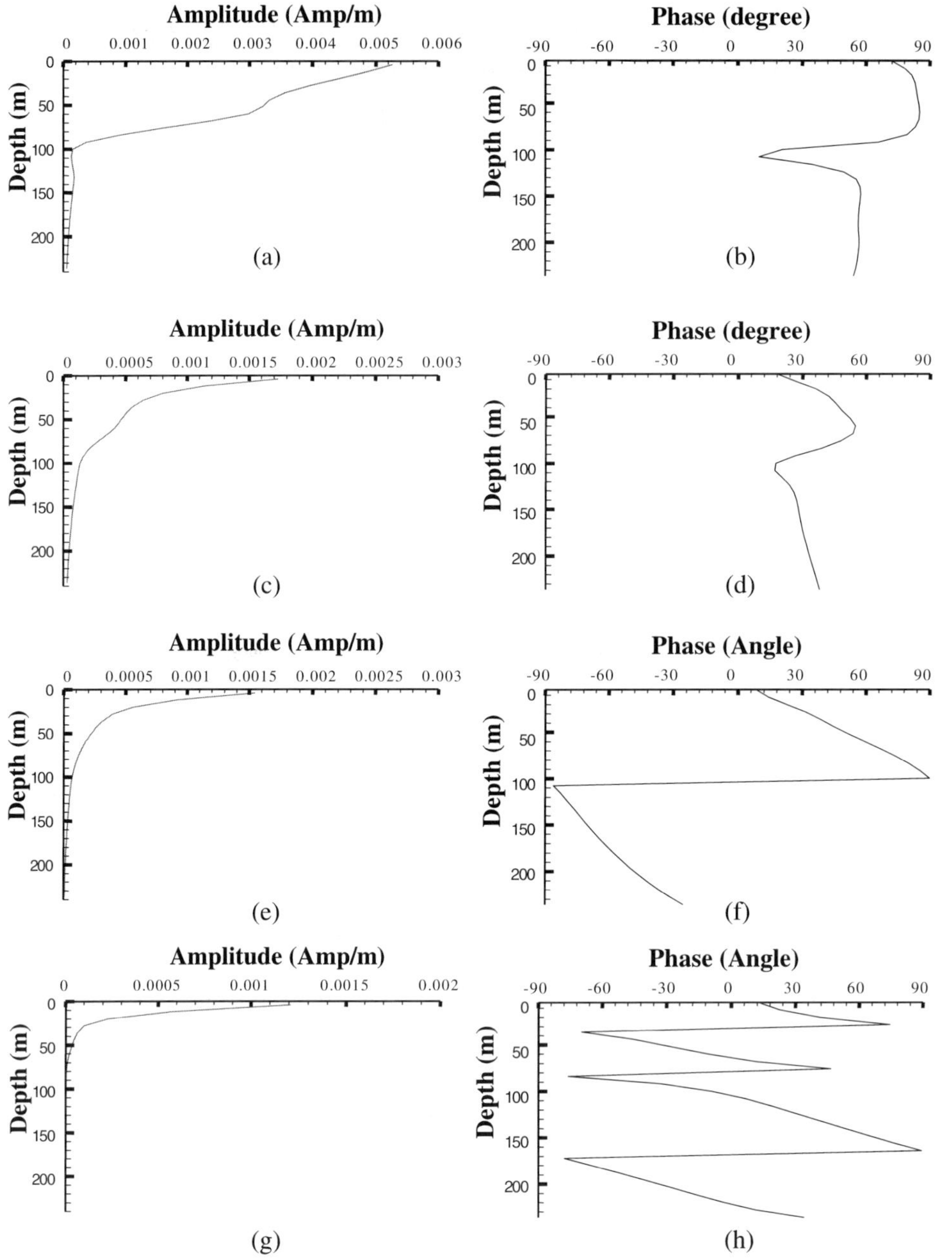

Figure 3. The vertical magnetic field (Hz): when frequency is 50 Hz, (a) the amplitude, (b) the phase; when frequency is 500 Hz, (c) the amplitude, (d) the phase; when frequency is 5000 Hz, (e) the amplitude, (f) the phase; when frequency is 50,000 Hz, (g) the amplitude, (h) the phase.

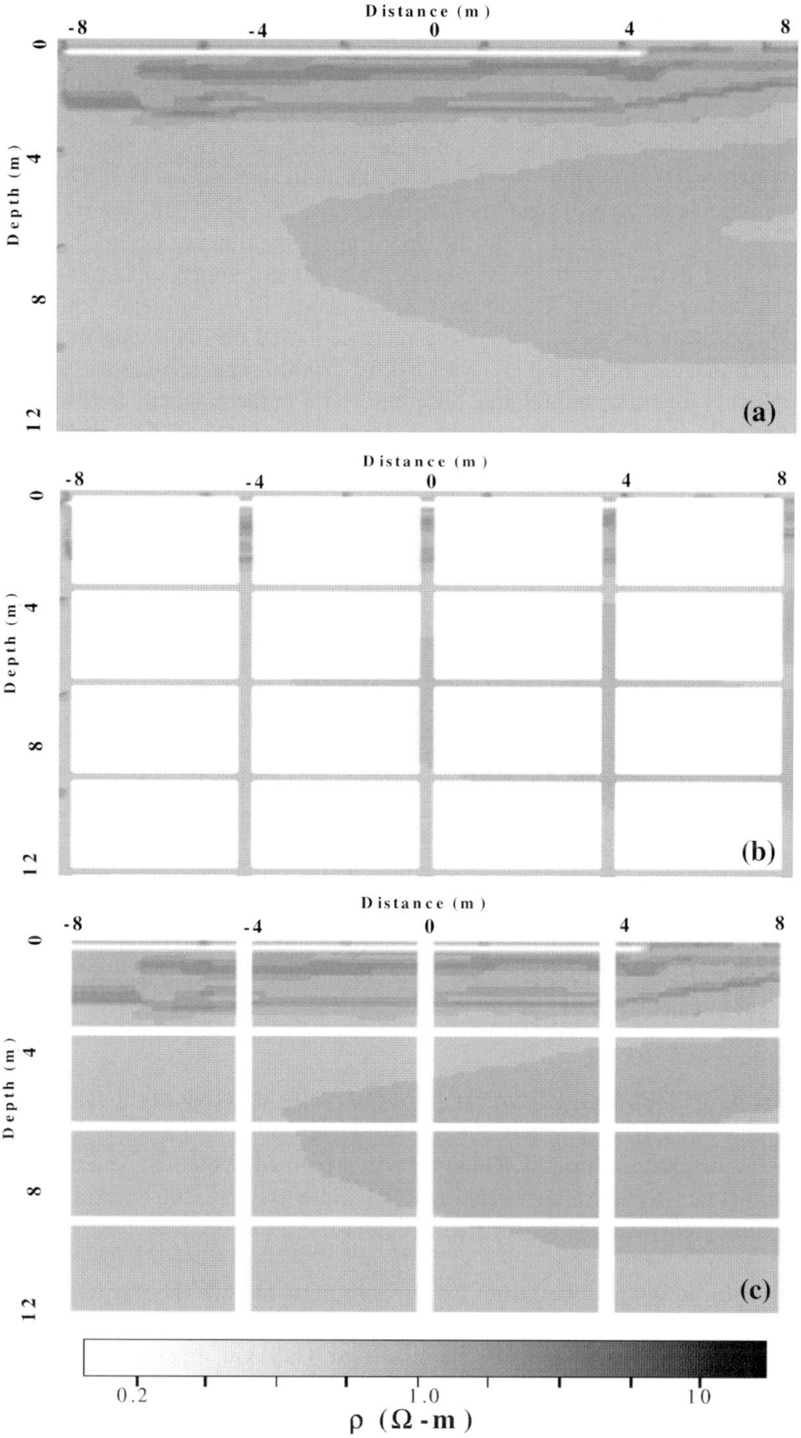

Figure 4. (a) Resistivity image of the field data using 3-D GILD EM inversion, (b) the resistivity image on the subdomain *CSI*, (c) the resistivity image on the subdomain *CSII*.

5. CONCLUSION

The stochastic inversion is an effective method to interpret data containing Gaussian noises. The SGILD inversion can be used to obtain a higher resolution ensemble mean resistivity image with the second order correction, cross covariance of the field and parameters, and standard deviations of the field and parameters. These statistics moments are useful to estimate the confidence of the field data interpretation. The integral equations are equivalent to the differential equations or Galerkin equations in the GILD and SGILD method. Xie and Li (2000) developed GILD-SOR modeling and an inversion for Earth, Ocean, and Atmosphere (EOA) strategic simulation. The new **XY** algorithm and new regularizing methods based on improving the constitutive law were proposed in the paper (Xie and Li, 2000). The continuous field, weakly singular kernel function, and relative difference term in the magnetic field and moments integral equations guarantee the equivalence and convergence of FEM in the GILD and SGILD. The GILD and SGILD methods can be useful for large-scale high resolution EM, seismic and hydrology image for geophysical and oil exploration, environmental remediation, and monitoring of the groundwater and vadose zone.

ACKNOWLEDGEMENTS

The authors thank Professors P.D. Lax, Zhonghua He, Y.M. Chen, Lu Ting, Qisu Zhou, and Drs. Feng Xie, Jinxing Cao, John Wu, Michael Hoversten, and Valeri Korneev for their suggestions and help. In particular, the authors are grateful to Dr. Michael Oristaglio, co-editor Dr. Philip E. Wannamaker and editor Dr. Michael Zhdanov for their help. Dr. Michael Oristaglio and Professor Wannamaker edited and improved this paper in many details. Support for this research was provided by the Director, Office of Science, Office of Basic Energy Science, Division of Chemical, Geoscience and Business of the U.S. Department of Energy under contract No. DE-AC03-76-SF00098. The algorithm has been validated on CRAY T3D and C90 computers operated by the LNBL National Energy Research Scientific Computer Center (NERSC) for the U.S. Department of Energy.

Appendix A. INTEGRAL EQUATIONS FOR THE MAGNETIC FIELD

A magnetic integral equation by Xie et al. was presented at the first Three Dimensional Electromagnetics Symposium held in 1995. Here we present several volume and boundary integral equations for the magnetic field.

A.1. The volume integral equation for the magnetic field in the frequency domain

The statistical parameter moment integral and Galerkin equations of the SGILD algorithm in this paper are derived from the following volume magnetic field integral equations (VIE),

$$H(r) = H_b(r) - \frac{\aleph}{i\omega\mu} \int_{V_s} \left\{ \frac{(\sigma - \sigma_b) + i\omega(\varepsilon - \varepsilon_b)}{(\sigma + i\omega\varepsilon)} E_b^M(r', r) \cdot (\nabla \times H(r')) \right\} dr'. \tag{11.A1}$$

This integral Equation (11.A1) is derived from the magnetic field differential equation,

$$\nabla \times \left(\frac{1}{\sigma + i\omega\varepsilon} \nabla \times H \right) + i\omega\mu H = Q_s, \tag{11.A2}$$

and Green formula

$$\begin{aligned} &\int_{V_s} \left(\nabla \times \left(\frac{1}{\sigma_b + i\omega\varepsilon_b} \nabla \times G_b^M \right) + i\omega\mu G_b^M \right) \cdot H \, d\Omega \\ &\qquad - \int_{V_s} \left(\nabla \times \left(\frac{1}{\sigma + i\omega\varepsilon} \nabla \times H \right) + i\omega\mu H \right) \cdot G_b^M \, d\Omega \\ &= \int_{V_s} \frac{(\sigma - \sigma_b) + i\omega(\varepsilon - \varepsilon_b)}{\sigma + i\omega\varepsilon} \frac{\nabla \times G_b^M}{\sigma_b + i\omega\varepsilon_b} \cdot \nabla \times H \, d\Omega, \end{aligned} \tag{11.A3}$$

where H is the magnetic field, H_b, background magnetic field, σ, the electric conductivity, σ_b, the background conductivity ε, the dielectric permittivity, ε_b, background permittivity, μ, the magnetic permeability, ω, the angular frequency, i is the imaginary unit, ∇, the gradient operator, $\times$, the vector cross product. $\aleph$, $\{[\text{magnetic unit}][\text{space unit}]\}^{-1}$ that gives the integral quantity magnetic field unit on the right-hand side of the equation, $E_b^M(r', r) = (\nabla \times G_b^M)/(\sigma_b + i\omega\varepsilon_b)$, G_b^M, the magnetic field Green's tensor excited by a magnetic dipole source in the background medium, Q_s is the source term.

A.2. The boundary integral equation for the magnetic field in the frequency domain

In GILD and SGILD modeling, we use a volume integral equation on the boundary and a differential equation in the internal domain. Another version of the GILD algorithm uses a boundary integral–differential equation (BIE) on the boundary.

The boundary integral–differential equation for the magnetic field is

$$\begin{aligned} MH(r) = H_b(r) &- \frac{\aleph}{i\omega\mu} \int_{\partial V_{s-}} \left(\frac{1}{\sigma + i\omega\varepsilon} \nabla_r \times H \times G_b^M(r', r) \right) \cdot d\vec{S}(r') \\ &+ \frac{\aleph}{i\omega\mu} \int_{\partial V_{s+}} \left(\frac{1}{\sigma_b + i\omega\varepsilon_b} \nabla_r \times G_b^M(r', r) \times H \right) \cdot d\vec{S}(r'), \end{aligned} \tag{11.A4}$$

where r and r' are points on the boundary. $\nabla_{r'}$ is gradient operator with respect to r', ∂V_{s+} is the exterior side of the boundary of the domain, and ∂V_{s-} is the interior side of the boundary of the domain. The leading coefficient matrix M is given by

$$\underset{(r \text{ is in plane})}{M} = \begin{bmatrix} 0.5 & 0 & 0 \\ 0 & 0.5 & 0 \\ 0 & 0 & 0.5 \end{bmatrix}, \qquad \underset{(r \text{ is on } x_1)}{M} = \begin{bmatrix} 0.75 & 0 & 0 \\ 0 & 0.75 & c \\ 0 & c & 0.83 \end{bmatrix}, \tag{11.A5}$$

$$\underset{(r \text{ is on } y_1)}{M} = \begin{bmatrix} 0.75 & 0 & c \\ 0 & 0.75 & 0 \\ c & 0 & 0.83 \end{bmatrix}, \qquad \underset{(r \text{ is on } z_1)}{M} = \begin{bmatrix} 0.75 & c & 0 \\ c & 0.75 & 0 \\ 0 & 0 & 0.83 \end{bmatrix}, \tag{11.A6}$$

$$\underset{(r \text{ in corner } P_1)}{M} = \begin{bmatrix} 0.875 & -d & d \\ -d & 0.875 & d \\ d & d & 0.917 \end{bmatrix}, \underset{(r \text{ in } P_8)}{M} = \begin{bmatrix} 0.875 & d & -d \\ d & 0.917 & d \\ -d & d & 0.875 \end{bmatrix}, \tag{11.A7}$$

were $c = \frac{1}{6\pi}$, $d = \frac{1}{12\pi}$, the edge $\underline{x}_1$ is the line connecting the corner points p1 and p2, $\underline{y}_1$ is the line connecting the corner points p1 and p4, etc. which is shown in figure 1 of Xie and Li (1998). By comparison, the computational cost of BIE (11.A4) is less than VIE (11.A1), the singularity of the kernel of (11.A4) is stronger than (11.A1), moreover, the leading coefficient matrix of (11.A4) is complicated.

A.3. The volume integral equation for the magnetic field in the time domain

Extending the GILD and SGILD algorithms in the time domain requires the integral equation and differential equation for the magnetic field in the time domain.

Using the inverse Fourier transform on two sides of the integral Equation (11.A1), we obtain a magnetic field integral equation in the time domain

$$\frac{\partial H}{\partial t} = \frac{\partial H_b}{\partial t} - \frac{\aleph}{\mu} \int_{V_s} \left(\frac{1}{\varepsilon} e^{-\frac{\sigma}{\varepsilon}t} - \frac{1}{\varepsilon_b} e^{-\frac{\sigma_b}{\varepsilon_b}t} \right) *_t \nabla \times H *_t \left(\nabla \times G_b^M(r',r) \right) dr', \tag{11.A8}$$

where $*_t$ is the time convolution.

A.4. The boundary integral equation for the magnetic field in the time domain

Using the inverse Fourier transform, we have the new boundary integral equation for the magnetic field in the time domain,

$$M \frac{\partial H}{\partial t}(r,t) = \frac{\partial H_b}{\partial t}(r,t) + \frac{\aleph}{\mu} \int_{\partial V_{s-}} \left(\frac{1}{\varepsilon} e^{-\frac{\sigma}{\varepsilon}t} *_t \nabla_{r'} \times H \times_{*t} G_b^M(r'r) \right) \cdot d\vec{S}(r')$$
$$- \frac{\aleph}{\mu} \int_{\partial V_{s+}} \left(\frac{1}{\varepsilon_b} e^{-\frac{\sigma_b}{\varepsilon_b}t} *_t \nabla_{r'} \times G_b^M(r',r) \times_{*t} H \right) \cdot d\vec{S}(r'), \tag{11.A9}$$

where $*_t$ is time convolution, $\times_{*t}$ is the vector cross product with time convolution.

A.5. The differential equation for the magnetic field in the time domain

The integral Equations (11.A8) and (11.A9) can be derived directly from the following differential equation for the magnetic differential equation in the time domain,

$$\nabla \times \left(\frac{1}{\varepsilon} e^{-\frac{\sigma}{\varepsilon} t} * t \nabla \times H \right) + \mu \frac{\partial H}{\partial t} = -\mu M \delta(t) \delta(r, r_s). \tag{11.A10}$$

Appendix B. THEORETICAL ANALYSIS

In the GILD and SGILD algorithms, we use integral equations in a subdomain, differential equations in the remanding subdomain. In this section, we prove the equivalence between the magnetic field integral equation and magnetic differential equation; moreover, the equivalence between their finite element equations is shown. The equivalence is the theoretical foundation of the convergence of the GILD and SGILD algorithms. This means the GILD and SGILD are consistent algorithms. This is different from the inconsistent traditional hybrid modeling where the strong singularity of the divergent kernel function in the electric integral equation (Lee et al., 1981). In particular, our GILD and SGILD uses the integral and differential equations for inversion that is obvious first in the world.

B.1. Equivalence between integral and differential equations

The Maxwell magnetic field differential Equation (11.A2) is equivalent to the Galerkin integral differential equation,

$$\int_{\Omega_\phi} \frac{1}{\sigma + i\omega\varepsilon} \nabla \times H \cdot \nabla \times \phi \, \mathrm{d}r + \oint_{\partial\Omega_\phi} \frac{1}{\sigma + i\omega\varepsilon} \nabla \times H \times \phi \cdot \mathrm{d}S + i\omega \int_{\Omega_\phi} \mu H \phi \, \mathrm{d}r$$
$$= \int_{\Omega_\phi} Q_s \cdot \phi \, \mathrm{d}r, \tag{11.B1}$$

where ϕ is an arbitrary basic test magnetic field, Ω_ϕ, support domain of the virtual field ϕ, $\partial\Omega_\phi$ is the boundary of the Ω_ϕ.

The Galerkin Equation (11.B1) is the basic equation for generating the Finite Element Method (FEM). We use it to derive the magnetic field integral Equation (11.A1) as follows.

Step 1: Upon substituting $\phi = G_b^M$ in Equation (11.B1) and change prime location, we yield

$$\int_{\Omega} \frac{1}{\sigma + i\omega\varepsilon} \nabla \times H \cdot \nabla \times G_b^M(r', r) \, \mathrm{d}r' + \oint_{\partial\Omega} \frac{1}{\sigma + i\omega\varepsilon} \nabla \times H \times G_b^M(r', r) \cdot \mathrm{d}S(r')$$
$$+ i\omega \int_{\Omega} \mu H G_b^M(r', r) \, \mathrm{d}r' = \int_{\Omega} Q_s \cdot G_b^M(r', r) \, \mathrm{d}r' = \frac{-i\omega\mu}{\aleph} H_b(r) \tag{11.B2}$$

Step 2: The magnetic field Green tensor satisfies the following Galerkin equation when $\phi = H$,

$$\int_{\Omega} \frac{1}{\sigma_b + i\omega\varepsilon_b} \nabla \times G_b^M(r',r) \cdot \nabla \times H \, dr' + \oint_{\partial\Omega} \frac{1}{\sigma_b + i\omega\varepsilon_b} \nabla \times G_b^M(r',r) \times H \cdot dS(r')$$

$$+ i\omega \int_{\Omega} \mu G_b^M(r',r) H \, dr' = \frac{1}{\aleph} \int_{\Omega} \delta(r',r) \cdot H(r') \, dr' = \frac{-i\omega\mu}{\aleph} H(r), \tag{11.B3}$$

where Ω is the domain.

Step 3: Subtracting Equation (11.B2) from Equation (11.B1), we have

$$H(r) = H_b(r) - \frac{\aleph}{i\omega\mu} \int_{V_s} \left\{ \frac{(\sigma - \sigma_b) + i\omega(\varepsilon - \varepsilon_b)}{(\sigma + i\omega\varepsilon)} E_b^M(r',r) \cdot (\nabla \times H(r')) \right\} dr'$$

$$- \frac{1}{i\omega\mu} \left\{ \oint_{\partial\Omega} \frac{1}{\sigma_b + i\omega\varepsilon_b} \nabla \times G_b^M(r',r) \times H \cdot dS(r') \right.$$

$$\left. - \oint_{\partial\Omega} \frac{1}{\sigma + i\omega\varepsilon} \nabla \times H \times G_b^M(r',r) \cdot dS(r') \right\} \tag{11.B4}$$

Because the interface condition of the internal conductivity and permittivity discontinuous interface and radiation decay condition on the infinite point, we have

$$\int_{\partial\Omega} \left(\frac{1}{\sigma_b + i\omega\varepsilon_b} \nabla \times G_b^M \times H \right) \cdot dS - \int_{\partial\Omega} \left(\frac{1}{\sigma + i\omega\varepsilon} \nabla \times H \times G_b^M \right) \cdot dS = 0, \tag{11.B5}$$

Step 4: Upon substituting (11.B5) in Equation (11.B4), we obtain a volume magnetic field integral representation Formula (11.A1), Finally, let r go to the domain V_s, we obtain the volume integral–differential Equation (11.A1) for the magnetic field.

Inversely, if the magnetic field, $H(r)$, satisfies the integral Equation (11.A1), we can prove that

$$H_b(r) = -\frac{\aleph}{i\omega\mu} \int_{\Omega} \frac{1}{\sigma + i\omega\varepsilon} \nabla \times H \cdot \nabla \times G_b^M(r',r) \, dr'$$

$$- \frac{\aleph}{i\omega\mu} \oint_{\partial\Omega} \frac{1}{\sigma + i\omega\varepsilon} \nabla \times H \times G_b^M(r',r) \cdot dS(r') + i\omega \int_{\Omega} \mu G_b^M(r',r) H \, dr', \tag{11.B6}$$

by Green's theorem. Because $H_b(r)$ is the background magnetic field, it satisfies Equation (11.B1) when $\sigma = \sigma_b$ and $\varepsilon = \varepsilon_b$, i.e.

$$\int_{\Omega_\phi} \frac{1}{\sigma_b + i\omega\varepsilon_b} \nabla \times H_b \cdot \nabla \times \phi \, dr + \oint_{\partial\Omega_\phi} \frac{1}{\sigma_b + i\omega\varepsilon_b} \nabla \times H_b \times \phi \cdot dS + i\omega \int_{\Omega_\phi} \mu H_b \phi \, dr$$

$$= \int_{\Omega_\phi} Q_s \cdot \phi \, dr \tag{11.B7}$$

Upon substituting $H_b(r)$ in (11.B6) into (11.B7) and using Equation (11.A2), we prove that $H(r)$ satisfies the Galerkin Equation (11.B1). Similarly, we can prove that $H(r)$ satisfies the interface conditions if it is the solution of integral Equation (11.A1). Therefore, we have proved that the volume integral Equation (11.A1) is equivalent to the Galerkin integro-differential Equation (11.B1) or differential Equation (11.A2). Similarly, we can prove that the volume integral equation in the time domain (11.A8) is equivalent to the differential Equation (11.A10) or its Galerkin equation.

B.2. Equivalence between collection and Galerkin finite element equations

We have proved the equivalence between the magnetic field integral equations and Galerkin in the last subsection. A similar approach can be used to prove the equivalence between the statistical parameter moment integral and Galerkin equations in the SGILD algorithm. In this subsection, we describe the equivalence between the finite element equations for the integral Equation (11.A1) and for the Galerkin integral differential Equation (11.B1). A similar conclusion will be valid for the collection and the Galerkin finite element equations in the SGILD.

When the magnetic permeability is continuous or constant, the magnetic field is continuous regardless of the discontinuity of the conductivity and permittivity. As is well known, the tri-linear finite element shape functions are also continuous. Therefore we chose the tri-linear finite element shape function to approximate the magnetic field in the integral Equation (11.A1) and the Galerkin Equation (11.B1).

The domain is divided into a set of finite cubic-block elements. There are eight vertex nodes in each element. The tri-linear element base function and the finite element approximation are described by Xie et al. (2000). Substituting the finite element approximate field into the magnetic field integral Equation (11.A1), we discretize the integral equation into the full matrix equation. We call this approach collocation FEM for integral equation. Because the Green function kernel of the integral Equation (11.A1) is an integrative singularity, the entries of the matrix and right-hand term are calculated by accurately integrating. On the other hand, substituting the finite element approximate field and base functions into the Galerkin Equation (11.B1), we discretize the magnetic field Galerkin equation into the sparse matrix equation. We call this approach the Galerkin FEM equation.

Because the magnetic field integral Equation (11.A1) is equivalent to the Galerkin Equation (11.B1), the tri-linear finite element function and magnetic field are continuous, and the entries of the matrices and right-hand term in (11.A1) and (11.B1) are calculated accurately. Therefore, we have the equivalence between the collocation FEM for (11.A1) and Galerkin FEM for (11.B1). The theory shows that the collocation FEM and Galerkin FEM in the GILD and SGILD are consistent equations.

Let

$$KH = Q \tag{11.B8}$$

be the Galerkin FEM matrix equation from the Galerkin Equation (11.B1). K is an infinite sparse matrix because we have discrete equations in an infinite domain without boundary conditions. We decompose the matrix K into the K_b and K_s,

$$K = K_b + K_s \tag{11.B9}$$

K_b is the background FEM matrix which is an infinite sparse matrix, K_s is the scattering FEM matrix which is a finite sparse matrix. Substituting (11.B9) into (11.B8) and multiplying K_b^{-1} on both sides of the resulting equation, we have

$$H + K_b^{-1} K_s H = K_b^{-1} Q, \tag{11.B10}$$

Let

$$K_b^{-1} Q = H_b$$

we have

$$H = H_b - K_b^{-1} K_s H. \tag{11.B11}$$

Matrix (11.B11) is the collocation FEM matrix equation from the integral Equation (11.A1). Because the kernel Green tensor of the magnetic field integral Equation (11.A1) is a weaker singularity, K_b^{-1} is the discrete FEM matrix of the integral of the background kernel Green tensor function in (11.A1). We have proved that the collocation FEM and Galerkin FEM for the magnetic field in the GILD and SGILD methods are equivalent.

REFERENCES

Ellis, R.G., 1995. Joint 3D EM inversion. Proceeding of International Symposium on 3D EM, SDR, pp. 307–323.

Habashy, T.M., Oristaglio, M.L. and de Hoop, A.T., 1994. Simulation nonlinear reconstruction of 2D permittivity and conductivity. Radio Sci., 29, 1101–1118.

Hohmann, G.W., 1975. Three-dimensional induced polarization and electromagnetic modeling. Geophysics, 40, 309–324.

Hoversten, G.M., Morrison, H.F. and Constable, S., 1998. Marine magnetotellurics for petroleum exploration Part 2: Numerical analysis of subsalt resolution. Geophysics, 63, 826–840.

Lee, K.H. and Xie, G., 1993. A new approach to imaging with low-frequency electromagnetic fields. Geophysics, 58, 780–796.

Lee, K.H., Pridmore, D.F. and Morrison, H.F., 1981. A hybrid three-dimensional electromagnetic modeling scheme. Geophysics, 46, 796–805.

Lee, K.H., Xie, G., Hoversten, M. and Pellerin, L., 1995. EM imaging for environmental site characterization. Proceedings of International Symposium on Three-Dimensional Electromagnetics. Oct. 4–6, Schlumberger–Doll Research, Ridgefield, CT, pp. 483–490.

Li, J. and Xie, G., 1995. A 3-D cubic-hole element and its application in resistivity imaging. 3-D Electromagnetic Methods. Proceedings of the International Symposium on Three-Dimensional Electromagnetics in Schlumberger–Doll Research, 1995, pp. 415–419.

Li, J., Lee, K.H., Javandel, I. and Xie, G., 1995. Nonlinear three-dimensional inverse imaging for direct current data: 65th Annu. Int. Mtg., Soc. Explor. Geophys., Expanded Abstracts, pp. 250–253.

Mackie, R.L. and Madden, T.R., 1993. Three-dimensional megnetotelluric inversion using conjugate gradients. Geophysics, 58, 215–229.

Newman, G.A., 1995. Crosswell EM inversion using integral and differential equations. Geophysics, 60, 899–911.

Oristaglio, M.L. and Habashy, T.M., 1996. Some uses (and abuses) of reciprocity in wavefield inversion. In: P.M. van den Berg, H. Blok and J. Fokkema (Eds.), Wavefields and Reciprocity. Delft University Press, pp. 1–22.

Oristaglio, M. and Spies, B. (Eds.), 1999. Geophysical Developments of SEGBOOK, Vol. 7, pp. 591–599.
Pellerin, L., Johnston, J.M. and Hohmann, G.W., 1993. Three-dimensional inversion of electromagnetic data. 63rd Annu. Int. Mtg., Soc. Explor. Geophys. Expanded Abstracts. pp. 360–363.
Raiche, A.P., 1974. An integral equation approach to 3-D modeling. Geophys. J. R. Astron. Soc., 36, 363–376.
Spichak, V., Menvielle, M. and Roussignol, M., 1995. Three-dimensional inversion of the magnetotelluric field using Bayesian statistics. Proceedings of the International Symposium on Three-Dimensional Electromagnetics in Schlumberger–Doll Research, 1995, pp. 347–353.
Wannamaker, P.E., Hohmann, G.W. and Dan Filipo, W.A., 1984. Electromagnetic modeling of three-dimensional bodies in layered earths using integral equations. Geophysics, 49, 60–74.
Xie, Ganquan, 1975. The 3-D finite element method in the elastic structure. J. Math. Practice Recogn., 3, 28–41 (in Chinese).
Xie, Ganquan, 1980a. Analysis of finite element method for minimum of convex functional. Acta Math. Appl. Sin., 3(1) (in Chinese).
Xie, Ganquan, 1980b. Finite element method for computation of nonlinear magnetic induction and its applications. J. Comput. Math, 2(8) (in Chinese).
Xie, Ganquan, 1981. A fast convergent finite element method for computation of nonlinear magnetic induction and corresponding generalizations. Advances in Computer Method for PDE, 4, IMACS.
Xie, G. and Li, J., 1972. Three-dimensional Finite Element Method on the elastic mechanics for displacement and stress analysis. Technology report of Hunan Computer Institute (in Chinese).
Xie, G. and Li, J., 1997. 3-D extrapolating electromagnetic imaging. Proceedings of Progress in Electromagnetics Research Symposium, Cambridge, MA, p. 756.
Xie, G. and Li, J., 1999a. New parallel stochastic global integral and local differential equation modeling and inversion. Physica D, 133, 477–487.
Xie, G. and Li, J., 1999b. A new algorithm for 3-D nonlinear electromagnetic inversion. In: M. Oristaglio and B. Spies (Eds.), Three-Dimensional Electromagnetics. Geophysical Developments of SEGBOOK, 1999, 7, 193–207.
Xie, G. and Li, J., 2000. GILD-SOR modeling and inversion for EOA strategic simulation. IMACS Ser. Comput. Appl. Math., 5, 135–148.
Xie, Ganquan, Li, J. and Wu, C., 1979. Finite element method of the electromagnetic field in the large scale electric motor. Transform, 36(3) (in Chinese).
Xie, G., Li, J. and Lee, K.H., 1995. New 3-D nonlinear electromagnetic inversion. Proceedings of International Symposium on Three-Dimensional Electromagnetics. Oct. 4–6, Schlumberger–Doll Research, Ridgefield, CT, pp. 405–414.
Xie, G., Li, J. and Zuo, D., 1997. 3-D electromagnetic imaging using a new parallel GILD nonlinear inversion. 67th Annu. Int. Mtg., Soc. Explor. Geophys., Expanded Abstracts, pp. 414–417.
Xie, G., Li, J., Moridis, G. and Majer, E., 1998. Three dimensional boundary-domain finite element method for electromagnetic modeling. 68th Annu. Int. Mtg., Soc Explor. Geophys. Expanded Abstracts. pp. 417–420.
Xie, G., Li, J., Majer, E., Zuo, D. and Oristaglio, M., 2000. New 3D electromagnetic modeling and inversion. Geophysics, 65, 804–822.
Xiong, Z., 1992. Electromagnetic modeling of three-dimensional structures by the method of system iterations using integral equations. Geophysics, 57, 1556–1561.
Zhdanov, M.S. and Fang, S., 1996. 3-D quasi-linear electromagnetic inversion. Radio Sci., 31, 741–754.
Zhou, Q., Becker, A. and Morrison, H.F., 1993. Audio-frequency electromagnetic tomography in 2-D. Geophysics, 58, 284–495.

Chapter 12

APPLICATION OF A NON-ORTHOGONAL COORDINATE SYSTEM TO THE INVERSE SINGLE-WELL ELECTROMAGNETIC LOGGING PROBLEM

Aria Abubakar and Peter M. van den Berg

Laboratory of Electromagnetic Research, Delft University of Technology, Mekelweg 4, 2628 CD Delft, The Netherlands

Abstract: In this study, we develop an inversion scheme to interpret low-frequency electromagnetic well log data collected in dipping layered formation with invasion. The logging problem is formulated in terms of an integral equation using an oblique (non-orthogonal) coordinate system. By using this coordinate system, discretization errors can be reduced, and a coarse grid can be employed to obtain the results with the same degree of accuracy as with the problem formulated using a Cartesian coordinate system. For the inverse problem this oblique coordinate system also allows inclusion of a priori information about the dip-angle and about the nature of the conductivity distribution. Furthermore, we have avoided the necessity of solving a full-forward problem in each step of the iterative inversion process; this makes a full nonlinear inversion feasible with moderate computer power. Numerical examples show that the approach can be used to interpret single-well logging measurements.

1. INTRODUCTION

Electromagnetic logging tools, either electrode or induction, are the most frequently run logs to characterize the region surrounding the well bore after the drilling of an oil well. The physical principle underlying the method is to measure the difference in conductivity between different zones by applying and measuring electromagnetic fields. When logging tools are used in a formation with dipping layers and zones of drilling mud invasion as shown in Figure 1, a full vector field analysis is required to interpret the measurements. For modeling the induction logging, there are already several efficient numerical schemes available. Eigenfunction solutions to dipping-bed induction response are given by Anderson et al. (1986), Hardman and Shen (1986) and Gianzero and Su (1990). Forward 3D modeling methods of the induction response are given by Howard and Chew (1992), Anderson et al. (1996), Druskin et al. (1999), and van der Horst et al. (1999). However, the algorithms to solve the inverse problem for such configurations are still rare. A 1D inversion method with dip is given by Minerbo (1989). For 3D parametric inversion where in each iterative step the finite difference code has been used to solve the forward problem has been used by Koelman et al. (1996). In their

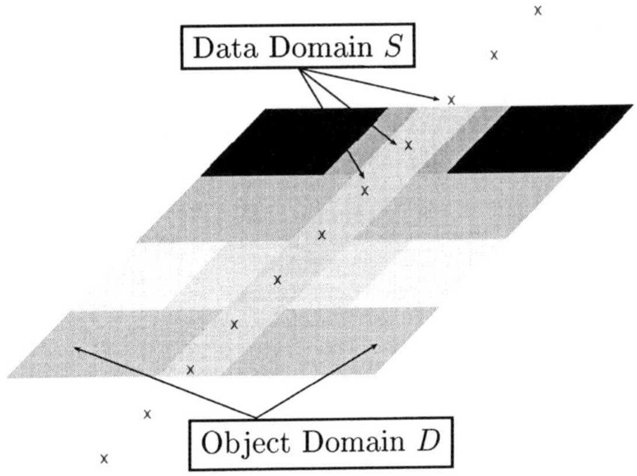

Figure 1. A theoretical model of a single-well electromagnetic problem. This figure gives a cross-section of the 3D formation at the (x_2, x_3) plane for fixed position of $x_1 = 0$. The crosses denote the source positions in the numerical experimentation described in this paper.

approach, the dipping bed environment is approximated using a staircase discretization grid.

Recently, we have developed an inversion scheme for electrode logging data in a highly deviated well with invasion zones (Abubakar and van den Berg, 2000). In this paper we extend the approach to handle the induction logging problem. We formulate the logging problem in terms of an integral equation that employs an oblique coordinate system. The vertical axis of the oblique coordinate system is defined to make an angle with the vertical Cartesian axis and to coincide with the borehole axis. The dip angle is assumed to be known and constant. By using this approach we have certain advantages over the common approach. First discretization errors inherent with the staircasing approximation for the dipping-bed environment can be avoided. Second, a priori information such as that the formation is horizontally symmetric around the borehole and the positivity of the conductivity can be included in simple fashions. Moreover, the dip angle is included explicitly in the algorithm. After the discretization procedure, the forward problem is solved with the conjugate gradient (CG) fast Fourier transform (FFT) method. The contrast source inversion (CSI) method introduced by van den Berg and Kleinman (1997) is extended and applied to solve the inverse problem. Unlike most nonlinear inversion methods, the CSI method does not use Tikhonov-type regularization techniques to deal with the problems of the non-uniqueness in inversion of data. It attempts to overcome this problem by recasting the problem as an optimization problem of minimizing a cost functional. Further discussions of the relation of this cost functional with the common regularization technique is beyond the scope of this paper. The cost functional consists of two terms. The first term is the misfit in the data (data equation) and the second is the misfit of the total field in D in satisfying the domain integral equation (object equation). In each iteration step, we alternately update the contrast source (the product of the total fields and the conductivity contrasts) using the CG direction so as to minimize the cost functional, and then the conductivity contrast is updated to minimize again the cost functional using the updated contrast sources with a CG direction. In this manner, we have avoided the necessity of solving full forward problems in each iteration of the iterative process.

As test examples, we consider the inversion of electrode and induction logging data. Moreover, we will show that we can improve the reconstruction results by using joint inversion of the electrode and induction logging data with an experiment hopping method; the so-called frequency hopping in wave scattering problems. In all of our numerical examples the borehole effect has not been taken into account such that we can study the formation related-effect without further complication. In a later study, the borehole effect can be included in the algorithm by replacing the homogeneous Green's function with a two-media borehole Green's function. Even further, if some a priori knowledge of the layered formation is known, we can take this as the background and construct the Green's function for this inhomogeneous background. The computation time of this Green's function will be very large, but can be computed beforehand. However, the convolution structure of the Green's function is lost, hence the computation time of the inversion scheme will increase. In this paper we suffice to take the homogeneous background.

2. NOTATION AND PROBLEM STATEMENT

The theoretical model of the single-well logging problem is shown in Figure 1. We define the unknown formation as the object domain D with conductivity distribution $\sigma(\boldsymbol{x})$ embedded in a homogeneous background medium of conductivity σ_0. The measurements are made along the axis of the borehole (data domain S) outside the object domain D. The inverse problem can now be formulated as the determination of the conductivity distribution $\sigma(\boldsymbol{x})$ in D from the measurements made in S. In our analysis we employ the 'scattered fields' as the data quantities by assuming that these fields can be computed from the total fields by subtracting off the incident fields. The incident fields are the fields generated by the given sources in the model without the inhomogeneity.

Consider a Cartesian coordinate frame for which a given location $\boldsymbol{x}$ is expressed in terms of the unit vectors as $\boldsymbol{x} = x_1\boldsymbol{i}_1 + x_2\boldsymbol{i}_2 + x_3\boldsymbol{i}_3$. We assume a time harmonic dependence $\exp(-\mathrm{i}\omega t)$, where $\mathrm{i}^2 = -1$, ω is angular frequency, and t is time. The induction logging problem can be formulated in terms of an integral representation (Hohmann, 1975) for the scattered magnetic field vector $\boldsymbol{H}^{\mathrm{sct}}$ as follows

$$\boldsymbol{H}^{\mathrm{sct}}(\boldsymbol{x}) = \sigma_0 \nabla \times \boldsymbol{A}(\boldsymbol{x}), \quad \boldsymbol{x} \in S, \tag{12.1}$$

where $\nabla = (\partial_1, \partial_2, \partial_3)$ denotes the spatial differentiation with respect to the position vector $\boldsymbol{x}$. Note that in the inversion, we use all three-components of the scattered magnetic field as suggested by Spies and Habashy (1995). The normalized vector potential $\boldsymbol{A}$ is the spatial convolution of the Green's function G, and the product of the conductivity contrast χ and the total electric field vector $\boldsymbol{E}$, i.e.,

$$\boldsymbol{A}(\boldsymbol{x}) = \int_D G(\boldsymbol{x} - \boldsymbol{x}')\chi(\boldsymbol{x}')\boldsymbol{E}(\boldsymbol{x}')\,\mathrm{d}v(\boldsymbol{x}'), \tag{12.2}$$

where

$$G(\boldsymbol{x}) = \frac{\exp[\mathrm{i}k_0|\boldsymbol{x}|]}{4\pi|\boldsymbol{x}|}, \qquad k_0 = (\mathrm{i}\omega\mu_0\sigma_0)^{\frac{1}{2}}. \tag{12.3}$$

The conductivity contrast χ is given by

$$\chi(\boldsymbol{x}) = \frac{\sigma(\boldsymbol{x}) - \sigma_0}{\sigma_0}. \tag{12.4}$$

The total electric fields $\boldsymbol{E}$ in D have to satisfy the domain integral equation

$$\boldsymbol{E}(\boldsymbol{x}) - k_0^2\boldsymbol{A}(\boldsymbol{x}) - \nabla\nabla\cdot\boldsymbol{A}(\boldsymbol{x}) = \boldsymbol{E}^{\mathrm{inc}}(\boldsymbol{x}), \qquad \boldsymbol{x} \in D. \tag{12.5}$$

The incident field $\boldsymbol{E}^{\mathrm{inc}}$ is the field in the background configuration due to the presence of point magnetic dipoles directed in the borehole direction. This simplified source model has been employed so that the approach presented in this paper can be tested without further complications.

3. OBLIQUE COORDINATE SYSTEM

As shown in Figure 1, the angle at which the borehole penetrates the formation is arbitrary. In our analysis, the borehole axis is defined to make an angle of θ with respect to the vertical Cartesian axis (x_3-axis) and the rotation is about the x_1-axis. The dip angle θ is assumed to be known and constant. We therefore introduce a new coordinate system such that the borehole axis is defined to coincide with the vertical axis of the new coordinate system ($\hat{x}^3$-axis). Figure 2 shows the oblique and the orthogonal Cartesian axis. With this coordinate system, we can avoid the use of a 'staircase' grid to model the boundary of the borehole and the intersecting layers as shown in the top plot of Figure 3. Note that the staircase approximation will always require more discretization points in the borehole direction than the model formulated in the oblique coordinate system. This aspect plays a crucial role in the application of these techniques to logging data as logging interval can be very large.

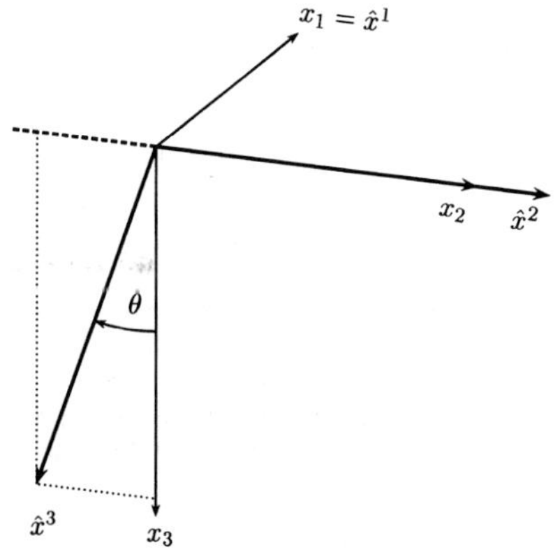

Figure 2. The oblique and the orthogonal Cartesian coordinate system.

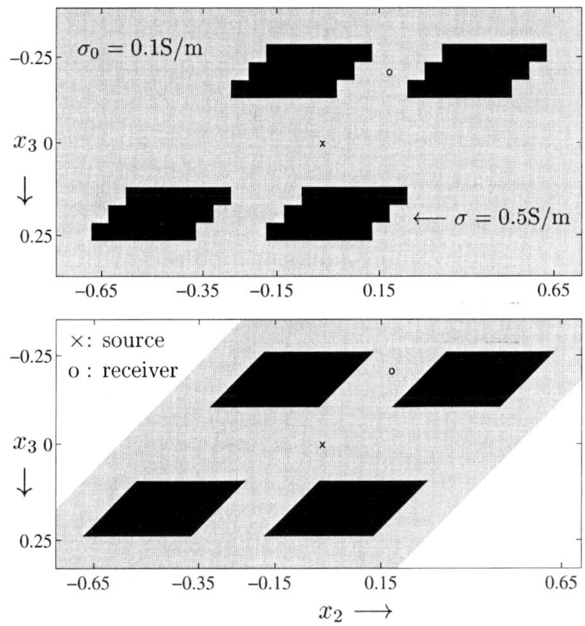

Figure 3. The dipping borehole model using the Cartesian coordinate system (top plot) and the oblique coordinate system (bottom plot). Both figures are cross-sections of 3D conductivity distributions in the (x_2, x_3)-plane at $x_1 = 0$.

In this oblique coordinate system, the volume, the surface, or the position is defined by independent coordinates, $\hat{x}^1$, $\hat{x}^2$, and $\hat{x}^3$. The relationship between these coordinates and the Cartesian coordinates is given by

$$\hat{x}^1 = x_1, \quad \hat{x}^2 = x_2 + \tan(\theta)x_3, \quad \hat{x}^3 = \cos^{-1}(\theta)x_3. \tag{12.6}$$

The vectors associated with these curves form a non-orthogonal system, such that the covariant and contravariant components of a vector must be introduced. Following the same procedure as Stratton (1941), any arbitrary vector $\boldsymbol{A}$ can be expanded in the contravariant or the covariant projections on either the unitary vectors $\hat{\boldsymbol{e}}_\kappa$, or the reciprocal unitary vectors $\hat{\boldsymbol{e}}^\kappa$;

$$\boldsymbol{A} = \sum_{\kappa=1}^{3} A_\kappa \boldsymbol{i}_\kappa = \sum_{\kappa=1}^{3} \hat{A}^\kappa \hat{\boldsymbol{e}}_\kappa = \sum_{\kappa=1}^{3} \hat{A}_\kappa \hat{\boldsymbol{e}}^\kappa, \tag{12.7}$$

where $\hat{A}^\kappa$ is the contravariant component of the vector $\boldsymbol{A}$, and $\hat{A}_\kappa$ is the covariant component. The contravariant component $\hat{A}^\kappa$ is related to the covariant component $\hat{A}_\kappa$ as follows:

$$\begin{bmatrix} \hat{A}^1 \\ \hat{A}^2 \\ \hat{A}^3 \end{bmatrix} = \frac{1}{\cos^2(\theta)} \begin{bmatrix} \cos^2(\theta) & 0 & 0 \\ 0 & 1 & \sin(\theta) \\ 0 & \sin(\theta) & 1 \end{bmatrix} \begin{bmatrix} \hat{A}_1 \\ \hat{A}_2 \\ \hat{A}_3 \end{bmatrix}. \tag{12.8}$$

Using (12.6) and the definition of the gradient, divergence, and curl operators given in Stratton (1941), we arrive at

$$\nabla V = \hat{\partial}_1 V \hat{\boldsymbol{e}}^1 + \hat{\partial}_2 V \hat{\boldsymbol{e}}^2 + \hat{\partial}_3 V \hat{\boldsymbol{e}}^3, \tag{12.9}$$

$$\nabla \cdot \boldsymbol{A} = \hat{\partial}_1 \hat{A}^1 + \hat{\partial}_2 \hat{A}^2 + \hat{\partial}_3 \hat{A}^3, \tag{12.10}$$

$$\nabla \times \boldsymbol{A} = \frac{1}{\cos(\theta)} \left[(\hat{\partial}_2 \hat{A}_3 - \hat{\partial}_3 \hat{A}_2)\hat{\boldsymbol{e}}_1 + (\hat{\partial}_3 \hat{A}_1 - \hat{\partial}_1 \hat{A}_3)\hat{\boldsymbol{e}}_2 + (\hat{\partial}_1 \hat{A}_2 - \hat{\partial}_2 \hat{A}_1)\hat{\boldsymbol{e}}_3 \right], \tag{12.11}$$

in which $\hat{\partial}_\kappa$ denotes the spatial differentiation with respect to $\hat{x}^\kappa$ with $\kappa \in \{1,2,3\}$. Note that the curl operator in (12.11) working on the covariant components of a vector field, and the result is a vector field represented by its contravariant components. Similar complication also happens with the gradient-divergence operator (gradient-divergence operator working on the contravariant components of a vector field, and the results are vector fields represented by its covariant components). As pointed out by Holland (1983) this alternation between the covariant and contravariant field components is the primary complication when using the oblique coordinate system. Note that, although the transformation procedure described in this section is used for integral equation modeling, this procedure can also be applied for finite difference modeling.

The forward and inverse problem can now be formulated in the oblique coordinate system given in Figure 2. In this work, we have chosen the covariant components to be the main field quantities. Using the results in (12.6)–(12.11), the integral representation in (12.1) becomes

$$H_1^{\text{sct}} = \hat{H}^{\text{sct};1}, \quad H_2^{\text{sct}} = \hat{H}^{\text{sct};2} - \sin(\theta)\,\hat{H}^{\text{sct};3}, \quad H_3^{\text{sct}} = \cos(\theta)\,\hat{H}^{\text{sct};3}, \tag{12.12}$$

where

$$\begin{bmatrix} \hat{H}^{\text{sct};1} \\ \hat{H}^{\text{sct};2} \\ \hat{H}^{\text{sct};3} \end{bmatrix} = \frac{\sigma_0}{\cos(\theta)} \begin{bmatrix} 0 & -\hat{\partial}_3 & \hat{\partial}_2 \\ \hat{\partial}_3 & 0 & -\hat{\partial}_1 \\ -\hat{\partial}_2 & \hat{\partial}_1 & 0 \end{bmatrix} \begin{bmatrix} \hat{A}_1 \\ \hat{A}_2 \\ \hat{A}_3 \end{bmatrix}, \quad \{\hat{x}^1, \hat{x}^2, \hat{x}^3\} \in S. \tag{12.13}$$

The covariant components of $\boldsymbol{A}$ are given by

$$\hat{A}_\kappa(\hat{x}^1,\hat{x}^2,\hat{x}^3) = \int_D G(\hat{x}^1 - \hat{x}^{1'}, \hat{x}^2 - \hat{x}^{2'}, \hat{x}^3 - \hat{x}^{3'})\chi(\hat{x}^{1'},\hat{x}^{2'},\hat{x}^{3'})\hat{E}_\kappa(\hat{x}^{1'},\hat{x}^{2'},\hat{x}^{3'})\,\mathrm{d}v(\boldsymbol{x}), \tag{12.14}$$

where

$$\mathrm{d}v(\boldsymbol{x}) = \cos(\theta)\,\mathrm{d}\hat{x}^1 \mathrm{d}\hat{x}^2 \mathrm{d}\hat{x}^3. \tag{12.15}$$

The domain integral equation in (12.5) becomes:

$$\hat{E}_\kappa - k_0^2 \hat{A}_\kappa - \hat{\partial}_\kappa \left(\hat{\partial}_1 \hat{A}^1 + \hat{\partial}_2 \hat{A}^2 + \hat{\partial}_3 \hat{A}^3 \right) = \hat{E}_\kappa^{\text{inc}}, \quad (\hat{x}^1,\hat{x}^2,\hat{x}^3) \in D, \ \kappa \in (1,2,3). \tag{12.16}$$

We observe that (12.16) together with (12.14) is a singular integral equation (see that for (12.16) the point of observation $(\hat{x}^1,\hat{x}^2,\hat{x}^3)$ in (12.15) is in D). This singular integral equation is discretized by a technique successfully used by Abubakar and van den Berg (2000). In this technique, the normalized vector potential is weakened by taking its spherical mean. This is the same as replacing the Green's function by its spherical mean (see Zwamborn and van den Berg (1994)). After that the weak normalized vector potential is computed using the trapezoidal rule, and the gradient-divergence operator working on the normalized vector potential is replaced by its finite difference operator.

In order to discuss our solution of the forward and inverse scattering problem we write our equations in an operator form. The integral representation defined by (12.12), (12.13) and (12.14) is called the data equation, and written symbolically as

$$\boldsymbol{H}^{\mathrm{sct}} = \mathcal{K}_S \chi \boldsymbol{E}, \quad \boldsymbol{x} \in S. \tag{12.17}$$

The domain integral equation defined by (12.14) and (12.16) is called the object equation, and written symbolically as

$$\boldsymbol{E} - \mathcal{K}_D \chi \boldsymbol{E} = \boldsymbol{E}^{\mathrm{inc}}, \quad \boldsymbol{x} \in D. \tag{12.18}$$

4. FORWARD PROBLEM

Assuming that χ is known and that the model has been properly discretized, the forward problem for each excitation may be formulated as a linear system of equations given by (12.18). Since the matrix operator $\mathcal{K}_D$ is composed of spatial convolutions, we can use advantageously FFT routines. However, we then need an iterative solution, and the conjugate gradient (CG) method seems to be one of the most efficient methods. With this so-called CGFFT (see Zwamborn and van den Berg (1994)) technique we are able to solve complex 3D problems efficiently. Furthermore, it gives also the fundament of our solution of the inverse problem.

4.1. Algorithm

The CG method consists of an algorithm to determine the vector $\boldsymbol{E}$ iteratively by reducing the value of the cost functional

$$F_D(\boldsymbol{E}) = \frac{\|\boldsymbol{E}^{\mathrm{inc}} - \boldsymbol{E} + \mathcal{K}_D \chi \boldsymbol{E}\|_D^2}{\|\boldsymbol{E}_j^{\mathrm{inc}}\|_D^2}, \tag{12.19}$$

where the norm on D is given by

$$\|\boldsymbol{E}\|_D^2 = \langle \boldsymbol{E}, \boldsymbol{E} \rangle_D = \int_D \sum_{\kappa=1}^{3} \hat{E}_\kappa(\hat{x}^1,\hat{x}^2,\hat{x}^3)\hat{E}^\kappa(\hat{x}^1,\hat{x}^2,\hat{x}^3)\cos(\theta)\,\mathrm{d}\hat{x}^1\mathrm{d}\hat{x}^2\mathrm{d}\hat{x}^3. \tag{12.20}$$

We construct sequences $\{\boldsymbol{E}_n\}$, for $n = 1,2,\ldots$, in the following manner. Define the object error at the nth iteration to be

$$\boldsymbol{r}_n = \boldsymbol{E}^{\mathrm{inc}} - \boldsymbol{E}_n + \mathcal{K}_D \chi \boldsymbol{E}_n. \tag{12.21}$$

We update $\boldsymbol{E}$ as follows:

$$\boldsymbol{E}_0 = \boldsymbol{0}, \quad \boldsymbol{E}_n = \boldsymbol{E}_{n-1} + \alpha_n \boldsymbol{e}_n, \quad n \geq 1, \tag{12.22}$$

where α_n is a constant parameter and the update direction $\boldsymbol{e}_n$ is a function of position. The update direction is chosen to be the Fletcher–Reeves gradient direction

$$\boldsymbol{e}_0 = \boldsymbol{0}, \quad \boldsymbol{e}_n = \partial \boldsymbol{e}_n + \frac{\|\partial \boldsymbol{e}_n\|_D^2}{\|\partial \boldsymbol{e}_{n-1}\|_D^2} \boldsymbol{e}_{n-1}, \quad n \geq 1. \tag{12.23}$$

In (12.23), $\partial \boldsymbol{e}_n$ is gradient of the cost functional F_D with respect to $\boldsymbol{E}$ evaluated at $\boldsymbol{E}_{n-1}$. Explicitly this is found to be

$$\partial \boldsymbol{e}_n = \frac{\boldsymbol{r}_{n-1} - \overline{\chi} \mathcal{K}_D^* \boldsymbol{r}_{j,n-1}}{\|\boldsymbol{E}^{\text{inc}}\|_D^2}, \tag{12.24}$$

where $\mathcal{K}_D^*$ is the adjoint operator of $\mathcal{K}_D$ and can be found explicitly in Abubakar and van den Berg (2000). After the update direction has been determined, the constant α_n is calculated via

$$\alpha_n = \frac{\langle \boldsymbol{r}_{n-1}, \boldsymbol{e}_n - \mathcal{K}_D \chi \boldsymbol{e}_n \rangle_D}{\|\boldsymbol{e}_n - \mathcal{K}_D \chi \boldsymbol{e}_n\|_D^2}, \tag{12.25}$$

which minimizes $F_D(\boldsymbol{E}_n)$ in (12.19).

4.2. Numerical example

In this example, the forward scheme using the oblique coordinate system will be compared to the one using the Cartesian coordinate system developed by Abubakar and van den Berg (1998). The exact model in both systems is given in Figure 3. Two layers with a conductivity 0.5 S/m are embedded in a homogeneous background of conductivity $\sigma_0 = 0.1$ S/m. The borehole of radius 0.15 m is filled with the same conductivity as the background, and is deviated with a dip angle of 45°.

The configuration using the Cartesian coordinate system is given in the upper plot of Figure 3. Here, the object domain is $1.5 \times 1.5 \times 0.7$ m^3. This domain is discretized into equal-sized subdomains with side lengths of $\Delta x_1 = \Delta x_2 = \Delta x_3 = 0.05$ m. Hence, we use 30 discretization points in the x_1 and x_2 directions and 14 points in the x_3 direction, yield a total of 12,600 discretization points. Note that as shown in top plot of Figure 3, the dipping bed environment (the boundary of the borehole and the intersecting layers) is approximated by using the 'staircased' discretization grid.

For the oblique coordinate system the object domain is $1.1 \times 1.1 \times 0.99$ m^3 as shown in the bottom plot of Figure 3. The object domain was partitioned into equal-sized subdomains with side lengths $\Delta \hat{x}^1 = \Delta \hat{x}^2 = 0.05$ m and $\Delta \hat{x}^3 = 0.07$ m. Hence, 22 discretization points are employed in the $\hat{x}_1$ and $\hat{x}_2$ directions, and 14 points in the $\hat{x}_3$ direction. Thus in total we have 6776 discretization points.

In Figure 4, for a given source position at $\hat{x}^{\mathrm{S};3} = 0$, we present the real and imaginary part of the vertical Cartesian component of the scattered magnetic field H_3^{sct} for different receiver positions $\hat{x}^{\mathrm{R};3}$ along the borehole axis. The results using the oblique coordinate system with 6776 discretization points are given by the solid lines, and the one using the Cartesian coordinate system with 12,600 discretization points by the dashed lines. We observe that the difference between these two forward solutions is large in the region near the source, while further away from the source the agreement is very good.

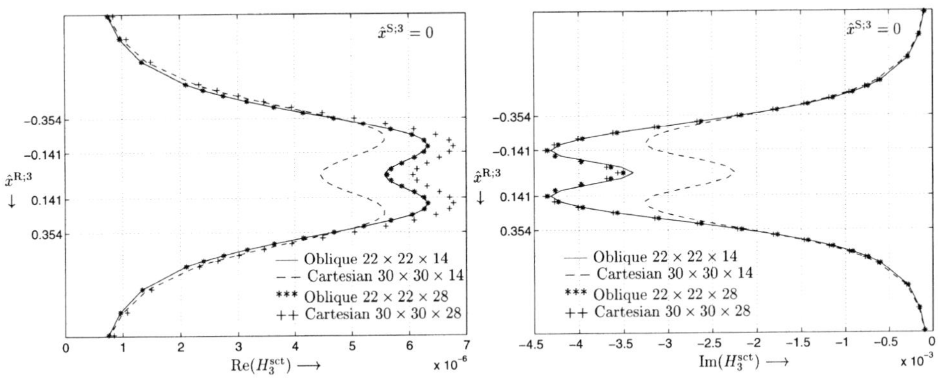

Figure 4. The vertical Cartesian component of the scattered magnetic field H_3^{sct} as a function of receiver position $\hat{x}^{R;3}$ in the borehole axis for a given source position at $\hat{x}^{S;3} = 0$.

Next, we employ a finer discretization grid in the x_3 direction for Cartesian solution. Here, the side length in the x_3 direction becomes $\Delta x_3 = 0.025$ m. Hence, we have 28 points in the x_3 direction, and in total have 25,200 discretization points. The results for this model are given by the plus-signs in Figure 4. We observe that this Cartesian solution using 25,200 discretization points tends towards the one using the oblique system. In Figure 4, we show also the forward solution using the oblique coordinate system with 28 discretization points in the $\hat{x}^3$ direction ($\Delta\hat{x}^3 = 0.035$). Hence, we are employing 13,552 discretization points. These results are given by the star-signs in Figure 4. We observe that the results calculated via the oblique coordinate system do not change significantly. Thus, it is clear that by using the oblique coordinate system we can suffice with less discretization points to obtain results to the accuracy as that of the Cartesian coordinate system.

5. INVERSE METHOD

In the inverse problem, we assume that the unknown formation (object domain D) is irradiated successively by a number ($J = 1, 2, \ldots$) of known incident electric fields $\boldsymbol{E}_j^{\mathrm{inc}}$. The inverse scattering problem can now be formulated as follows: find χ in the object domain D, given $\boldsymbol{H}_j^{\mathrm{sct}}$ within the data domain S; or solving the data equation (12.17) for χ, subject to the additional and necessary condition that χ and $\boldsymbol{E}_j$ satisfy (12.18).

5.1. Algorithm

A major observation is that the data equation contains both the unknown field and the unknown contrast in the form of a product; it can be written as a single quantity, viz. the contrast source

$$\boldsymbol{W}_j(\boldsymbol{x}) = \chi(\boldsymbol{x})\boldsymbol{E}_j(\boldsymbol{x}), \tag{12.26}$$

which can be considered as an equivalent source that produces the measured scattered field. Using (12.26) in (12.17), the data equation becomes

$$\boldsymbol{H}_j = \mathcal{K}_S \boldsymbol{W}_j, \quad \boldsymbol{x} \in S, \tag{12.27}$$

while the object equation in (12.18) becomes

$$\boldsymbol{E}_j = \boldsymbol{E}_j^{\text{inc}} + \mathcal{K}_D \boldsymbol{W}_j, \quad \boldsymbol{x} \in D. \tag{12.28}$$

Substituting (12.28) into (12.26), we obtain an object equation for the contrast source rather than for the field, viz.,

$$\chi \boldsymbol{E}_j^{\text{inc}} = \boldsymbol{W}_j - \chi \mathcal{K}_D \boldsymbol{W}_j, \quad \boldsymbol{x} \in D. \tag{12.29}$$

Although the data equation in (12.27) is linear in the contrast source, it is a classic ill-posed equation. It was shown that there exist non-trivial solutions of the homogeneous form of (12.27), although it was argued by some that uniqueness could be restored from physical considerations (a priori information). A good summary of the debate is given by Deveney and Sherman (1982) and the responses by Bojarski (1982) and Stone (1982). It is not our intention to renew this controversy since it is now well accepted that non-trivial solutions of Equation (12.27) exist. Moreover, it also has been shown that the minimum norm solution of (12.27), the solution produced, for example, by the CG method described in the previous section, is not the appropriate physical solution (Habashy et al., 1994). Therefore, we recast the problem as an optimization problem in which not only the contrast sources but also the conductivity contrast itself are sought to minimize a cost functional. The method updates in each iterative step the contrast sources $\boldsymbol{W}_{j,n}$ using a CG direction (only one step) and the conductivity contrast χ_n also using a CG direction (only one step) in order to reduce the value of the cost functional. This iterative process is repeated until a prescribed error level is achieved. In this way, we have avoided the necessity of solving one or more full forward problems in each iteration of our inversion process.

The cost functional F is a superposition of the normalized errors in the data equation and object equation, i.e.,

$$F(\boldsymbol{W},\chi) = \frac{\sum_j \|\boldsymbol{H}_j^{\text{sct}} - \mathcal{K}_S \boldsymbol{W}_j\|_S^2}{\sum \|\boldsymbol{H}_j^{\text{sct}}\|_S^2} + \frac{\sum_j \|\chi \boldsymbol{E}_j^{\text{inc}} - \boldsymbol{W}_j + \chi \mathcal{K}_D \boldsymbol{W}_j\|_D^2}{\sum_j \|\chi \boldsymbol{E}^{\text{inc}}\|_D^2}, \tag{12.30}$$

where the norm on S is given by

$$\|\boldsymbol{H}_j\|_S^2 = \langle \boldsymbol{H}, \boldsymbol{H} \rangle_S = \sum_{\forall \boldsymbol{x}^{\text{R}}} \sum_{\kappa=1}^{3} \hat{H}_{\kappa;j}(\boldsymbol{x}^{\text{R}}) \hat{H}_j^{\kappa}(\boldsymbol{x}^{\text{R}}). \tag{12.31}$$

The normalizations in Equation (12.30) are chosen in such a way that both terms are equal to one if $\boldsymbol{W}_j = \boldsymbol{0}$. This is a quadratic functional in $\boldsymbol{W}_j$, but highly nonlinear in χ. Note that this cost functional is introduced by van den Berg and Kleinman (1997), and is very similar to the ones used by Kleinman and van den Berg (1992) and Zhdanov and Fang (1999). In this cost functional we have not included an extra regularization technique to stabilize the inversion process. If one wants to think in terms of regularization, it is remarked that the object equation acts as a regularizer for the data equation. Further discussions of the relation of this cost functional with the common regularization technique is beyond the scope of this paper. The algorithm involves the construction of sequences $\{\chi_n\}$ and $\{\boldsymbol{W}_{j,n}\}$, for $n = 1, 2, \ldots$ by reducing the value of F in the following manner.

5.1.1. Updating the contrast sources

Define the data error and the object error at the n-th step to be

$$\boldsymbol{\rho}_{j,n} = \boldsymbol{H}^{\text{sct}}_{j,n} - \boldsymbol{\mathcal{K}}_S \boldsymbol{W}_{j,n}, \quad \boldsymbol{r}_{j,n} = \chi_n \boldsymbol{E}_{j,n} - \boldsymbol{W}_{j,n}, \tag{12.32}$$

where the total field $\boldsymbol{E}_{j,n}$ follows from (12.28). Now suppose $\boldsymbol{W}_{j,n-1}$ and χ_{n-1} are known. We update the contrast sources ($\boldsymbol{W}_j$) by

$$\boldsymbol{W}_{j,n} = \boldsymbol{W}_{j,n-1} + \alpha^W_{j,n} \boldsymbol{w}_{j,n}, \tag{12.33}$$

where $\alpha^W_{j,n}$ are constant parameters and the update directions $\boldsymbol{w}_{j,n}$ are functions of position.

The update directions are chosen to be the Polak–Ribière conjugate gradient directions (Brodlie, 1977), which search for improved directions when a change with respect to the directions of the last iteration occurs, and restart the optimization when the changes are small. These update directions are obtained as

$$\boldsymbol{w}_{j,0} = \boldsymbol{0}, \quad \boldsymbol{w}_{j,n} = \partial \boldsymbol{w}_{j,n} + \frac{\langle \partial \boldsymbol{w}_{j,n}, \partial \boldsymbol{w}_{j,n} - \partial \boldsymbol{w}_{j,n-1} \rangle_D}{\| \partial \boldsymbol{w}_{j,n-1} \|^2_D} \boldsymbol{w}_{j,n-1}, \quad n \geq 1, \tag{12.34}$$

where $\partial \boldsymbol{w}_{j,n}$ is the gradient (Fréchet derivative) of the cost functional F with respect to $\boldsymbol{W}_j$ evaluated at $\boldsymbol{W}_{j,n-1}$ and χ_{n-1}. Explicitly, the gradient is found to be

$$\partial \boldsymbol{w}_{j,n} = -\eta_S \boldsymbol{\mathcal{K}}^*_S \boldsymbol{\rho}_{j,n-1} - \eta_{D;n-1} \left(\boldsymbol{r}_{j,n-1} - \boldsymbol{\mathcal{K}}^*_D \overline{\chi}_{n-1} \boldsymbol{r}_{j,n-1} \right), \tag{12.35}$$

where

$$\eta_S = \left(\textstyle\sum_j \| \boldsymbol{H}^{\text{sct}}_j \|^2_S \right)^{-1}, \quad \eta_{D;n-1} = \left(\textstyle\sum_j \| \chi_{n-1} \boldsymbol{E}^{\text{inc}}_j \|^2_D \right)^{-1}, \tag{12.36}$$

while $\boldsymbol{\mathcal{K}}^*_S$ and $\boldsymbol{\mathcal{K}}^*_D$ denote the adjoint operators of $\boldsymbol{\mathcal{K}}_S$ and $\boldsymbol{\mathcal{K}}_D$ mapping $L_2(S)$ into $L_2(D)$ and $L_2(D)$ into $L_2(D)$, respectively.

With the update directions completely specified the constant parameters $\alpha^W_{j,n}$ are determine by minimizing $F(\boldsymbol{W}_{j,n}, \chi_{n-1})$ in (12.30). The parameters $\alpha^W_{j,n}$ are found explicitly to be

$$\alpha^W_{j,n} = \frac{\eta_S \langle \boldsymbol{\rho}^{(i)}_{k-1}, \boldsymbol{\mathcal{K}}_S \boldsymbol{w}_{j,n} \rangle_S + \eta_{D;n-1} \langle \boldsymbol{r}^{(i)}_{n-1}, \boldsymbol{w}_{j,n} - \chi_{n-1} \boldsymbol{\mathcal{K}}_D \boldsymbol{w}_{j,n} \rangle_D}{\eta_S \| \boldsymbol{\mathcal{K}}_S \boldsymbol{w}_{j,n} \|^2_S + \eta_{D;n-1} \| \boldsymbol{w}_{j,n} - \chi_{n-1} \boldsymbol{\mathcal{K}}_D \boldsymbol{w}_{j,n} \|^2_D}. \tag{12.37}$$

5.1.2. Updating the contrast

We observe that the conductivity contrast χ is only present in the second term of the cost functional F in (12.30), which we write as

$$F_D(\chi) = \frac{\sum_j \| \chi \boldsymbol{E}_{j,n} - \boldsymbol{W}_{j,n} \|^2_D}{\sum_j \| \chi \boldsymbol{E}^{\text{inc}}_j \|^2_D}. \tag{12.38}$$

where

$$\boldsymbol{E}_{j,n} = \boldsymbol{E}_j^{\text{inc}} + \mathcal{K}_D \boldsymbol{W}_{j,n}, \tag{12.39}$$

and $\boldsymbol{W}_{j,n}$ are the updated contrast sources. We observe that the numerator of F_D is minimized by taking

$$\widetilde{\chi}_n := \frac{\sum_j \boldsymbol{W}_{j,n} \cdot \overline{\boldsymbol{E}}_{j,n}}{\sum_j |\boldsymbol{E}_{j,n}|^2} = \chi_{n-1} - \frac{\sum_j \boldsymbol{r}'_{j,n} \cdot \overline{\boldsymbol{E}}_{j,n}}{\sum_j |\boldsymbol{E}_{j,n}|^2}, \tag{12.40}$$

where the object error $\boldsymbol{r}'_{j,n}$ is given by

$$\boldsymbol{r}'_{j,n} = \chi_{n-1}\boldsymbol{E}_{j,n} - \boldsymbol{W}_{j,n}. \tag{12.41}$$

However, because of the presence of χ in the denominator of F_D, this choice will not necessarily reduce the normalized error quantity F_D. To ensure this error is reduced by the CG update, we update the conductivity contrast (χ) by

$$\chi_n = \chi_{n-1} + \alpha_n^\chi d_n, \tag{12.42}$$

where α_n^χ is a constant parameter and the update direction d_n is a function of position.

The update direction is chosen to be the Polak–Ribière conjugate gradient direction,

$$d_0 = 0, \quad d_n = g_n^\chi + \frac{\langle g_n^\chi, g_n^\chi - g_{n-1}^\chi \rangle_D}{\|g_{n-1}^\chi\|_D^2} d_{n-1}, \quad n \geq 1, \tag{12.43}$$

where g_n^χ is the gradient of F with respect to χ evaluated at χ_{n-1}. Explicitly the gradient is found to be

$$g_n^\chi = g_n^D, \tag{12.44}$$

where g_n^D is chosen to be the gradient of the numerator of F_D

$$g_n^D = -\sum_j \boldsymbol{r}'_{j,n} \cdot \overline{\boldsymbol{E}}_{j,n}. \tag{12.45}$$

The update in (12.40) is the optimum updating, if changes of the denominator of F_D are neglected. Comparing (12.40), (12.42) and (12.45), we observe that we can devise a preconditioned gradient as

$$g_n^\chi = \frac{g_n^D}{\sum_j |\boldsymbol{E}_{j,n}|^2} = -\frac{\sum_j \boldsymbol{r}'_{j,n} \cdot \overline{\boldsymbol{E}}_{j,n}}{\sum_j |\boldsymbol{E}_{j,n}|^2}. \tag{12.46}$$

With the update direction completely specified, if we choose α_n^χ to be a real parameter that minimizes $F_D(\chi_n)$ in (12.38), this parameter is found explicitly by

$$\alpha_n^\chi = \frac{-(aC - Ac) + \sqrt{(aC - Ac)^2 - 4(aB - Ab)(bC - Bc)}}{2(aB - Ab)}, \tag{12.47}$$

where the coefficients are defined as

$$\begin{aligned} a &= \textstyle\sum_j \|d_n \boldsymbol{E}_{j,n}\|_D^2, & A &= \textstyle\sum_j \|d_n \boldsymbol{E}_j^{\text{inc}}\|_D^2, \\ b &= \operatorname{Re}\textstyle\sum_j \langle \boldsymbol{r}'_{j,n}, d_n \boldsymbol{E}_{j,n}\rangle_D, & B &= \operatorname{Re}\textstyle\sum_j \langle \chi_{n-1}\boldsymbol{E}_j^{\text{inc}}, d_n \boldsymbol{E}_j^{\text{inc}}\rangle_D, \\ c &= \textstyle\sum_j \|\boldsymbol{r}'_{j,n}\|_D^2, & C &= \textstyle\sum_j \|\chi_{n-1}\boldsymbol{E}_j^{\text{inc}}\|_D^2. \end{aligned} \tag{12.48}$$

This completes the updating of χ if there is no a priori information. However, because σ is positive, in each iteration, σ (can be obtained from the updated χ by using Equation (12.4)) is enforced to be zero if it attains a negative value. Note that another way to incorporate positivity as in Arikan (1994) can also be used. Furthermore, in the practical applications we have a priori information about the conductivity distribution such as that around the borehole the conductivity distribution can not change abruptly. This means that the formation is horizontally symmetrical around the well-bore. We can enforce this constraint by assigning, in this second step of each iteration, the mean value of χ_n, in (12.42), to all constant points which have the same horizontal distance to the borehole axis ($\hat{x}^3$-axis) in a given horizontal plane.

5.1.3. Starting values

We are not able to start with initial estimates of zero for the contrast (contrast sources), since then the cost functional F of (12.30) is undefined. Therefore, we start with finding the contrast sources that minimize only the normalized data error,

$$F_S(\boldsymbol{W}_{j,0}) = \frac{\sum_j \|\boldsymbol{H}_j^{\text{sct}} - \mathcal{K}_S \boldsymbol{W}_{j,0}\|_S^2}{\sum_j \|\boldsymbol{H}_j^{\text{sct}}\|_S^2}. \tag{12.49}$$

Using the gradient method, we arrive at

$$\boldsymbol{W}_{j,0} = \frac{\|\mathcal{K}_S^{\star} \boldsymbol{H}_j^{\text{sct}}\|_D^2}{\|\mathcal{K}_S \mathcal{K}_S^{\star} \boldsymbol{H}_j^{\text{sct}}\|_S^2} \mathcal{K}_S^{\star} \boldsymbol{H}_j^{\text{sct}}. \tag{12.50}$$

The operator $\mathcal{K}_S^{\star} \boldsymbol{H}_j^{\text{sct}}$ is the back projection of the data from the data domain S into the object domain D. With $\boldsymbol{W}_{j,0}$, the initial field and contrast estimates are obtained by

$$\boldsymbol{E}_{j,0} = \boldsymbol{E}_j^{\text{inc}} + \mathcal{K}_D \boldsymbol{W}_{j,0}, \qquad \chi_0 = \frac{\sum_j \boldsymbol{W}_{j,0} \overline{\boldsymbol{E}}_{j,0}}{\sum_j |\boldsymbol{E}_{j,0}|^2}. \tag{12.51}$$

5.2. Numerical example

As a numerical test, we consider a formation consisting of four beds (two with invasion) and a borehole deviated 45° with respect to the intersecting layers as shown in the top plot of Figure 5. The plots in Figure 5 give the cross-sections of 3D conductivity (σ) distributions in the (x_2, x_3) plane for a fixed $x_1 = 0$. The conductivity of the layers of the actual profile given in the top plot in Figure 5 from top to bottom are 1, 0.45, 0.01, and 0.45 S/m. The conductivity of the invasion zone of the first (top) layer is 0.6 S/m, and of the third layer (from the top) is 0.2 S/m. The dimension of each layer in the logging direction ($\hat{x}^3$ dimension) is 0.3 m. The radius of the borehole is 0.1 m and filled with conductivity 0.3 S/m. This formation is assumed to be embedded in background medium with conductivity $\sigma_0 = 0.3$ S/m. The synthetic data are generated by solving a forward problem numerically using a two-times finer discretization grid than the one used in the inversion. In this example, we use 15 sources and 14 receivers located at the borehole axis ($\hat{x}^3$-axis) with spacings of 0.2 m. The source positions are given by $\hat{x}^{S;3} = -1.4, -1.2, \ldots, 1.4$ m, respectively, and the receiver positions by

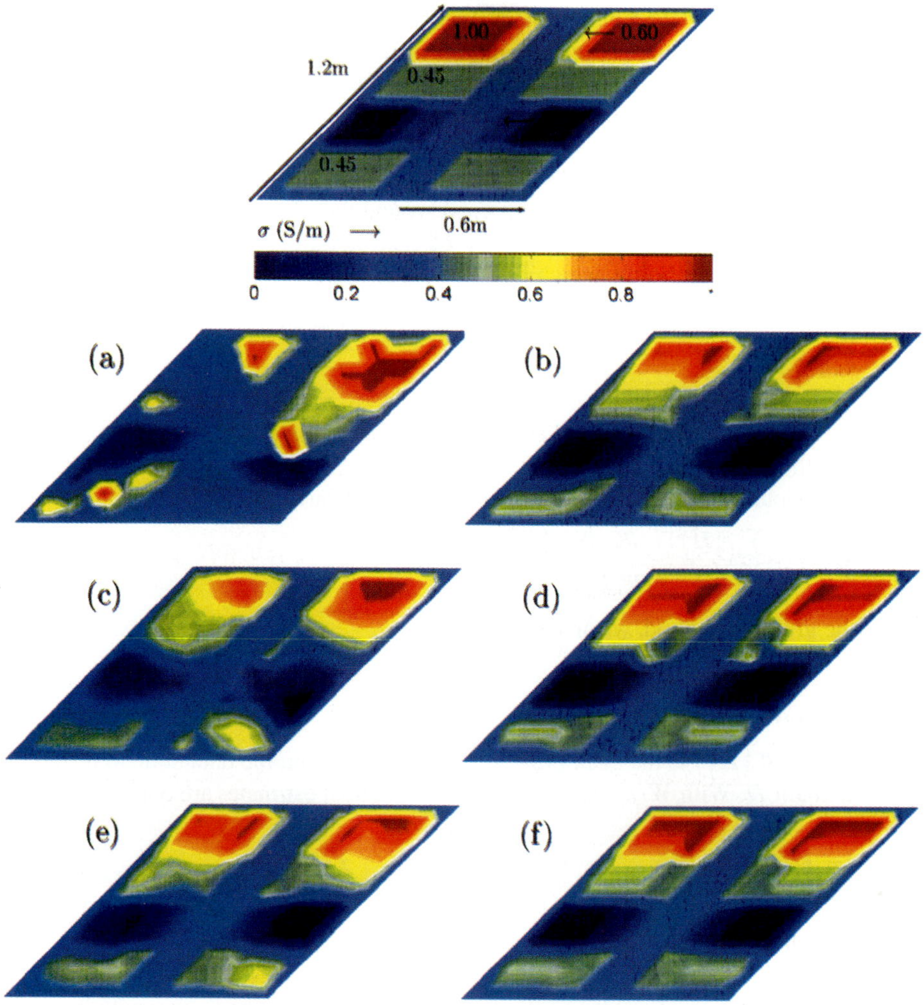

Figure 5. Plot of the actual formation (top plot) and the conductivity distribution inverted from electrode logging data (a, b), induction logging data (c, d), and both data sequentially (e, f). The horizontal symmetry constraint has been used in (b), (d) and (f).

$\hat{x}^{R;3} = -1.3, -1.1, \ldots, 1.3$ m, respectively. The source positions are indicated by the crosses in Figure 1. Thus we have 210 data points for the electrode logging and 630 data points for the induction logging.

In the inversion, we assume that the unknown formation was located entirely within a test domain of $1.4 \times 1.4 \times 1.4$ m^3. This test domain was partitioned into equal-sized subdomains with side lengths $\Delta\hat{x}^1 = \Delta\hat{x}^2 = \Delta\hat{x}^3 = 0.1$ m. Thus, the total number of conductivity contrast unknowns is equal to 2744.

The reconstruction results after 1024 iterations of the inversion procedure from the electrode logging data without and with using horizontally symmetry constraints are given in Figure 5a and b, respectively. Note that although the number of iterations is large, we do not solve a full forward problem in each iteration of our inversion scheme. One iteration of the inversion procedure takes only 8 s on a personal computer with a 400 MHz Pentium II processor. In Figure 5c,d we present the reconstructed profile from the induction logging data at 20 kHz. Note that for the induction logging data we have used all three components of the scattered vector magnetic fields as the data quantity. Now, one iteration of the inversion procedure takes 14 s. We observe that the reconstructed results from induction logging data give more accurate results for the region further away from the borehole than the one from the electrode logging data. Furthermore, the results in Figure 5a indicate that it is impossible for electrode logging data to reconstruct this type of formation without using a priori information. As seen in Figure 5b–d the results are quite reasonable despite the large variation of the conductivity distribution in the unknown formation ($0.01 \leq \sigma \leq 1$).

We give also in Figure 5e,f the results of the joint inversion of electrode and induction logging data using a 'hopping method' (the so-called frequency hopping in the wave problem). Here, we use the reconstructed profile from the electrode logging data (given in Figure 5a,b) as the initial estimate of contrast sources $\boldsymbol{W}_{j,0}$ for the inversion of induction logging data. Note that contrast sources $\boldsymbol{W}_{j,0}$ are obtained by solving a forward problem using the reconstructed profile from electrode logging data. We observe that the resolution of each layer of the reconstructed profile in the borehole direction is improved.

Next, in order to simulate a more realistic experiment, we add noise to the synthetic data. Note that the noise has been added to the total fields (the scattered plus the incident fields), and then after the addition of noise the scattered fields can be calculated by subtracting off the incident fields. Both electrode and induction data are contaminated with 10% additive random white noise. In Figure 6, we show five plots of the conductivity distributions as functions of measurement depth ($\hat{x}^3$) for fixed lateral positions ($\hat{x}_1, \hat{x}_2$) from the borehole axis ($\hat{x}^3$ axis). The fixed positions from the borehole axis from the left to the right in Figure 6 are 0.15, 0.25, 0.35, 0.45, and 0.55 m. The solid lines represent the original profile which has been used to generate the synthetic data. The inverted conductivity distributions from data without noise are given by the dashed lines (same as the results in Figure 5f) and from data with noise by dotted lines. From the results we observe that the algorithm is stable, and we can still determine the different layers in the formation.

6. CONCLUSIONS

We have presented a new approach to reconstruct a 3D conductivity distribution in a formation with dip, simultaneous layering, and invasion zones using both the electrode and induction logging data. The approach employs an integral equation modeling using an oblique coordinate system. By using the oblique (non-orthogonal) coordinate system we have gained advantages above the usual approach. First, the use of the staircasing approximation for the dipping bed environment can be avoided. This results in the reduction of the discretization errors, and we can suffice with less discretization points

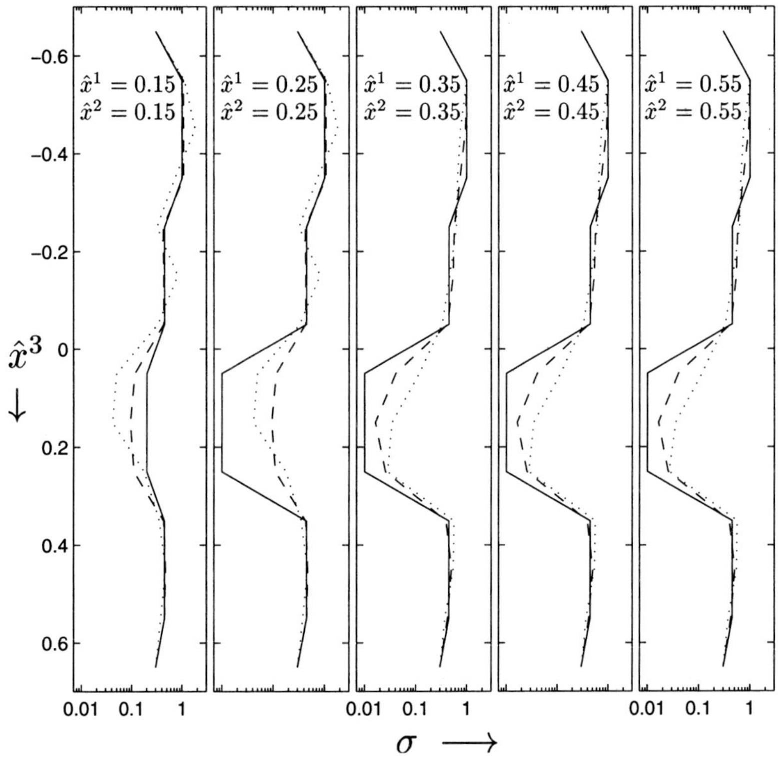

Figure 6. Plots of the conductivity distribution inverted sequentially from electrode and induction logging data. The solid lines give the original profiles, the dashed lines give the results from noiseless data, and the dotted lines give the results from data with 10% additive random white noise.

to obtain the results with the same degree of accuracy of the problem formulated in the Cartesian coordinate system. Second, within the inversion, a priori information about the dip angle is included explicitly, and the horizontal symmetry and positivity constraints of the conductivity distribution are included in a simple fashion. Furthermore, in our inversion algorithm, we have avoided solving a full forward problem in each iteration of the iterative process. The results of the numerical examples have shown that the present inversion method may be used as an interpretation technique for single-well logging measurements.

REFERENCES

Abubakar, A. and van den Berg, P.M., 1998. Three-dimensional nonlinear inversion in crosswell electrode logging. Radio Sci., 4, 989–1004.

Abubakar, A. and van den Berg, P.M., 2000. Nonlinear inversion in electrode logging in a highly deviated formation with invasion using an oblique coordinate system. IEEE Trans. Geosci. Remote Sensing, 38, 25–39.

Anderson, B., Safinya, K. and Habashy, T.M., 1986. Effect of dipping beds on the response of induction tools. 61st Annu. Tech. Conf. Soc. Pet. Eng., Paper SPE 15488.
Anderson, B., Barber, T., Druskin, V., Lee, P., Dussan, V.E., Knizhnerman, L. and Davydycheva, S., 1996. The response of multiarray induction tools in highly dipping formations with invasion and in arbitrary 3D geometries. 37th Annu. Log. Symp. June 16–19, Proc. SPWLA.
Arikan, O., 1994. Regularized inversion of a two-dimensional integral equation with applications in borehole induction measurements. Radio. Sci., 29, 519–538.
Bojarski, N.N., 1982. Comments on 'Nonuniqueness in inverse source and scattering problems'. IEEE Trans. Antennas Propag., 30, 1037–1038.
Brodlie, K.W., 1977. Unconstrained minimization. In: D.A.H. Jacobs (Ed.), The State of Art in Numerical Analysis. Academic Press, New York, NY, pp. 255–257.
Deveney, A.J. and Sherman, G.C., 1982. Nonuniqueness in inverse source and scattering problems. IEEE Trans. Antennas Propag., 30, 1034–1037.
Druskin, V., Knizhnerman, L.A. and Lee, P., 1999. A finite difference scheme for elliptic equations with rough coefficients using a Cartesian grid nonconforming to interfaces. Siam J. Numer. Anal., 36, 442–464.
Gianzero, S. and Su, S.M., 1990. The response of induction dipmeter and standard induction tools to dipping beds. Geophysics, 55, 1128–1147.
Habashy, T.M., Oristaglio, M.L. and De Hoop, A.T., 1994. Simultaneous nonlinear reconstruction of two-dimensional permittivity and conductivity, Radio Science, 29, 1101–1118.
Hardman, R.H. and Shen, L.C., 1986. Theory of induction sonde in dipping beds. Geophysics, 51, 800–809.
Hohmann, G.W., 1975. Three-dimensional induced polarization and electromagnetic modeling. Geophysics, 40, 309–324.
Holland, R., 1983. Finite-difference solution of Maxwell's equations in generalized nonorthogonal coordinates. IEEE Trans. Nucl. Sci., 6, 4589–4591.
Howard, A.Q., Jr. and Chew, W.C., 1992. Electromagnetic borehole fields in a layered dipping-bed environment with invasion. Geophysics, 37, 451–465.
Kleinman, R.E. and van den Berg, P.M., 1992. A modified gradient method for twodimensional problems in tomography. J. Comput. Appl. Math., 42, 17–35.
Koelman, J.M.V.A., Van der Horst, M., Lomas, A.T., Koelmij, A.T., Bonnie, J.H.M., 1996. Interpretation of resistivity logs in horizontal logs an application to complex reservoirs from Oman. 37th Annu. Log. Symp. June 16–19, Proc. SPWLA.
Minerbo, G.N., 1989. Inversion of induction logs in dipping beds. PIERS Conf. Proc., 293–294.
Spies, B.R. and Habashy, T.M., 1995. Sensitivity analysis of crosswell electromagnetics. Geophysics, 60, 834–845.
Stone, W., 1982. Comments on 'Nonuniqueness in inverse source and scattering problems'. IEEE Trans. Antennas Propag., 30, 1037–1038.
Stratton, J.H., 1941. Electromagnetic Theory. McGraw-Hill, New York, NY.
Van den Berg, P.M. and Kleinman, R.E., 1997. Contrast source inversion method. Inverse Problems, 13, 1607–1620.
Van der Horst, M., Druskin, V. and Knizhnerman, L., 1999. Modeling induction logs in 3-D geometries. In: M. Oristaglio and B. Spies (Eds.), Three-Dimensional Electromagnetics. Society of Exploration Geophysicists Press, Tulsa, OK, pp. 611–622.
Zhdanov, M.S. and Fang, S., 1999. Three-dimensional quasi-linear electromagnetic modeling and inversion. In: M. Oristaglio and B. Spies (Ed.), Three-Dimensional Electromagnetics. Society of Exploration Geophysicists Press, Tulsa, OK, pp. 233–255.
Zwamborn, A.P.M. and van den Berg, P.M., 1994. Computation of electromagnetic fields inside strongly inhomogeneous objects by the weak-conjugate gradient FFT method. J. Opt. Soc., 11, 1414–1421.

Part III

3-D EM in Practice

Chapter 13

DECOMPOSITION OF THREE-DIMENSIONAL MAGNETOTELLURIC DATA

Xavier Garcia and Alan G. Jones

Geological Survey of Canada, 615 Booth Street, Ottawa, ON, K1A 0E9, Canada

Abstract: Decomposition of magnetotelluric data into a local galvanic 3D distortion matrix and a regional 2D Earth caused a quantum leap in our understanding of complex data and our ability to handle those data. The Groom–Bailey method is the most widely adopted tensor decomposition approach, and rightly so given its physical basis and its separation of distortion parameters into determinable and indeterminable parts. However, on occasion the 3D over 2D (3D/2D) decomposition fails in that the misfit of the model to the data is far greater than the data errors permit, and this failure is due to either the distortion model being invalid or to inappropriately small error estimates for the data. In this paper we describe and demonstrate our attempts to extend MT tensor decomposition to local galvanic 3D distortion of regional 3D data (3D/3D). There are insufficient data to accomplish this uniquely for a single MT site, so some approximations must be made. The approach we use is to assume that two neighboring sites sense the same regional structure if they are sufficiently close compared to the skin depth to the structure, but that the two sites have differing galvanic distortion matrices. We use a decomposition method similar to the Groom–Bailey one, but with a different parameterization, and we solve the problem using a Newton method. We demonstrate the method to a synthetic data set, and highlight the difficulties that result as a consequence of inherent parameter-resolution instabilities.

1. INTRODUCTION

The presence of electrical charges near electrical conductivity transitions in the Earth cause galvanic distortions of magnetotelluric (MT) data to varying degrees. When such charges occur associated with local, near-surface inhomogeneities there is an inherent spatial aliasing problem which must be addressed prior to interpretation. The basic formulation of galvanic distortion consists of the tensor decomposition of the measured (superscript D) distorted electromagnetic fields ($\boldsymbol{E}$, $\boldsymbol{H}$) into a series of regional fields and distortion matrices (Wannamaker et al., 1984; Habashy et al., 1993; Chave and Smith, 1994):

$$\begin{aligned}
\boldsymbol{E}^{\mathrm{D}} &= \boldsymbol{E}\boldsymbol{P}, \\
\boldsymbol{H}^{\mathrm{D}} &= \boldsymbol{E}\boldsymbol{Q}_{\mathrm{h}} = \boldsymbol{I} + \boldsymbol{Q}_{\mathrm{h}}\boldsymbol{Z}, \\
\boldsymbol{H}_z^{\mathrm{D}} &= \left[(\boldsymbol{A},\boldsymbol{B}) + \boldsymbol{Q}_z\boldsymbol{Z}\right]\boldsymbol{H},
\end{aligned} \tag{13.1}$$

where $\boldsymbol{P}$, $\boldsymbol{Q}_{\rm h}$ and $\boldsymbol{Q}_z$ are real valued and frequency-independent distortion matrices (Jiracek, 1990):

$$\boldsymbol{P} = \begin{pmatrix} P_{xx} & P_{xy} \\ P_{yx} & P_{yy} \end{pmatrix}, \quad \boldsymbol{Q}_{\rm h} = \begin{pmatrix} Q_{xx} & Q_{xy} \\ Q_{yx} & Q_{yy} \end{pmatrix} \quad \text{and} \quad \boldsymbol{Q}_z = \begin{pmatrix} Q_{zx} & Q_{zy} \end{pmatrix}, \tag{13.2}$$

and $(\boldsymbol{A}, \boldsymbol{B})$ is the vertical magnetic transfer function or tipper. Thus, the distortion of the regional MT transfer function can be written as:

$$\boldsymbol{Z}^{\rm D} = (\boldsymbol{I} + \boldsymbol{P})\boldsymbol{Z}\left(\boldsymbol{I} + \boldsymbol{Q}_{\rm h}\boldsymbol{Z}\right)^{-1}, \tag{13.3}$$

where $\boldsymbol{Z}$ is the general regional impedance, $\boldsymbol{Z}^{\rm D}$ is the observed impedance, and $\boldsymbol{I}$ is the identity matrix. Similarly, the regional vertical field transfer function is distorted

$$(\boldsymbol{A}, \boldsymbol{B})^{\rm D} = \left[(\boldsymbol{A}, \boldsymbol{B}) + \boldsymbol{Q}_z\boldsymbol{Z}\right](\boldsymbol{I} + \boldsymbol{Q}\boldsymbol{Z})^{-1}, \tag{13.4}$$

where $(\boldsymbol{A}, \boldsymbol{B})$ is the regional transfer function and $(\boldsymbol{A}, \boldsymbol{B})^{\rm D}$ is the observed one.

When the EM fields are observed in an arbitrary coordinate system that is not aligned with the strike angle of the regional structures, the observed impedance (Equation 13.3) becomes

$$\boldsymbol{Z}^{\rm D} = \boldsymbol{R}(\theta)(\boldsymbol{I} + \boldsymbol{P})\boldsymbol{Z}\left(\boldsymbol{I} + \boldsymbol{Q}_{\rm h}\boldsymbol{Z}\right)^{-1}\boldsymbol{R}^{\rm T}(\theta), \tag{13.5}$$

where θ is the azimuth angle respect to the regional strike, $\boldsymbol{R}$ is the Cartesian rotation matrix, and superscript T denotes transpose.

Following this basic tensor decomposition, several authors developed different approaches to extract the regional impedances given the observed ones (Larsen, 1977; Groom and Bailey, 1989; Bahr, 1991; Zhang et al., 1993; Chave and Smith, 1994; Smith, 1997).

Most of the authors neglect galvanic magnetic distortion. The reason for this lies in the fact that the magnetic distortion in Equation (13.3) is $\boldsymbol{Q}_{\rm h} \cdot \boldsymbol{Z}$, rather than $\boldsymbol{Q}_{\rm h}$ alone which describes the scattering. This product term is frequency-dependent, due to the frequency dependence of $\boldsymbol{Z}$, and vanishes for low frequencies. For this reason magnetic galvanic distortion is usually ignored. Jones and Groom (1993), Chave and Smith (1994) and Smith (1997) account for the magnetic term in their galvanic distortion of MT data, and Ritter and Banks (1998) consider the effects on the vertical field transfer functions. For distortion of controlled-source data there is the theoretical work of Qian and Pedersen (1992), while Li et al. (2000) study the anisotropy case, and Garcia et al. (2000) provide a case study of controlled-source data in which the magnetic distortion is taken into account.

The most widely used decomposition technique is the one proposed by Groom and Bailey, 1989 (called GB). Two angles, *twist* and *shear*, describe the galvanic distortion of the regional electric field. There are two other distortion parameters, *local anisotropy* and *site gain*, that are unresolvable amplitude scaling effects and are included in the regional impedance matrix. The anisotropy can be approximated by assuming that the asymptote of the apparent resistivity curves for both off-diagonal terms for the high frequency should be coincident. The remaining distortion parameter, site gain, is usually small and can be addressed using either additional geophysical or geological information or determined as part of the inversion procedure.

When applying MT for addressing mining scale problems, the frequencies of interest range from approximately 1 Hz to 20,000 Hz. Within the depth of penetration corresponding to this frequency band most mineralized structures will show 3D inductive effects, and existing decomposition schemes are not valid as they assume regional 1D or 2D Earth models. The first attempt to deal with distortion over 3D regional structures was that of Ledo et al. (1998), who assumed a 3D/2D/3D Earth, with the upper 3D structures causing the galvanic distortion. Following their approach, conventional Groom–Bailey decomposition is performed at high frequencies, where the regional structures can be considered 2D. Then they apply the derived distortion parameters to the data over the whole observational frequency range. Recently, Utada and Munekane (2000) presented a 3D/3D decomposition method that makes use of the spatial derivatives of the magnetic fields to estimate galvanic distortion. Their method requires a large number of stations located in a 2D array, and measurement of the vertical magnetic field at each site. Both of these are not common practice in mineral exploration MT surveys.

In the present work we discuss our attempts to extend galvanic decomposition of the electric field to 3D regional Earth models. Two theoretical data sets are examined and decomposed using our approach, with varying degrees of success due to the inherent instabilities of the problem as posed. We conclude that this approach may prove fruitful, but that further work is required.

2. 3D DECOMPOSITION

Figure 1 shows a generalized model of galvanic and inductive interactions at a variety of depths and scales. The electromagnetic spectrum can be separated into four different bands depending on the effects observed. At the highest frequencies the shallow surficial

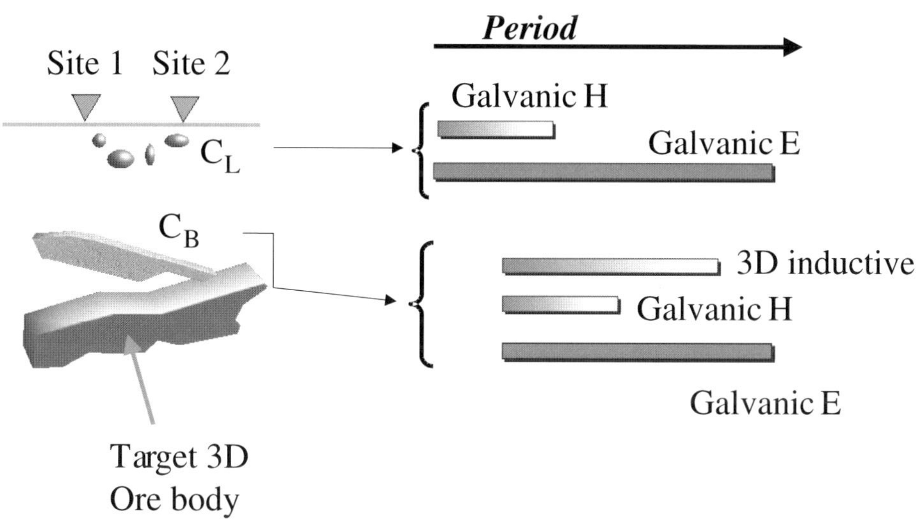

Figure 1. Cartoon of the different inductive and galvanic effects that appear in a MT mining scale survey.

inhomogeneities are the only source of galvanic distortion effects, and these can be described by a distortion tensor, $\boldsymbol{C}_{\mathrm{L}}$. At lower frequencies the 3D distorting structures, if sufficiently elongate, will have a response that can be characterized as 2D. As the frequency decreases (and penetration depth increases) the 3D inhomogeneity causes 3D inductive effects. At the lowest frequencies only the galvanic distortion of both the surficial ($\boldsymbol{C}_{\mathrm{L}}$) and distorting body ($\boldsymbol{C}_{\mathrm{B}}$) of the electric fields will remain, and a classical GB scheme could be applied to retrieve the combined distortion parameters $\boldsymbol{C}$. For even lower frequencies the 3D inductive and galvanic effects from the 3D regional structure, the mineralized ore body, appear in the response curves. In this frequency range of interest, the distorted impedance can be decomposed into two inseparable distortion matrices and a regional 3D response:

$$\boldsymbol{Z}^{\mathrm{D}} = \boldsymbol{C}\boldsymbol{Z} = \boldsymbol{C}_{\mathrm{L}}\boldsymbol{C}_{\mathrm{B}}\boldsymbol{Z}, \tag{13.6}$$

where $\boldsymbol{C}_{\mathrm{L}}$ is the distortion caused by the shallow structures, $\boldsymbol{C}_{\mathrm{B}}$ the distortions caused by the 3D structure, $\boldsymbol{C}$ is their combined effect, and $\boldsymbol{Z}$ the regional 3D response. We wish to determine the 3D impedance of the mineralized body, $\boldsymbol{Z}$, given the observations $\boldsymbol{Z}^{\mathrm{D}}$, and therefore need to extract the parameters of $\boldsymbol{C}$.

Given that only eight equations are obtained for each frequency from the observed impedance tensor for each station, and that the 3D distortion problem described by Equation (13.6) has 12 unknowns (4 complex impedances plus 4 distortion parameters), an approximation is required in order to solve the problem. We assume that two adjacent stations respond to the same 3D regional response $\boldsymbol{Z}^{\mathrm{D}}$ but different local galvanic effects, C^1 and C^2. This assumption reduces the problem to $16 \times N$ equations and $8 \times N + 8$ unknowns (4 impedances $\times$ 2 (complex) $\times$ N frequencies + 4 distortion parameters $\times$ 2 stations) for N frequencies, and the problem may have a solution.

This approach is limited to situations where the distortion parameters are different at both sites but the regional parameters are the same. In the case of both stations being affected by the same galvanic distortion their responses should be identical (to within statistical error), then the problem has no solution as the real number of equations will be smaller than the number of parameters.

3. DECOMPOSITION ALGORITHM

Two different approaches have been applied to address this problem. The first one was to consider the distortion parameters as described in Equation (13.6), then the impedances recorded at two nearby sites can be rewritten as:

$$Z^{\mathrm{D},1} = \begin{pmatrix} C_1^1 & C_2^1 \\ C_3^1 & C_4^1 \end{pmatrix} \begin{pmatrix} Z_{xx} & Z_{xy} \\ Z_{yx} & Z_{yy} \end{pmatrix},$$

$$Z^{\mathrm{D},2} = \begin{pmatrix} C_1^2 & C_2^2 \\ C_3^2 & C_4^2 \end{pmatrix} \begin{pmatrix} Z_{xx} & Z_{xy} \\ Z_{yx} & Z_{yy} \end{pmatrix}, \tag{13.7}$$

where the superscript index refers to each site.

For this form, two different least-squares algorithms have been tried; one based on a Marquardt–Levenbergs scheme (Pedersen and Rasmussen, 1989) and the second using a sequential quadratic programming method as implemented in the NAG [1] libraries. Both methods failed to solve the problem, and an inspection of the matrices revealed that the condition number was of the order of around 80, which is too big for the accuracy required to reach a minimum and invalidated the use of such algorithms.

The second description used was similar to the GB approach (Groom and Bailey, 1989), where the impedances for two neighboring sites can be decomposed and rewritten as:

$$
\begin{aligned}
Z^{\mathrm{D},i} &= g^i T^i S^i A^i Z \\
&= \begin{pmatrix} 1 & -t^i \\ t^i & 1 \end{pmatrix} \begin{pmatrix} 1 & s^i \\ s^i & 1 \end{pmatrix} \begin{pmatrix} g^i(1+a^i) & 0 \\ 0 & g^i(1-a^i) \end{pmatrix} \begin{pmatrix} Z_{xx} & Z_{xy} \\ Z_{yx} & Z_{yy} \end{pmatrix} \\
&= \begin{pmatrix} 1 & -t^i \\ t^i & 1 \end{pmatrix} \begin{pmatrix} 1 & s^i \\ s^i & 1 \end{pmatrix} \begin{pmatrix} g_1^i & 0 \\ 0 & g_2^i \end{pmatrix} \begin{pmatrix} Z_{xx} & Z_{xy} \\ Z_{yx} & Z_{yy} \end{pmatrix}, \qquad i = 1,2. \qquad (13.8)
\end{aligned}
$$

We studied various parameterizations of this equation, including those of Groom and Bailey (1989) and of Chave and Smith (1994). The classical distortion decomposition parameterization of Groom and Bailey (1989) was based on the Pauli matrices, and is a natural one for decomposition of a tensor with only elements on the diagonals or off-diagonals. When testing with the GB parameterization we found the algorithm to be highly sensitive to start model and to deviate rapidly to local minima.

Of the forms that we tested, the parameterization that yielded the greatest stability was to solve for a new set of complex functions α_i defined in the following way:

$$
\begin{aligned}
\alpha_0^i &= \frac{Z_{xx}^{\mathrm{D},i} - Z_{yx}^{\mathrm{D},i}}{2}, \qquad \alpha_1^i = \frac{Z_{xx}^{\mathrm{D},i} + Z_{yx}^{\mathrm{D},i}}{2}, \\
\alpha_2^i &= \frac{Z_{xy}^{\mathrm{D},i} - Z_{yy}^{\mathrm{D},i}}{2}, \qquad \alpha_3^i = \frac{Z_{xy}^{\mathrm{D},i} + Z_{yy}^{\mathrm{D},i}}{2}, \qquad i = 1,2. \qquad (13.9)
\end{aligned}
$$

In our new parameterization every α_i depends only on two regional impedance elements instead of four as required by the classical GB one. Each α_i is more sensitive to the parameters that form it, which accounts both for their sensitivity and for the stability.

To solve this system of equations we used a Newton algorithm to find the minimum of the square of the norm between the estimated impedance and the measured impedance:

$$
\mathrm{fn} = \sum_j \left(\alpha_j^0 - \alpha_j^{\mathrm{m}}\right)^2, \qquad (13.10)
$$

where α^0 are the data and α^m are the model parameters defined according to Equation (13.9). As we will see later, different tests performed on synthetic data were successful in retrieving the regional response together with the distortion parameters.

The misfit that we use is a χ-square misfit:

[1] NAG: Numerical Algorithms Group, Fortran 77 mathematical libraries, version 17. The specific library use in this work is E04KDF, that is a modified Newton algorithm.

$$\mathrm{RMS} = \sqrt{\sum_j \left(\frac{\alpha_j^0 - \alpha_j^{\mathrm{m}}}{\delta\alpha_j^0} \right)^2}, \qquad j = 0, \ldots, 3, \tag{13.11}$$

where α_j^0 are the data and α_j^{m} are the model responses, and $\delta\alpha_j^m$ are the data variances.

4. SYNTHETIC TESTS

We discuss below two tests using synthetic data that illustrate the reliability and limitations of our approach. We derived the theoretical responses at an MT station above a three-dimensional (3D) resistivity model, shown in Figure 2, using Randy Mackie's 3D forward modelling code (Mackie et al., 1994). The model consisted of a moderately conductive exposed medium (100 Ω m), a dipping resistive structure (2000 Ω m) at depth, and a conductive (5 Ω m) elongated structure that crosses the two media at an angle of 45°. The grid used was of 105 × 105 × 50 cells, with a minimum grid size in X and Y of 250 m and in Z of 90 m. Responses at a total of 11 frequencies were calculated from 10^{-3} to 10^2 Hz. The scaled impedances for the station located in the center of this model are shown in Figure 3. These synthetic responses were perturbed using two sets of distortion parameters (listed in Table 1), and Figure 4 shows the responses of the two stations obtained in this way. For both tests we assumed that the data had associated errors of 2% of their impedance amplitudes.

Previous to any three-dimensional decomposition we proceeded to decompose both stations using a 2D algorithm (McNeice and Jones, 2001). As both stations have the same response with different galvanic distortion, we expect that the final 2D decomposed data should be the same, but different twist and shears. The results corresponding to the decomposition of both sites simultaneously are presented in Figure 5a for station 1 and Figure 5b for station 2. The error in interpreting these data is clearly shown in these figures (left upper corner panel). Besides that the error misfit is too big to be considered a good fit, the regional responses (middle right panel) are totally different and the program could not retrieve the distortion parameters (bottom left panel).

The results we present below consist of the decomposition of an impedance tensor at four frequencies, both with the gains fixed and with them free. Different tests were undertaken using different numbers of frequencies and different distortion models, and the results were essentially the same.

Table 1. Distortion parameters used to obtain stations 1 and 2 used in the present work [a]

Station 1		Station 2	
$g1 = g_1^1$	0.9	$g3 = g_1^2$	1.43
$g2 = g_2^1$	0.9	$g4 = g_2^2$	1.43
t1	−20°	t2	−5°
s1	35°	s2	−17°

[a] If the anisotropy is zero, then the gain factors in each station are equal.

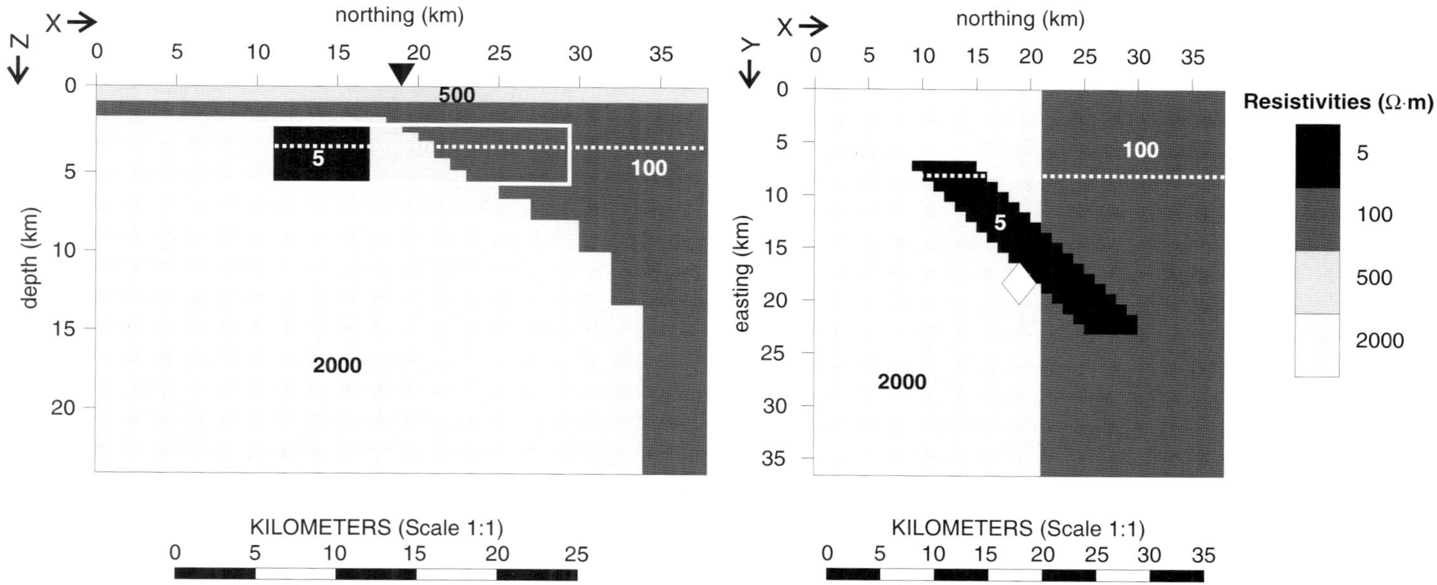

Figure 2. Theoretical regional model used in this study. It consists of a 100-Ω m background with an embedded resistive structure dipping towards the *X* direction, and a conductive structure crossing the model at 45°. We calculate the responses of this model using Randy Mackie's 3D forward modelling code. The white box on the right indicates the extent of the conductor. The white dotted lines indicate the position of the slices.

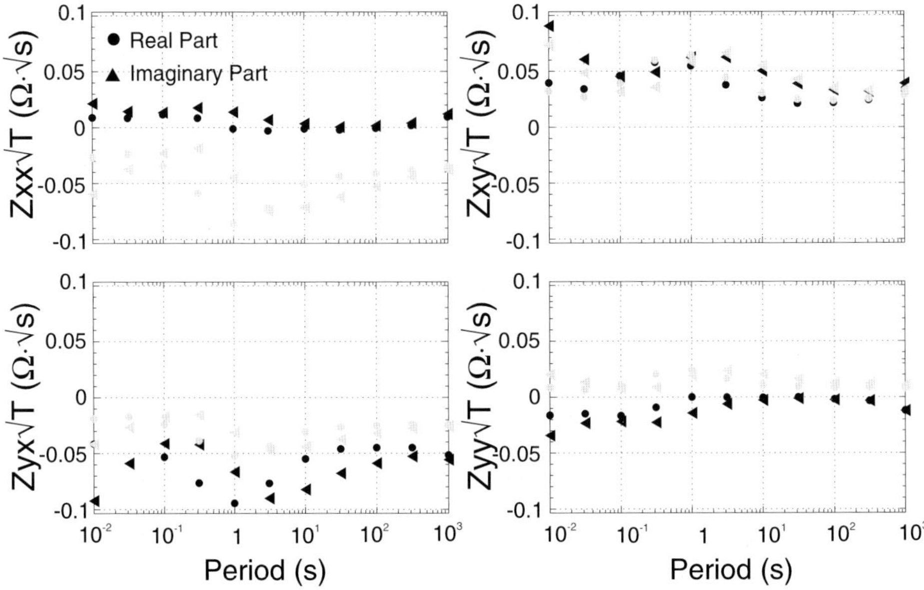

Figure 3. Impedance tensor scaled by the square of the period, obtained from the theoretical model in Figure 2. This is the regional response used in this study.

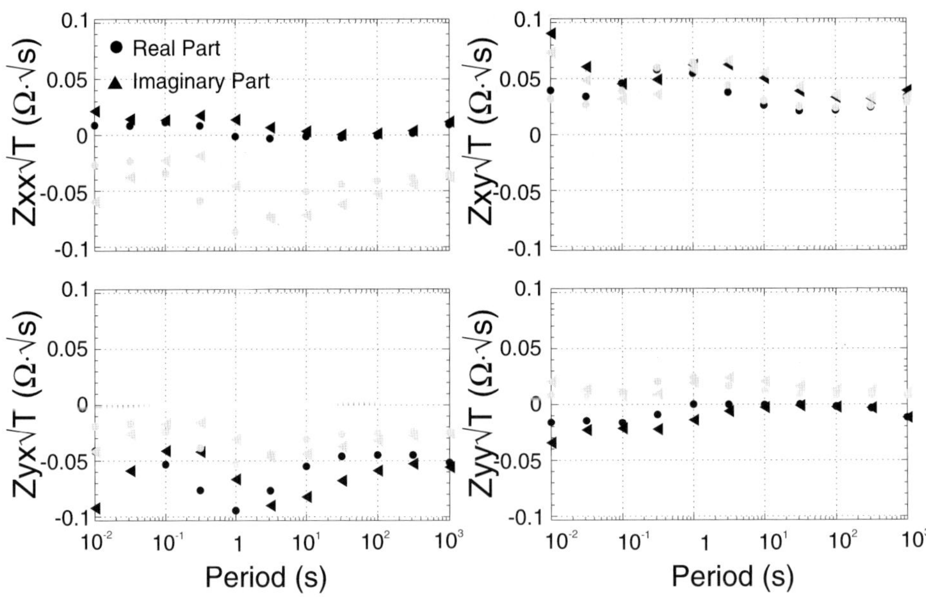

Figure 4. Impedances scaled by the square of the period corresponding to the two stations obtained applying distortions parameters (Table 1) to the station from the 3D model. In gray there is station 1 and the black responses are station 2.

4.1. Example 1: Gains fixed

In this first test we held the gain parameters fixed to their actual values. Figure 6a shows the decrease of RMS misfit with successive iterations. For this case convergence was rapid and within 10 iterations a satisfactory solution was achieved with an RMS misfit of 0.6972. As an example of convergence of the parameters to their correct values we show the evolution of the shear and twist angles (Figure 6b) and the impedance elements Z_{xy} and Z_{yx} (both real and imaginary parts) at a frequency of 122.1 Hz (Figure 6c) as the iterations increased. The correct values are indicated in Figure 6b,c, and an excellent agreement between the determined parameters and the actual ones is apparent. We can conclude that if we know a priori the gain parameters, we can recover the other distortion parameters to within statistical accuracy.

4.2. Example 2: All parameters free

For the second example we demonstrate decomposition of the same data but with the gain parameters free in an attempt to recover them. For the initial tests the twist and shear angles at both sites were held fixed to their true values. The final solutions recovered the correct impedances and the gain parameters after six iterations.

However, the real challenge is decomposition of the data with all parameters unconstrained. Figure 7a shows the evolution of the misfit function for successive iterations, with a final RMS misfit of 0.9265 after 10 iterations. The twist angles at both sites have been recovered after few iterations (Figure 7b) and the shear angles are close to their correct values (Figure 7b). However, the impedances (Figure 7d) and gains (Figure 7c) are significantly different from their correct values. Nevertheless the final decomposed model fits the data to within statistical tolerances (RMS misfit < 1.0), but the final parameters are far from the correct solution. Figure 8 shows the actual regional scaled impedances compared to those derived using the above final solution. As can be observed, the fit is better at high frequencies showing a constant shift at lower frequencies. As stated above in the Introduction, there is a strong equivalence problem that must be solved using additional geological and geophysical information, just as in the GB case for solving for the unknown site gain and anisotropy parameters.

The main problem associated with this method is that the gain parameters appear to be virtually impossible to recover from the distorted impedances. Including the gain factors into the decomposition, as undertaken in the Groom–Bailey method, results in two differently scaled regional parameters, and the parameters to obtain will be $16 \times N + 4$ instead of the prior $8 \times N + 8$, making the problem impossible to solve.

To study this equivalence problem we undertook an analysis consisting of a series of realizations obtained by adding 1% Gaussian noise and scatter to the impedance tensor elements of the two distorted stations. These individual realizations were each decomposed, and the results demonstrated that the problem is sensitive to small departures. A statistically satisfactory fit (RMS misfit < 1.0) can be achieved for each realization, but the 3D regional impedance parameters are not correctly recovered. As an example of the poor sensitivity of the individual parameters, we studied the variation of misfit as a function of two free parameters, the real part of the element Z_{yx} of

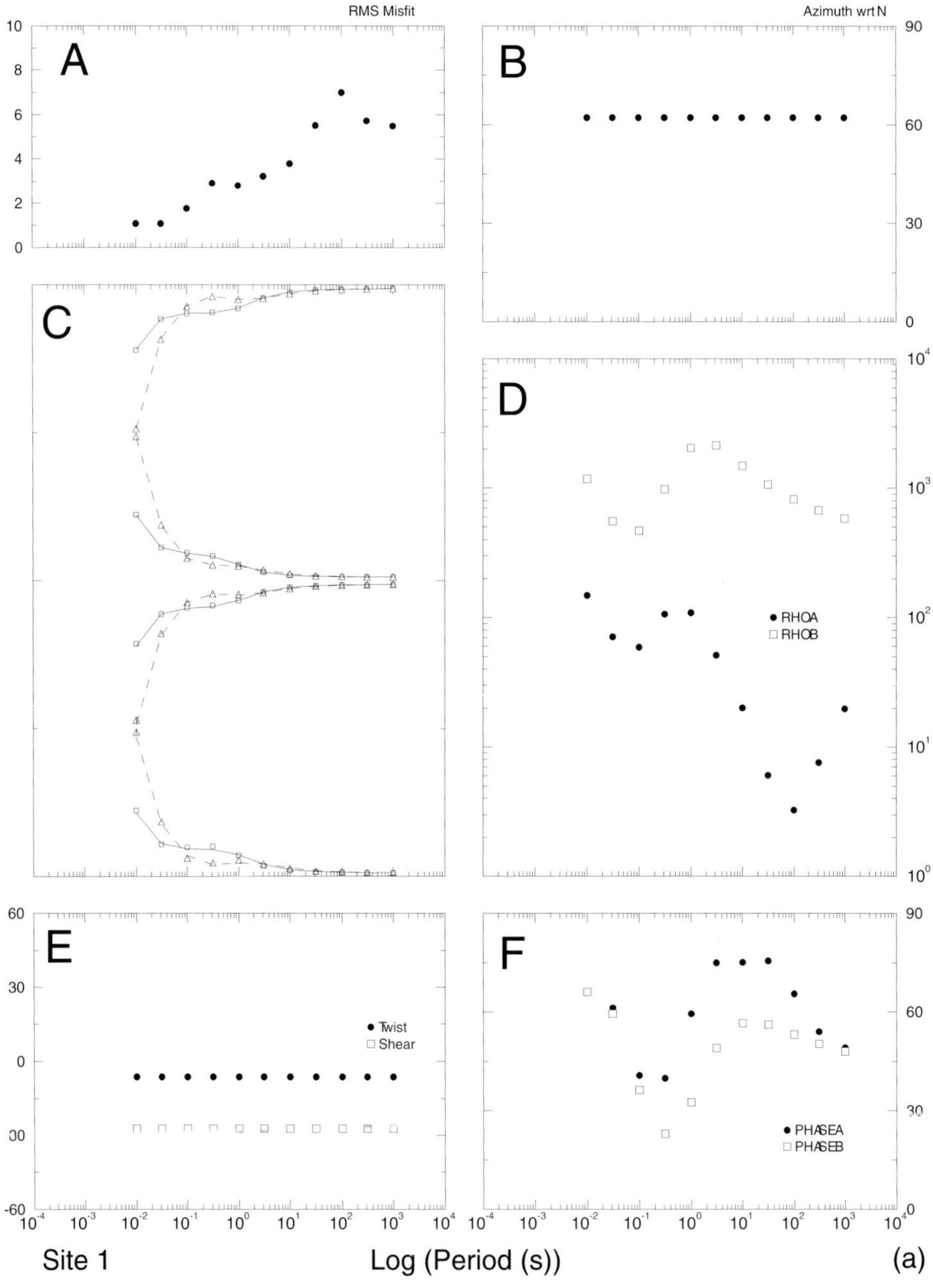
RMS Misfit
Azimuth wrt N
A
B
C
D
E
F
RHOA
RHOB
Twist
Shear
PHASEA
PHASEB
Site 1
Log (Period (s))
(a)

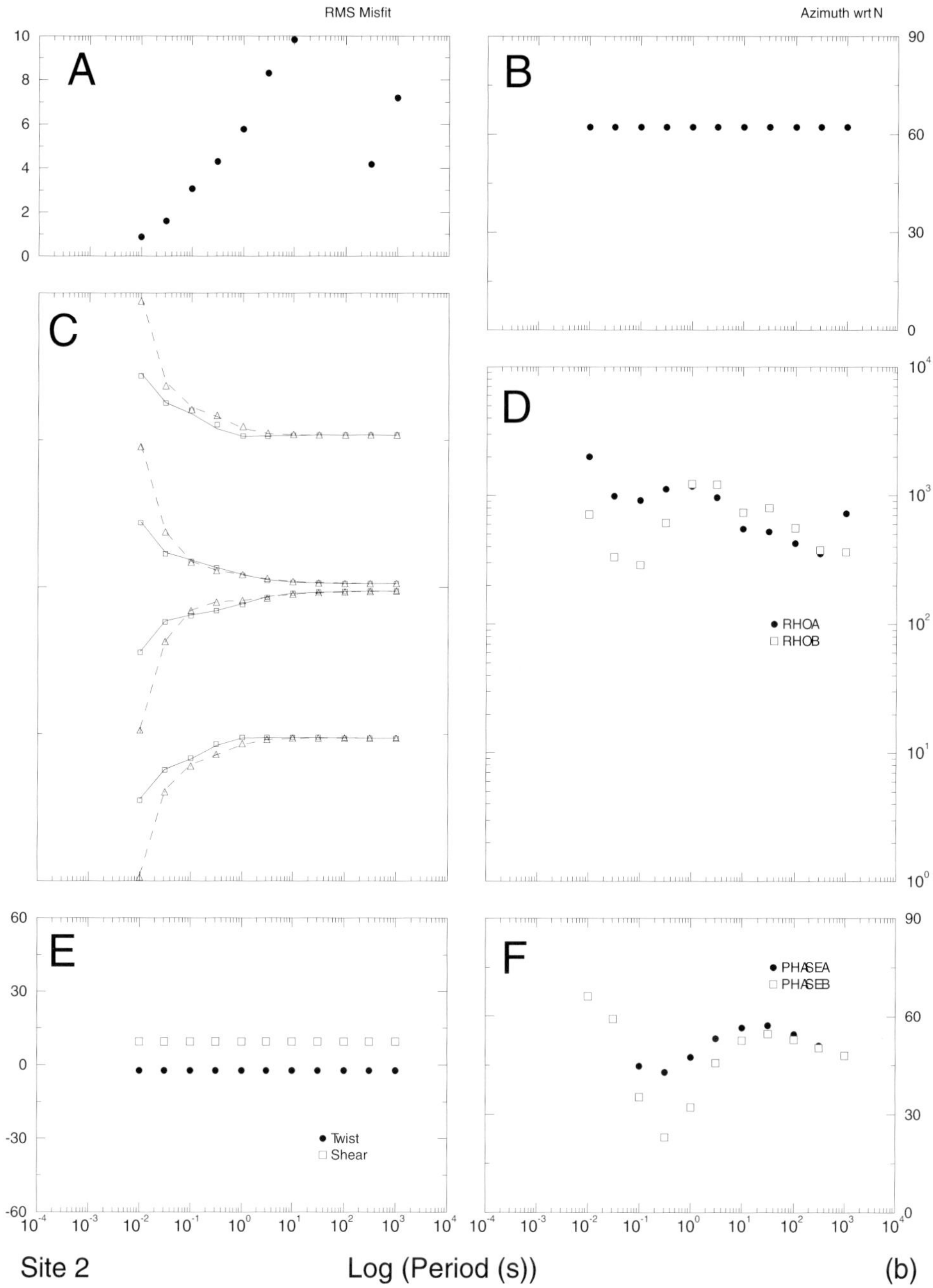

Figure 5. Results from the 2D galvanic decomposition of the stations depicted in Figure 4. (a) Station 1. (b) Station 2. Both figures: A = RMS misfit of the decomposition; B = strike angle; C = impedance fit of the final 2D model to the data; D = regional 2D apparent resistivity; E = twist and shear angles; F = regional 2D phase.

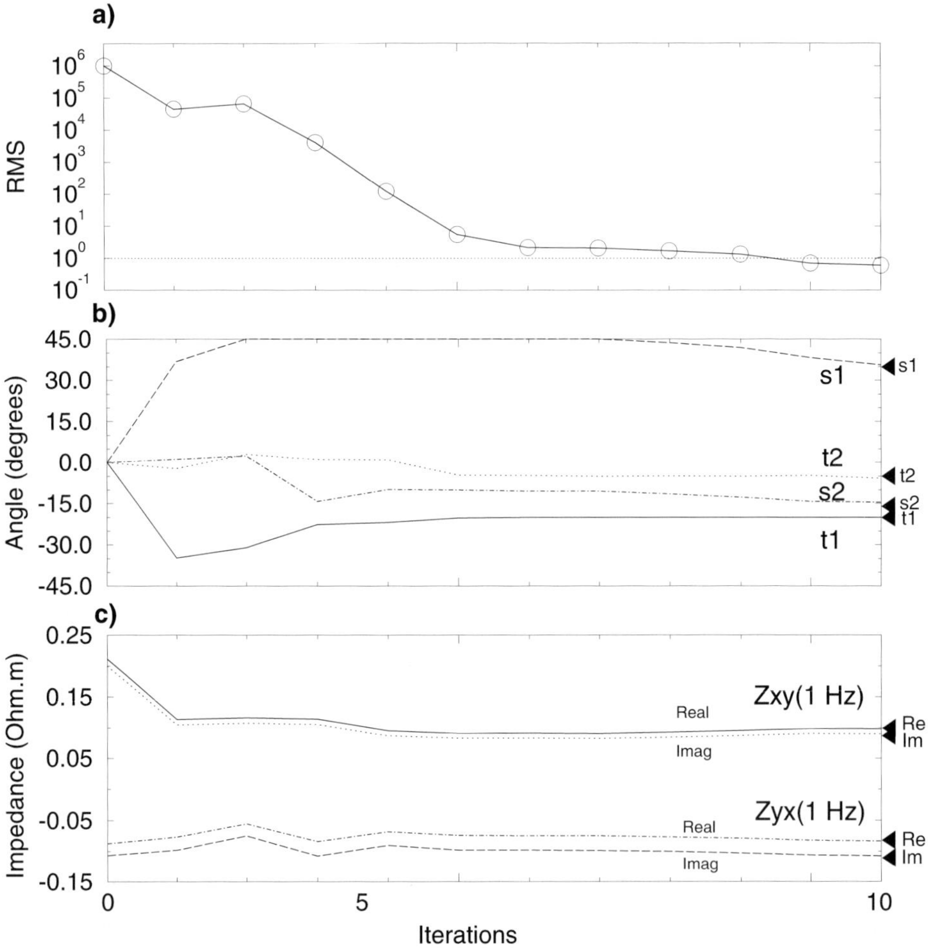

Figure 6. Results corresponding to example 1. (a) RMS misfit. (b) Evolution of the twist and shear angles from an initial value of 0 to their actual values. (c) Evolution of the Z_{xy} and Z_{yx} element of the impedance tensor at a frequency of 1 Hz: continuous line, real part Z_{xy}; dashed line, imaginary part Z_{xy}; dotted line, real part Z_{yx}; dotted–dashed line, imaginary part Z_{yx}. As can be observed, with 10 iterations the correct model is reached.

the regional impedance tensor (at 0.1 Hz) against the shear of the first site, as the other parameters are held constant. Figure 9a shows the variation of the RMS objective function (Equation 13.11) as the two free parameters are varied with all other parameters held at their *correct* values. As can be appreciated, the contoured minimum indicates the correct values for the two parameters (shown by the white cross). In Figure 9b we again contour the misfit against the two free parameters, but in this case the fixed parameters are held to the best-fit values found from our unconstrained procedure. As can be observed in this figure, the minimum is located far from the true solution (black cross), for a shear angle of 45°, but the RMS misfit is still statistically acceptable. Comparison

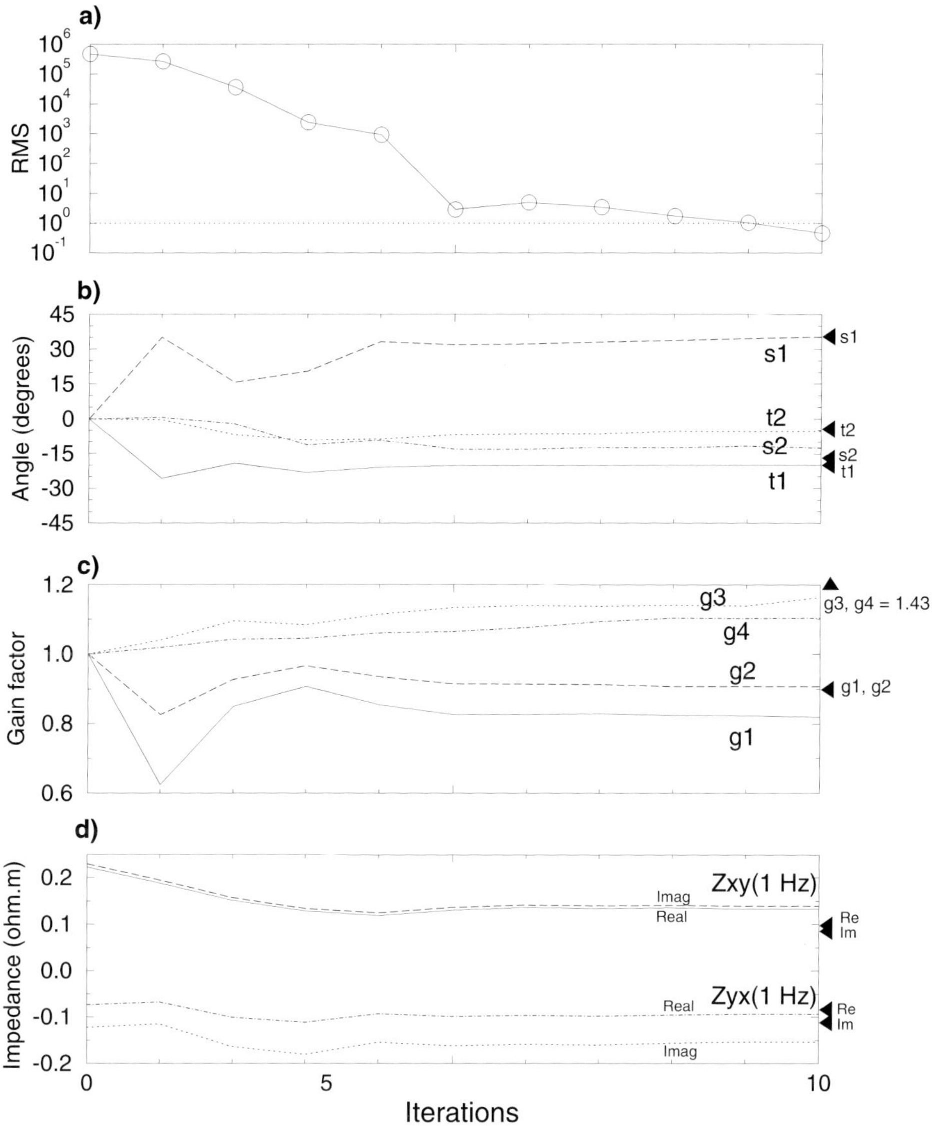

Figure 7. Results corresponding to example 2. (a) RMS misfit. (b) Evolution of the shear and twist angles: continuous line, twist site 1; dashed line, shear site 1; dotted line, twist site 2; dotted–dashed line, shear site 2. (c) Evolution of the gain parameters from an initial value of 1: continuous line, first gain factor site 1; dashed line, second gain factor site 1; dotted line, first gain factor site 2; dotted–dashed line, second gain factor site 2. (d) Evolution of the Z_{xy} and Z_{yx} impedance components at a frequency 1 Hz: continuous line, real part Z_{xy}; dashed line, imaginary part Z_{xy}; dotted line, real part Z_{yx}; dotted–dashed line, imaginary part Z_{yx}.

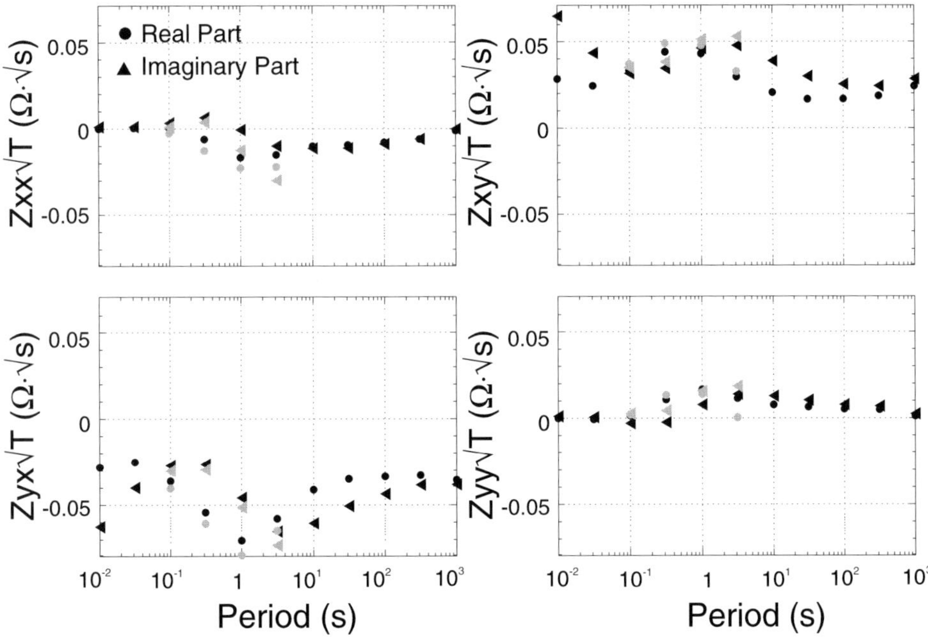

Figure 8. Comparison of the regional impedance tensor scaled by the square of the period from test 2 (in gray) and the theoretical regional values (black).

of Figure 9a and b also indicates that the regional impedances may be more stable and robustly determined than the distortion parameters. However, further work needs to be done to establish whether this is a generic feature, or only one resulting from our choice of impedance tensor and distortion parameters.

5. DISCUSSION

We have developed a 3D/3D decomposition approach for removing 3D galvanic distortion of a 3D regional response. For this purpose we use two adjacent stations and consider that both are sufficiently close to have the same regional 3D response but that they are affected differently by galvanic effects. The decomposition is undertaken in the frequency band where the inductive and magnetic galvanic effects due to the 3D scattering structure vanish.

Using a decomposition scheme similar to that of Groom and Bailey (1989), but with a different parameterization, we can solve the problem. As in the Groom and Bailey case the problem becomes unstable when we try to solve for the site gains, although they can be recovered in our case under some circumstances.

In this work we show that the 2D decomposition technique fails to retrieve the regional model from two three-dimensional sites affected by galvanic distortion. As a consequence, a 2D interpretation of these data would display erroneous features.

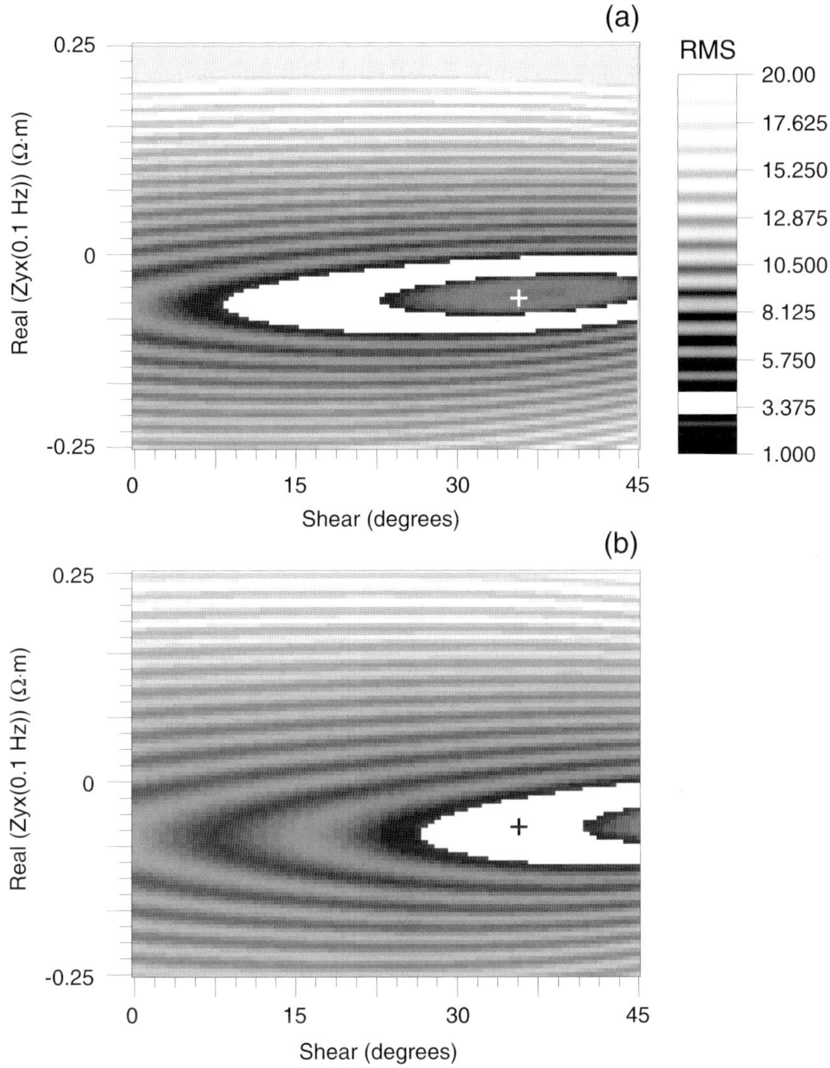

Figure 9. Analysis of the RMS misfit as a function of two parameters (real part of the impedance tensor Z_{yx} at 0.1 Hz and the shear angle from station 1) fixing the rest. (a) The fixed parameters are equal to their actual values. (b) The fixed parameters have been obtained from the inversion of the data. The white area shows the RMS values between 3 and 4. The crosses indicate the location of the true minimum.

The current work is focused on finding a more stable algorithm and a new parameterization of the decomposition equations that allow us to solve the total problem. The addition of the tipper vector to the problem is not advised, as this will add four more equations for each frequency to the problem, but at the same time it will add $4 \times N + 2$ more unknowns: two complex regional tipper per frequency and two vertical magnetic field distortion parameters Q_z to the number of unknowns.

ACKNOWLEDGEMENTS

XG is supported by an industry-GSC research fellowship, and thanks the GSC, Geosystem, Phoenix and Inco for his support. Comments by Jim Craven, Juanjo Ledo, John Weaver, Alan Chave, Phil Wannamaker, Martyn Unsworth and an anonymous reviewer are also acknowledged. Geological Survey of Canada contribution number 2000227.

REFERENCES

Bahr, K., 1991. Geological noise in magnetotelluric data: a classification of distortion types. Phys. Earth Planet. Int., 60, 119–127.

Chave, A.D. and Smith, J.T., 1994. On electric and magnetic galvanic distortion tensor decompositions. J. Geophys. Res., 94, 14,215–14,225.

Garcia, X., Boerner, D. and Pedersen, L.B., 2000. Full galvanic decomposition of tensor CSAMT data. Application to data from the Buchans mine (Newfoundland). Geophys. J. Int., submitted.

Groom, R.W. and Bailey, R.C., 1989. Decomposition of magnetotelluric impedance tensors in the presence of local three-dimensional galvanic distortion. J. Geophys. Res., 94, 1913–1925.

Habashy, T.M., Groom, R.W. and Spies, B.R., 1993. Beyond the Born and Rytov approximations: A nonlinear approach to electromagnetic scattering. J. Geophys. Res., 98, 1759–1776.

Jiracek, G.R., 1990. Near-surface and topographic distortions in EM induction. Surv. Geophys., 11, 163–203.

Jones, A.G. and Groom, R.W., 1993. Strike angle determination from the magnetotelluric tensor in the presence of noise and local distortion: rotate at your peril! Geophys. J. Int., 113, 524–534.

Larsen, J., 1977. Removal of local surface conductivity effects from low frequency mantle response curves. Acta Geodaet., Geophys., Montanist. Acad. Sci. Hung., 12, 183–186.

Ledo, J., Queralt, P. and Pous, J., 1998. Effects of galvanic distortion on magnetotelluric data over a three-dimensional regional structure. Geophys. J. Int., 132, 295–301.

Li, X., Oskooi, B. and Pedersen, L.B., 2000. Inversion of controlled-source tensor magnetotelluric data for a layered earth with azimuthal anisotropy. Geophysics, 65, 452–464.

Mackie, R.L., Smith, J.T. and Madden, T.R., 1994. Three-dimensional electromagnetic modeling using finite difference equations: the magnetotelluric example. Radio Sci., 29(4), 923–935.

McNeice, G.W. and Jones, A.G., 2001. Multi-site, multi-frequency tensor decomposition of magnetotelluric data. Geophysics, 66, 158–173.

Pedersen, L.B. and Rasmussen, T.M., 1989. Inversion of magnetotelluric data: a non-linear least-squares approach. Geophys. Prospect., 37, 669–695.

Qian, W. and Pedersen, L.B., 1992. Near-surface distortion effects on controlled source magnetotelluric transfer functions. Geophys. J. Int., 108, 833–847.

Ritter, P. and Banks, R.J., 1998. Separation of local and regional information in distorted GDS response functions by hypothetical event analysis. Geophys. J. Int., 135, 923–942

Smith, J.T., 1997. Estimating galvanic-distortion magnetic fields in magnetotellurics. Geophys. J. Int., 130, 65–72.

Utada, H. and Munekane, H., 2000. On galvanic distortion of regional three-dimensional magnetotelluric impedances. Geophys. J. Int., 140, 385–398.

Wannamaker, P.E., Hohmann, G.W. and Ward, S.H., 1984. Magnetotelluric responses of three dimensional bodies in layered earth. Geophysics, 49, 1517–1533.

Zhang, P., Pedersen, L.B., Mareschal, M. and Chouteau, M., 1993. Channelling contributions to tipper vectors: a magnetic equivalent to electrical distortion. Geophys. J. Int., 113, 693–700.

Chapter 14

ON THE DETERMINATION OF THE ELECTRICAL 3-D CONDUCTIVITY DISTRIBUTION BENEATH ICELAND WITH LONG-PERIOD MAGNETOTELLURICS

Anja Kreutzmann and Andreas Junge

Institute of Meteorology and Geophysics, J.W. Goethe University, Feldbergstrasse 47, D-60323 Frankfurt/M., Germany

Abstract: The cause for Iceland's existence is assumed to be the upwelling of a hot mantle plume. An important result of more than 200 magnetotelluric (MT) measurements over the last decades is a layer of increased conductivity at an average depth of 10 km within the Icelandic crust. Three-dimensional (3-D) model studies show the effects of various conductivity structures on the magnetotelluric transfer functions for periods between 10 and 100,000 s. In particular the resolution of the plume structure and the influence of the highly conductive seawater is discussed. The assumed plume can be detected for a plume–host conductivity contrast $> \sim 3$. Using these values, it is possible to differentiate between the highly conducting crust layer and the plume head as they show a characteristic pattern at clearly separated period ranges in the magnetotelluric transfer functions. Thus the influence of the crustal conductor does not mask the effect of an assumed plume structure. Also the influence of the seawater does not shield the effects of the mentioned conductive zones in crust and mantle; however, the coast effect can be detected up to 50 km inside the island. An assumed plume stem cannot be detected in the model results.

---

1. INTRODUCTION

Iceland's exceptional crust and mantle structure is a result of the coaction of different geodynamic processes: there is horizontal movement because of its position on the Mid-Atlantic Ridge, and up-doming crust as a result of a raising plume and subsequent volcanic and geothermal activities. For understanding the various processes the shape, size, depth, composition, and the exact centre of the assumed mantle plume are of great interest. For example, such a plume could contain a certain percentage of melt, possibly leading to a reduced resistivity. Hence, we examine whether it would be possible to see the plume in long-period magnetotelluric (LMT) measurements.

Between 1980 and 1993 more than 200 MT sites were collected on Iceland. They all measured in a period range of 2–2000 s or less. The most prominent result of these previous investigations was the determination of a low-resistivity layer within the Icelandic crust with increasing depth from 10 to 30 km for increasing distance from the trench structure crossing Iceland in a NE–SW direction (cf. Beblo and Björnsson,

1980; Beblo et al., 1983; Hersir et al., 1984; Eysteinsson, 1993). But the data showed no indication of a plume structure because of the limited period range. So 3-D model studies were performed to obtain an overview of the resolution of different conductivity structures beneath Iceland. Furthermore allowance was made for the influence of the seawater and of the crust conductor. Finally these model calculations will help to interpret the data of several new LMT sites on Iceland, which were installed by M. Beblo, Munich in 1999. These data will reach a period range of up to 50,000 s.

2. 3-D MODELLING

All presented 3-D models were computed using the finite difference forward algorithm released by Mackie et al. (1994) and modified by Booker and Handong (1999). The algorithm calculates the horizontal components of the magnetic and electric fields at the surface and additionally the vertical component of the magnetic field. This is done twice, for source field polarised in the north–south (x) direction and in the east–west (y) direction, applying a plane-wave source condition. These field values are used to calculate the transfer functions, i.e. apparent resistivities and phases for the MT interpretation. Preliminary 3-D conductivity models were constructed, which were deduced from various 1-D and 2-D results of previous MT measurements (periods up to 2000 s) and roughly reflect the bathymetry around Iceland. Figure 1 shows a plan view and a cross-section of the model geometry and the used resistivity values. The island covers a rectangular area of 460 km × 460 km. The background resistivity is 100 Ω m. The seawater depth increases in two steps: till 50 km offshore it is 200 m and in the outer area 1500 m with a constant resistivity of 0.25 Ω m. A crustal conductor of 10

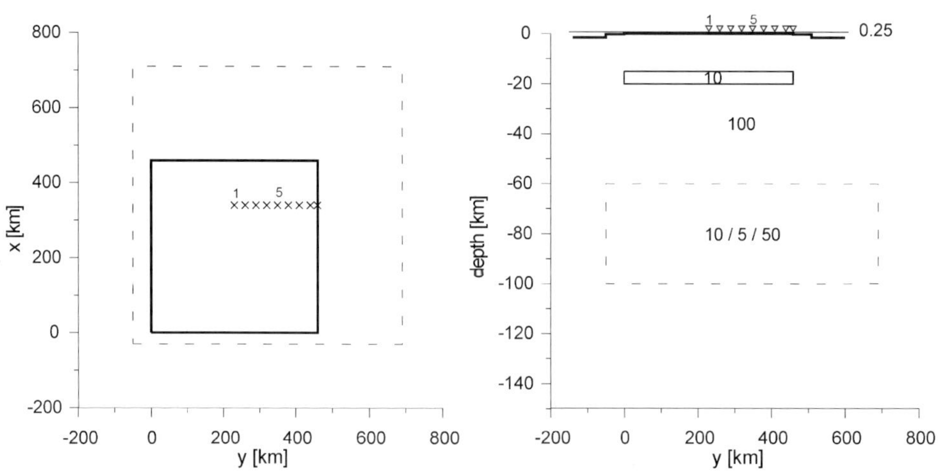

Figure 1. 3-D models. Grid size: 50 x 52 x 32 cells. Left: plan view. Right: cross-section through centre of plume (strongly superelevated). Numbers indicate resistivities in Ω m, for the plume head three alternative resistivities. Solid lines: island (thick), good conducting layer (thin). Dashed lines: plume head. Crosses and triangles: nine model site locations.

Table 1. Differences in the presented models

Model	Plume head	Crust conductor
1	10 Ω m	X
2	5 Ω m	X
3	50 Ω m	X
4	10 Ω m	No crust conductor
5	No plume	X

Ω m extends beneath the island at 15–20 km depth. A 10-Ω m half-space concludes the model beneath 550 km depth. While these parameters remain constant, the influence of the assumed plume head on the transfer functions is investigated. Its lateral extension is 740 km × 740 km, its vertical extension is 60–100 km, while the resistivity varies between 5 and 50 Ω m. Table 1 gives an overview on the different models. Models 1 to 3 differ in the resistivity of the plume head, whereas model 4 was considered without the good crust conductor and model 5 has no plume head. The electric and magnetic field values were computed for 13 periods between 10 and 100,000 s.

3. RESULTS

For all the models apparent resistivity and phase curves were calculated at nine sites along an east–west profile in the northeastern part of the island. Thereby the x-axis is positive to the north, y is positive to the east and z is positive downwards. The spacings between the site is 30 km, with site 1 in the centre and site 9 500 m from the coast.

The impedance tensor $\underline{\boldsymbol{Z}}$ is defined by $\underline{\boldsymbol{E}} = \underline{\boldsymbol{Z}}\,\underline{\boldsymbol{B}}$ with $\underline{\boldsymbol{E}} = (\boldsymbol{E}_x, \boldsymbol{E}_y)$ and $\underline{\boldsymbol{B}} = (\boldsymbol{B}_x, \boldsymbol{B}_y)$ as complex Fourier transforms of the electric and magnetic field in the frequency domain. In the following, the complex off-diagonal elements of $\underline{\boldsymbol{Z}}$, i.e. $\boldsymbol{Z}_{xy}$ and $\boldsymbol{Z}_{yx}$, are presented as apparent resistivities ρ_a^{xy} and ρ_a^{yx} and phases φ^{xy} and φ^{yx}.

To show the changes of the effects of the conductivity structures along the profile in Figure 2 the curves for model 1 at all sites are drawn whereas the different sites are characterized by numbers in the curves, respectively. In this model the crust conductor and the plume head are included with a plume resistivity of 10 Ω m.

For site 9 the coast effect is dominant for the xy-component at lower periods only and for the yx-component for the whole period range. This effect strongly decreases with increasing distance from the coast. It can be detected till 100 km inland (site 7), but is not dominant anymore at around 50 km (between sites 7 and 8).

Furthermore the pattern of the curves is characterized by the conductivity structures in the subsurface. The first slight minimum in the ρ_a curves and the maximum in the phase are caused by the crustal conductor. The second deeper and broader minimum in ρ_a and the respective phase maximum can be assigned to the plume head. The following steep rise in phase is caused by the 10-Ω m half space at the bottom of the model.

The ρ_a and phase of site 1 (centre of the island) for the models described in Table 1 are compared in Figure 3. Note that the numbers in the curves now indicate the different

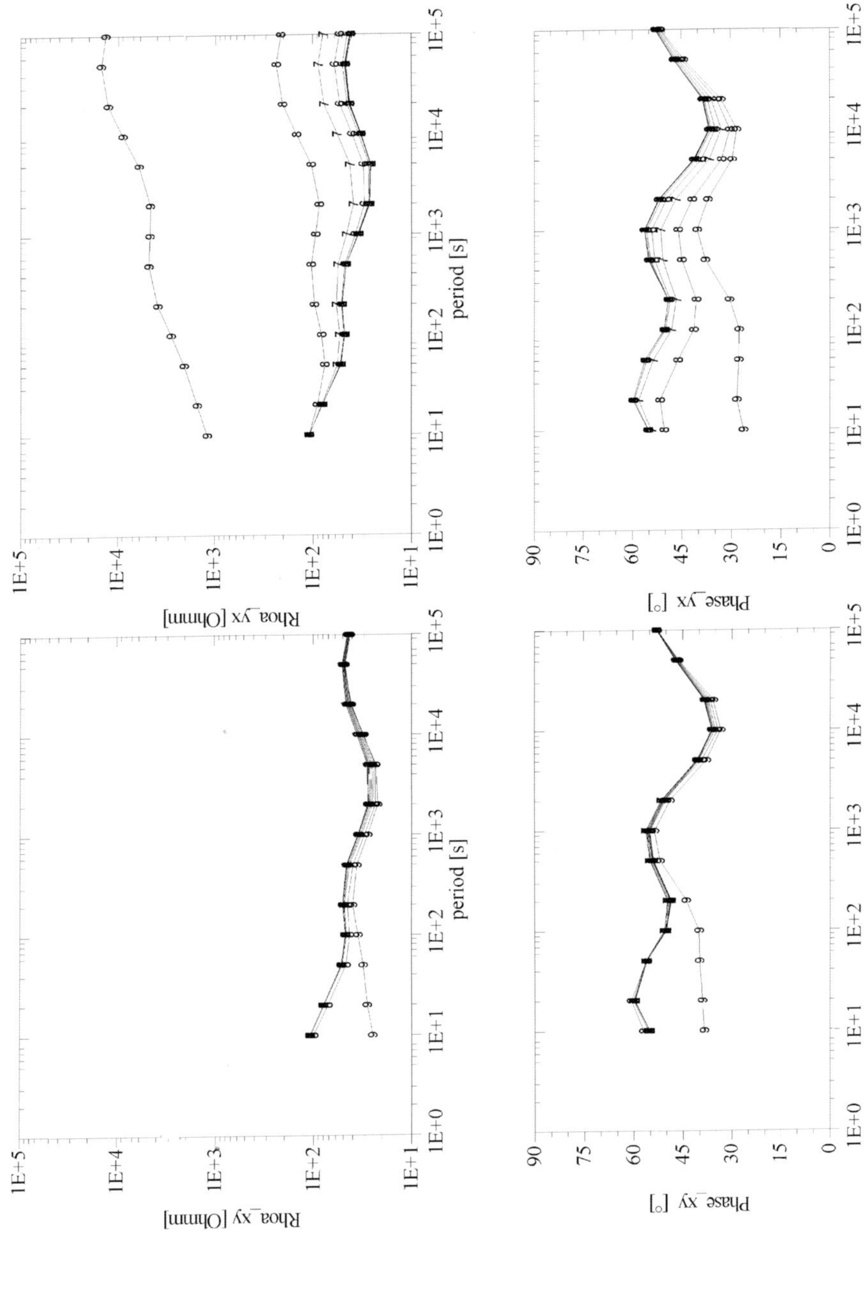

Figure 2. Apparent resistivities and phases for *xy*- and *yx*-component for model 1 and all nine sites. The crustal conductor is included, and the plume head resistivity is 10 Ω m.

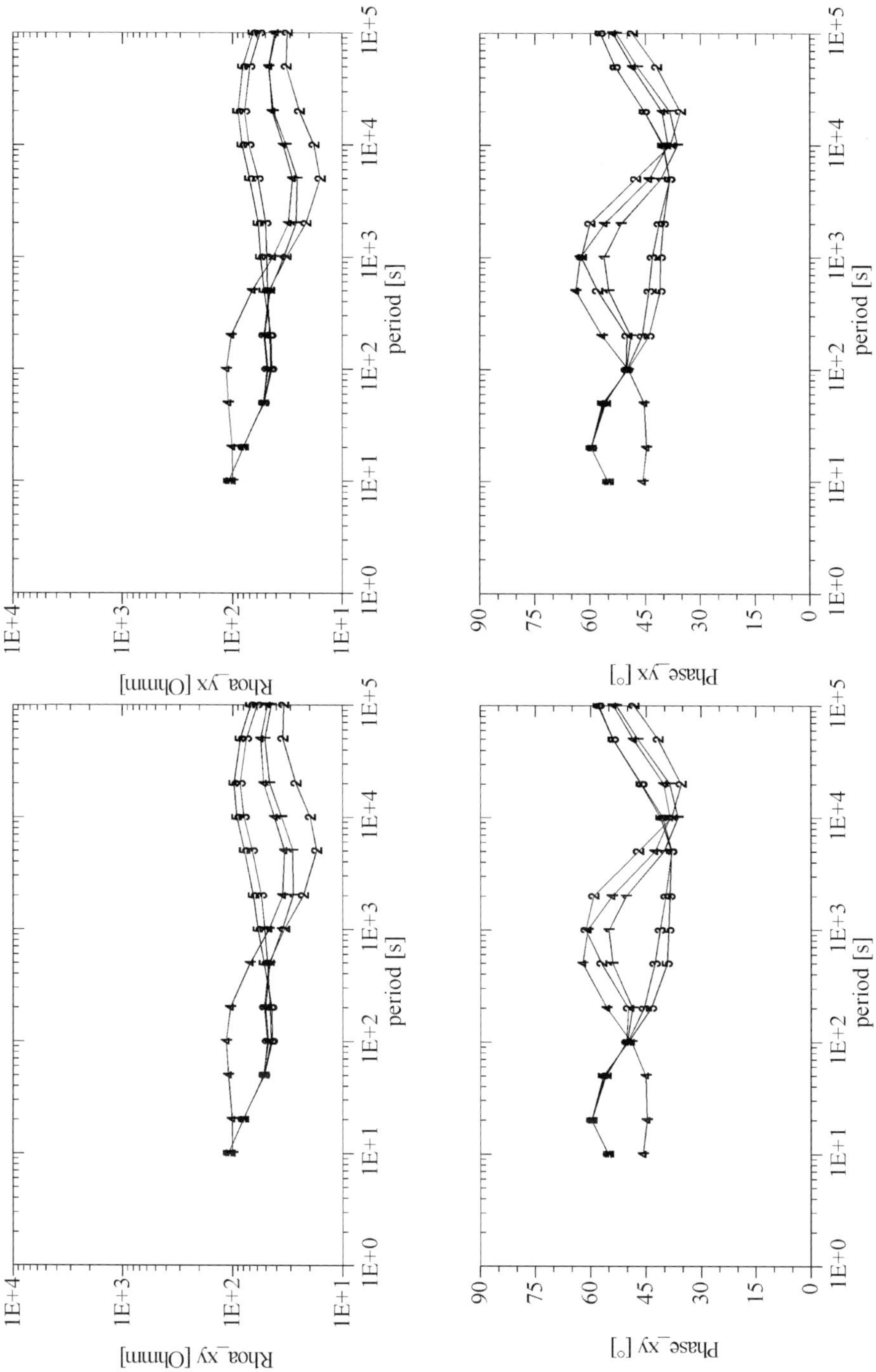

Figure 3. Comparison of all five models (cf. Table 1) for the *xy*-component at site 1 (island centre).

model numbers. In Figure 4 the same is done for a position 3 km away from the coast (2.5 km inland from site 9).

At site 1 the xy- and yx-components are almost identical (Figure 3). Model 5 serves as a reference since it has no plume structure. Compared to model 3, which considers a plume head with 50 Ω m, almost no differences can be found, in contrast to model 1 (10 Ω m), were clear differences occur for periods above 100 s in phase and above 500 s in ρ_a. A similar situation is found for model 2, in which the plume resistivity was set to 5 Ω m. In that case the second minimum in ρ_a becomes deeper and broader compared to model 1. Model 4 does not include the crustal conductor. Thus the first minimum in ρ_a disappears and the phase starts at 45°.

Approaching the coast (Figure 4) the xy- and yx-components of a site 3 km away from the east coast show pronounced differences: the xy-component is similar to that of site 1 for all models. However, the effect of the crustal conductor cannot be seen as distinctly as before. The minimum in ρ_a is very flat and the phase does not show a maximum at 20 s, due to the influence of the seawater which affects the curves already at lower periods. For the yx-component the coast effect completely overwhelms the influence of the crustal conductor whereas the plume head can still be detected.

In several additional models a plume tail was included. It had a diameter of 230 km × 230 km, a vertical extent down to 200 km and the resistivity of the respective plume head. In all cases the results were the same compared to the models without a plume tail, indicating that a plume tail such as this is not detectable with MT.

4. DISCUSSION AND CONCLUSIONS

The model studies show that the highly conducting seawater and the conducting layer of 10 Ω m between 15 and 20 km depth do not shield the deep conductivity structures, i.e. the plume head, for a conductivity contrast $> \sim 3$. In this case the MT transfer functions show separate minima and maxima in ρ_a and phase curves due to each structure (cf. Table 1, Figures 1–4). It is also possible to resolve the thickness of the plume for resistivities less than 10 Ω m. Furthermore one can conclude that the lateral extent of the plume head has no significant influence on the data as long as it is larger than the extent of the island. Thus MT neither gives sufficient information about the size of the plume nor resolves a plume tail as its effect on the transfer functions is completely covered by the upper structures.

Probably also the exact size of the crust conductor will be hard to specify due to the influence of the seawater. However, a model without such a conducting layer causes clear changes in ρ_a and phase curves compared to models including this zone. As nearly all measurements on Iceland, except sites in the southwest and some close to the coast, show a well distinct minimum and phases above 45° in a period range between 10 s and 200 s (cf. Beblo and Björnsson, 1980; Beblo et al., 1983; Hersir et al., 1984; Eysteinsson, 1993) the strong effect in the observed data must be explained with a conducting layer in the subsurface and cannot be an effect of the seawater. In general the coast effect can be seen up to a distance of 50 km from the coast, but it is not dominant further inland (cf. Figure 2, sites 8 and 7).

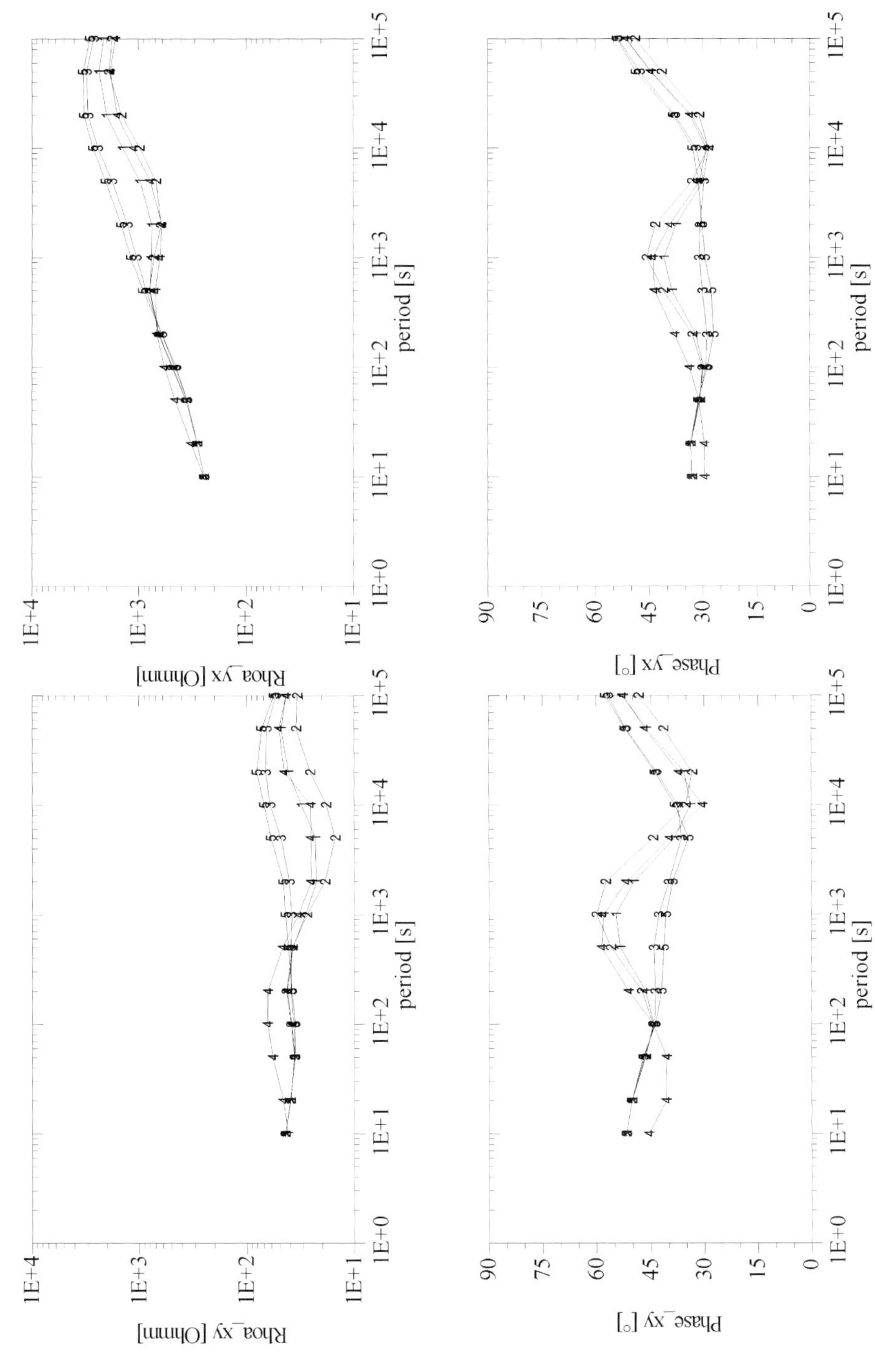

Figure 4. Comparison of all five models (cf. Table 1) for the *xy*- and *yx*-component at a site 3 km away from the east coast.

REFERENCES

Beblo, M. and Björnsson, A., 1980. A model of electrical resistivity beneath NE-Iceland, correlation with temperature. J. Geophys., 47, 184–190.

Beblo, M., Björnsson, A., Arnason, K., Stein, B. and Wolfgram, P., 1983. Electrical conductivity beneath Iceland — constraints imposed by magnetotelluric results on temperature, partial melt, crust- and mantle structure. J. Geophys., 53, 16–23.

Booker, J. and Handong, T., 1999. Description of changes and the 3-D modelling code can be found on internet site ftp://ftp.geophys.washington.edu/pub/out/booker/mackie3D

Eysteinsson, H., 1993. MT soundings in southwest Iceland. Phys. Iceland VI, 105–116.

Hersir, G.P., Björnsson, A. and Pedersen, L.B., 1984. Magnetotelluric survey across the active spreading zone in southwest Iceland. J. Volcanol. Geotherm. Res., 20, 253–265.

Mackie, R.L., Smith, J.T. and Madden, T.R., 1994. Three-dimensional electromagnetic modeling using finite difference equations: the magnetotelluric example. Radio Sci., 29, 923–935.

Chapter 15

THREE-DIMENSIONAL MONITORING OF VADOSE ZONE INFILTRATION USING ELECTRICAL RESISTIVITY TOMOGRAPHY AND CROSS-BOREHOLE GROUND-PENETRATING RADAR

Douglas LaBrecque [a], David L. Alumbaugh [b], Xianjin Yang [c], Lee Paprocki [d] and Jim Brainard [e]

[a] *Multi-Phase Technologies, 310 Rebecca Drive, Sparks, NV 89436, USA*
[b] *Department of Civil and Environmental Engineering/Geological Engineering Program, University of Wisconsin–Madison, Madison, WI 53706, USA*
[c] *Advanced Geosciences Inc., 12700 Volente Road, Building A, Austin, TX 78726, USA*
[d] *Formerly E&ES Department, New Mexico Institute of Mining and Technology, Socorro, NM 89502, USA*
[e] *Sandia National Laboratories, Albuquerque, NM 87185, USA*

Abstract: We discuss results from field experiments conducted at the Sandia–Tech Vadose Zone experimental facility on the New Mexico Tech campus at Socorro, New Mexico, as part of a project to develop a joint hydrological–geophysical method to characterize fluid flow properties of the vadose zone. The site contains dense arrays of tensiometers, access tubes for neutron moisture and ground-penetrating radar probes, and arrays of surface and subsurface electrical resistivity tomography electrodes installed in shallow clays, sands and gravels. We collected electrical resistivity tomography (ERT), cross-borehole ground-penetrating radar (XBGPR) and neutron data before and during a controlled infiltration of water at the site. Using local, empirical relations, we estimated subsurface moisture contents from images generated with the XBGPR and ERT data. The XBGPR images provided an excellent comparison to the neutron-derived moisture contents along a plane through the center of the experimental site. The ERT results were limited in terms of resolution by the coarse electrode spacing and inversion mesh used at the site, but provided a full three-dimensional picture of the wetting front as it progressed.

1. INTRODUCTION

The vadose zone, the hydrological region lying above the water table, is difficult to monitor and the hydrological processes within this region are still poorly understood. As a result, there are many documented cases of contaminants being found where they are not expected and/or not predicted. For example, recently at the Hanford Reservation, cesium was detected near the water table beneath one of the tank farms although hydrologic modeling indicated that contaminant transport rates would be much slower. Studies from the Yucca Mountain project provides a second example where Chlorine 36 from airborne nuclear detonations at the Nevada Test Site during the 1950s was found at much greater depths than thought possible. Thus, there is a need to develop methods for generating three-dimensional (3-D) quantitative estimates of hydrological parameters within the vadose zone. Electrical and electromagnetic geophysical methods have received much attention in this regard as (1) these geophysical imaging techniques are fairly nondestructive, (2) they are very sensitive to changes in subsurface moisture contents, and (3) they provide estimates of properties in regions between boreholes. In comparison, typical hydrologic measurements tend to be sensitive only to the region immediately surrounding the sensor. The work here focused on the comparison of two electrical geophysical methods, electrical resistivity tomography (ERT) and cross-borehole ground-penetrating radar (XBGPR), with a variety of subsurface hydrologic measurements.

To facilitate these measurements, the Sandia–Tech Vadose Zone (STVZ) test site was designed and built adjacent to the New Mexico Institute of Mining and Technology in Socorro, New Mexico. The test site is a cooperative venture between scientists at Sandia National Labs and New Mexico Tech, and is installed into unsaturated, unconsolidated, heterogeneous fluvial deposits. While most of the site consists of fine sands that contain a moderate amount of iron-oxide, there is a good deal of heterogeneity present, with the upper 2 m consisting of coarse sands and gravels, and two distinct silt and/or clay layers present between 4 and 6 m depth.

Dense arrays of tensiometers, surface and subsurface ERT electrodes were installed at the STVZ site to a depth of 12 m along with 13 polyvinylchloride (PVC)-cased boreholes. The tensiometers provide point measurements of the in-situ pore water pressure/tension. The PVC-cased boreholes provided subsurface access for XBGPR instruments, neutron moisture measurements, and other geophysical logging tools. A plan-view of the site showing the locations of various instruments is given in Figure 1.

Geophysical and hydrological characterization to determine initial subsurface conditions were initiated a year prior to beginning the experiment. Infiltration was started on March 11, 1999, by introducing water over a 3 m by 3 m infiltrometer at a rate of 2.7 cm/day. The infiltrometer was designed to produce a constant, uniform, known flux of water over its 3 m by 3 m area. The influx of water creates a region of increased moisture extending downward and outward from the infiltrometer. The boundaries of this region are known as the wetting front. Measurements were made with the various subsurface hydrologic and geophysical sensors to monitor the advance of the wetting front, and the conditions within the region of increased moisture content.

2. ERT METHODOLOGY

ERT data were collected using 8 vertical electrode arrays (VEA) (Figure 1) that each contained 17 copper electrodes spaced at 76 cm intervals. The uppermost electrode is just below the surface, while the lowest is 12 m deep. In addition to the 8 VEAs, 36 additional electrodes were placed on the surface (Figure 1). The ERT data presented in this paper were collected using combinations of the VEAs and surface electrodes, and a complete data set consisted of about 20,000 individual measurements.

A 3-D Occam's inverse algorithm described by LaBrecque et al. (1995) and Morelli

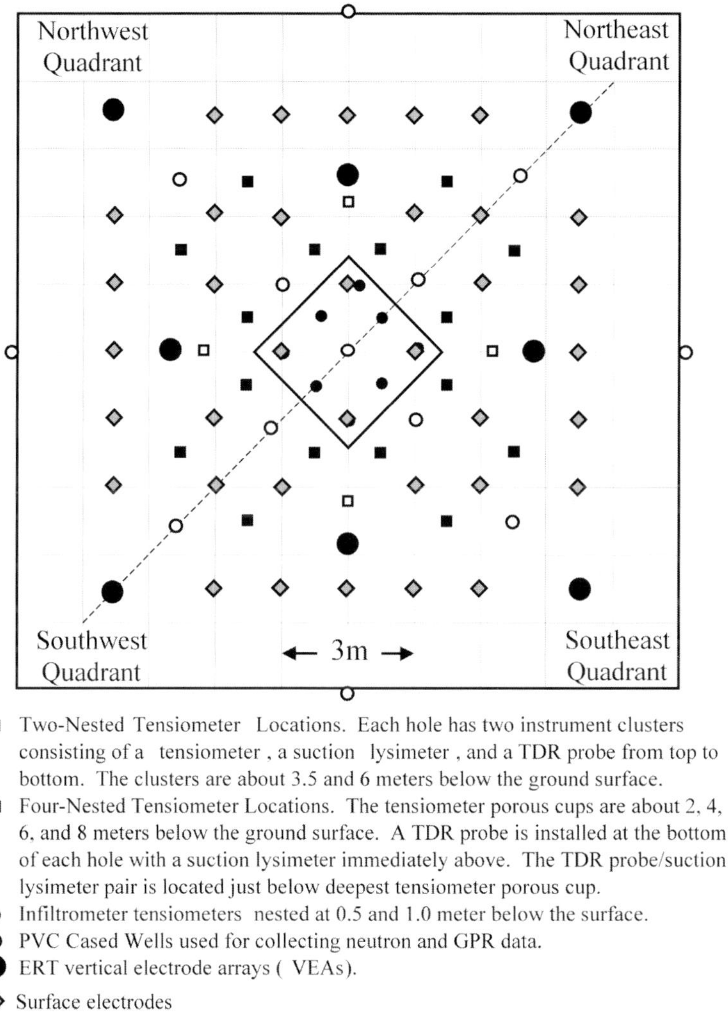

Figure 1. A plan-view schematic of the STVZ site showing the locations of the various instruments. The innermost square of 3 m by 3 m is the infiltrometer. The dashed line is the 2-D XBGPR image transect.

and LaBrecque (1996) was used to reconstruct the electrical conductivity image from the electric potential data. Occam's inversion finds the smoothest possible model whose response best fits the measured data to an a-priori, chi-square statistic (Constable et al., 1987; DeGroot-Hedlin and Constable, 1990). In the 3-D inversion routine, the subsurface is divided into rectangular–parallelepiped regions of constant conductivity called voxels. Smoothness of the model is enforced by minimizing the differences in log conductivity of adjacent voxels, in each of the three Cartesian coordinate directions. The conjugate-gradient method was employed to solve both the forward and inverse matrix systems, and a data-error reweighting scheme implemented to suppress the effects of data outliers (Morelli and LaBrecque, 1996). To improve the resolution in the center of 3-D volume, a roughness-weighting scheme was incorporated (Morelli and LaBrecque, 1996). The region to be imaged was discretized into a mesh of $40 \times 40 \times 39 = 62{,}400$ elements. Element size ranged from 0.38 m in the center of the mesh to 6 m along the boundaries. The inversion of a complete data set took about 45 h on a Pentium II 400 MHz PC.

2.1. Moisture content estimation using ERT images

Archie (1942) and Waxman and Smits (1968) developed empirical relations between resistivity and porosity, fluid salinity, moisture content, and the cation-exchange capacity (CEC) of the various mineral constituents. Thus moisture content can be estimated from resistivity if the other parameters (porosity, CEC etc.) are known. This was not practical for this application in which the available data were the ERT resistivity images coupled with spatially sparse hydrological measurements. Our solution was to determine a local, empirical relationship between resistivity and moisture content for the 'background' images generated from data collected before the infiltration began, and a second expression that relates changes in resistivity to changes in moisture content during infiltration.

In almost all sediments, there are two mechanisms for electric current flow: normal conduction due to the flow of mobile ions in the interstitial pore water and surface conduction due to the flow of semi-bound ions on mineral surfaces (Waxman and Smits, 1968). In typical unconsolidated sediments, the normal conduction is primarily a function of the pore-water salinity and the volume fraction of the sediment filled with pore water. Although these and a number of other factors affect surface conduction, this conduction mechanism is most strongly influenced by the amount of surface area of the mineral grains which of course increases as the mineral-grain sizes become smaller. In the vadose zone, water is retained by capillary force. Therefore, sediments with small grain sizes (high curvature and thus large capillary forces) retain a larger volume fraction of water than coarse-grained ones (small capillary forces). Thus, in unsaturated sediments, both the normal and surface conduction modes are strongly correlated to grain size, volumetric-water content, and each other. Because these factors are all strongly correlated, we assume that we can apply a single, power-law function to the moisture-content/resistivity relationship. However, the constants within this equation are very different from those normally employed in formulas such as that of Archie (1942), since they reflect not only the relationship between saturation and resistivity, but also the correlation between grain size and residual water content in the unsaturated sediments. We use a relationship of the form

$$\rho_t = a \cdot \theta^b \tag{15.1}$$

where ρ_t is the resistivity of partially saturated sediments and θ is moisture content. The variables, a and b, are empirical constants that were determined by comparing resistivity values to neutron-derived moisture contents. We found good comparisons between the ERT estimated moisture contents and those derived from the neutron data using $a = 0.021$ and $b = -3.2$. These relations were found from fitting neutron data at

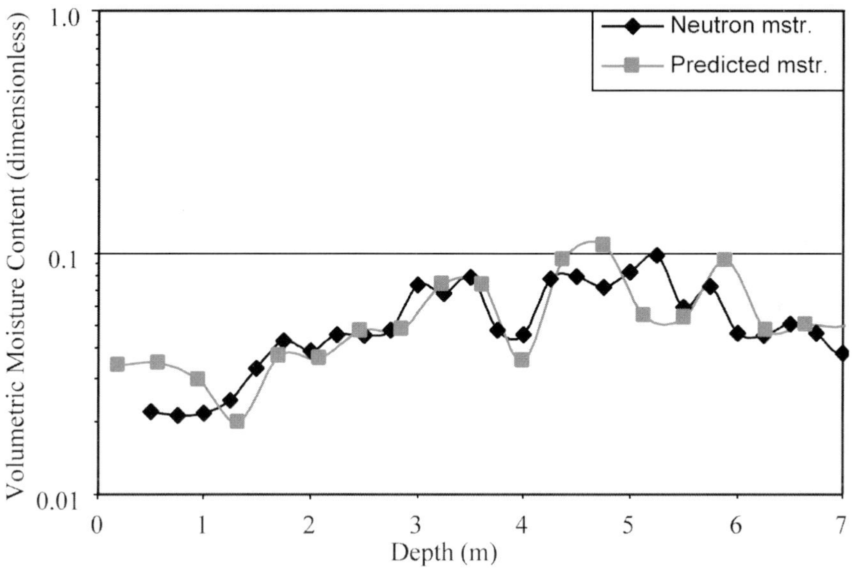

Figure 2. Comparison of neutron-derived moisture content with that predicted from ERT for the southwestern-most well shown in Figure 1.

one of the wells to the resistivities derived from ERT. Since there is a wide range of volumetric moisture content, ranging from as low as 2% to over 20%, these coefficients are determined to about two decimal places. However, it is important to keep in mind that these coefficients are accurate only for the specific conditions on this site. We cannot extrapolate the results from a single site to determine the range of variability of these coefficients at other sites.

The application of this relation also depends on a number of assumptions. The first assumption is that the system is undisturbed prior to infiltration such that the capillary forces have reached equilibrium and so that the distribution of pore water is mainly controlled by the pore size. A second assumption is that the water chemistry, most importantly the salinity of the pore water, is uniform throughout the site. Finally, we assume that the porosity is fairly constant. This is generally a good assumption for shallow, unconsolidated sediments. Figure 2 compares the pre-infiltration, neutron-derived moisture content with the moisture content estimated from the ERT inverse results via Equation (15.1).

Once we begin infiltrating water, we violate the assumption that the capillary forces are in equilibrium. The changes in resistivity caused by increased moisture content during infiltration can also follow the power-law relation given by Equation (15.1), but the constants, a and b, are more typical of those employed in Archie's relation (Archie, 1942). Using laboratory measurements of resistivity versus saturation that were made on 12 core samples collected at the site, we chose a value of $b = -1.5$, the geometric mean of the 12 samples. These laboratory measurements were made on the wetting/saturation phase using the same solution used for the field infiltration experiment. The coefficient a

varied substantially from sample to sample. To determine the value of a, the background moisture content is calculated using the background formula. Then the value of a in the post-infiltration formula was allowed to vary from cell to cell so that the two equations matched for the background data set.

3. XBGPR METHODOLOGY

At the Socorro site, XBGPR data were acquired using a Sensors and Software Pulse EKKO 100. The system consists of a transmitter antenna and a separate receiver antenna, both tied to a central acquisition computer via fiber optic cables. The Pulse EKKO 100 can transmit a pulse with a center frequency of 50, 100 or 200 MHz. In this experiment, a center frequency of 100 MHz was employed along with a source–receiver sampling interval of 0.25 m. Because the image resolution is dictated both by the wave-length of the transmitted pulse as well as the spatial sampling interval, the maximum resolution that is attainable from these data is 0.25 m. XBGPR surveys were conducted using the same PVC-cased boreholes that were employed for neutron logging.

Although the full XBGPR wave-field was acquired, the imaging only employed the travel time of the wave that arrived first at the receiver, and these first arrivals were picked using the routine supplied by Sensors and Software. The data were then inverted to produce a 2-D velocity image using the 'PRONTO' scheme of Aldridge and Oldenburg (1993). This is a nonlinear curved-ray inversion scheme that imposes constraints in the form of spatial derivatives that, like the ERT inversion algorithm described above, force the inverted model to be smoothly varying from one point to the next. For a more detailed description of the algorithm the reader is referred to Aldridge and Oldenburg (1993).

3.1. Estimating moisture content from XBGPR data

The first step in estimating the moisture content from the XBGPR velocity images is to convert velocity to apparent dielectric constant (ε_a). For materials of relatively low electrical conductivity (say electrical conductivities less than 0.03 S/m), the EM wave velocity is given by the expression,

$$v = \left(\frac{1}{\mu\varepsilon_a\varepsilon_0}\right)^{-1/2}, \tag{15.2}$$

where ε_0 is the dielectric permittivity of free space ($\varepsilon_0 = 8.85 \times 10^{-12}$ F/m), and μ is the magnetic permeability. Usually, μ is assumed to be that of free space ($\mu_0 = 4\pi \times 10^{-7}$ H/m), in which case the above relationship can be rearranged to solve for ε_a as

$$\varepsilon_a = \left(\frac{c}{v}\right)^2, \tag{15.3}$$

where c is the velocity of light in a vacuum ($\varepsilon_a = 1$) (Davis and Annan, 1989). Both c and v have dimensions of m/ns, and thus ε_a is dimensionless. The values of ε_a are often converted to moisture content, θ, via an empirical relationship known as Topp's equation (Topp et al., 1980),

$$\theta = 5.3 \times 10^{-2} + 2.92 \times 10^{-2}\varepsilon_a - 5.5 \times 10^{-5}\varepsilon_a^2 + 4.3 \times 10^{-6}\varepsilon_a^3. \quad (15.4)$$

At the Socorro site, we found that moisture estimates created by applying Topp's equation to XGPR images significantly overestimated in-situ moisture content determined by neutron logging. We determined that the problem resulted from a limitation in application of Topp's equation with velocity estimates made using Equation (15.3). We made laboratory time-domain-reflectrometry (TDR) measurements of ε_a versus moisture content on samples collected from the site. TDR uses a combination of equipment originally designed for testing cables and specially designed probes to estimate the electromagnetic velocity of a sample about a cubic decimeter in size. The TDR gave us an independent measure of velocity and showed that the problem of overestimation of moisture content was not due to problems with the XGPR data collection, the XGPR field instrumentation, the neutron measurements, or the XGPR reconstruction algorithm. When Equation (15.3) and Topp's equation were used to derive moisture contents, the results significantly overestimated the true moisture content of the laboratory samples.

The overestimation is due to the fact that Equation (15.3) is only an approximation and assumes that the magnetic permeability (μ) of the sediments is equal to that of free space, i.e., that there are no magnetic minerals present. However, most of the sediments deposited at the Socorro site were derived from volcanic terrain, and in fact samples collected from the site showed up to 5% magnetic minerals by weight. Magnetic mineralization in this amount causes the magnetic permeability of the sediments to be greater than that of free space by a factor of 1.2 to 1.5 depending on the actual mineral composition. Since the EM velocity is inversely proportional to the square root of μ, the velocity of these materials will be lower than materials that are absent of magnetic minerals, and thus the estimates of ε_a will be higher. In addition, because the moisture content increases with increasing ε_a, employing Topp's equation to convert values of ε_a that have been derived from Equation (15.3) in magnetic soils will lead to estimates of moisture content that are too high. Thus, to convert the images of ε_a to volumetric moisture content, we employed a site-specific empirical relationship derived from the aforementioned TDR experiments, and this conversion is given as

$$\theta = 0.0136\varepsilon_a - 0.033. \quad (15.5)$$

4. COMPARISON OF ERT AND RADAR MOISTURE ESTIMATES

As outlined above, multiple sets of neutron, XBGPR, and ERT data were collected to determine initial conditions at the site. XBGPR data that were collected using five PVC-cased boreholes along the southwest to northeast profile are shown in Figure 1. Four planes of data were collected between the four adjacent pairs of electrodes. Figure 3 shows the moisture content estimates that were derived from the inverted radar data. Note that this image was produced by inverting all four planes of data simultaneously rather than inverting each separately and then piecing the resulting images together. Superimposed on the radar-derived moisture contents are the calibrated neutron logs from the same five boreholes. The dark, high-moisture-content regions represent the

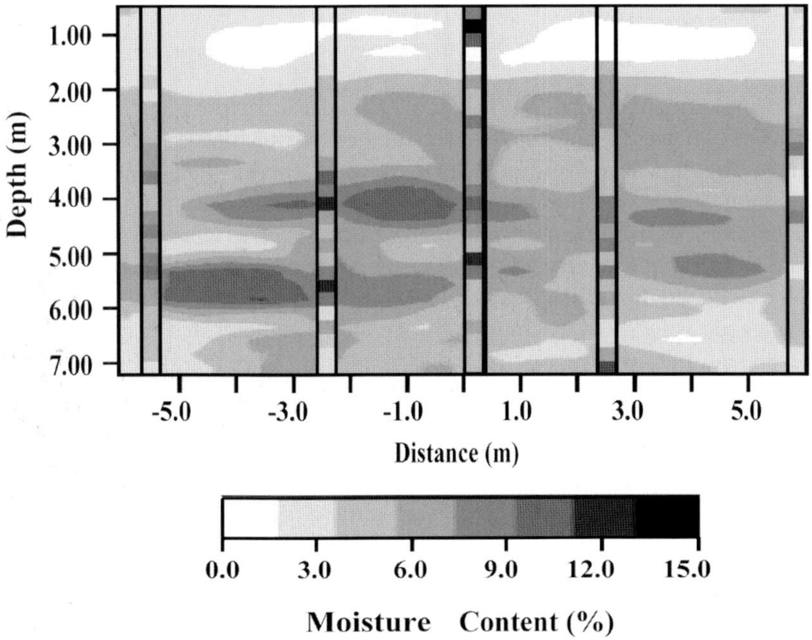

Figure 3. Comparison of moisture content from GPR with neutron data for background data prior to infiltration.

clay and silt layers, the lighter zones the sands and gravels. Notice that the XBGPR and neutron moisture contents results compare very well. Even at the very near surface, where the inversion routine had some difficulty reconstructing the image due to the presence of the high-velocity air layer, the agreement between neutron and radar is still very good. The one exception is the dark, high-moisture content region near the surface in the center borehole that is located within the infiltrometer. This anomalous region is likely the result of a bentonite seal that was emplaced around this particular borehole to prevent water from migrating down along the casing. This is a near borehole effect, a few centimeters in diameter. It is observed in the neutron data which are highly sensitive to borehole effects (the probe used was not compensated for borehole effects) but are not large enough to appear in the XBGPR image.

Figure 4 shows the moisture content estimated from ERT data along the same profile as the XBGPR results shown in Figure 3. The ERT data were collected using a 0.76-m electrode separation and inverted using elements 0.38 m thick. Thus the best possible resolution for the method is 0.38 m and that resolution would be approached only at or near the boreholes. In the center of the image, the resolution is probably closer to the 0.76-m electrode separation or likely poorer. Despite this, the ERT does a fairly good job of resolving a series of relatively thin layers. The comparison between ERT and GPR is particularly good on the southwest edge of the image where the image plane approaches one of the VEAs. In addition, toward the center of the image, the ERT shows similar features to the GPR and neutron. However, the depths of the layers within the

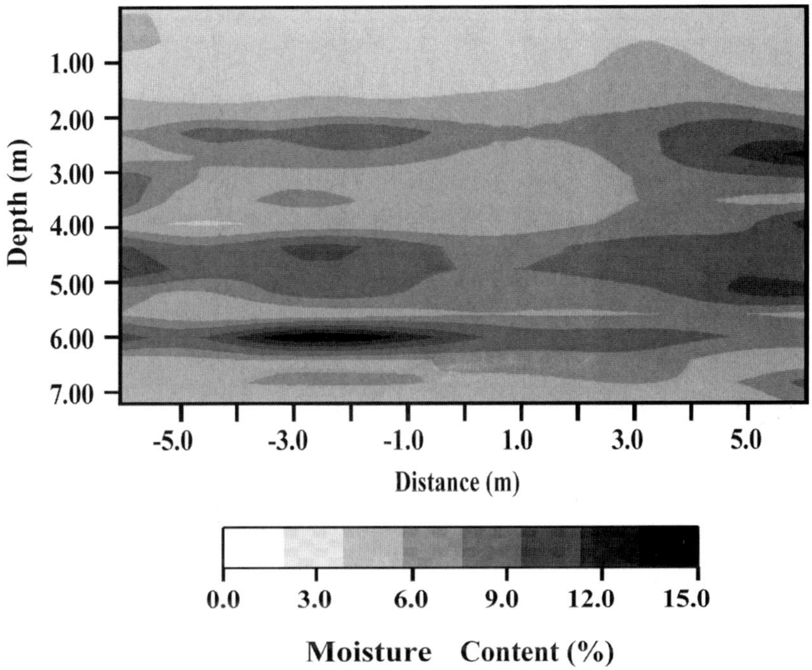

Figure 4. Background moisture content (prior to infiltration) estimated from ERT. Results are shown for the same plane as the XBGPR results in Figure 3.

ERT images are shifted by as much as a 1/2 m when compared to the other results. These shifts probably result from limited resolution, coarse discretization, and probably most importantly, the fact that there are no electrodes within this region. Regions away from the electrodes can be expected to be the most poorly resolved portions of ERT images.

Figure 5 shows the XBGPR-derived moisture content after eight days of infiltration and Figure 6 shows the ERT results for the following day. Note that both images as well as the neutron data indicate that by this point in time, the moisture front infiltration has reached a depth of 3 m. Again, the XBGPR and ERT results compare fairly well. The ERT image shows what is probably a more realistic view of the moisture content in the near surface, especially near the infiltration pad (center of image). Figures 7 and 8 show the XBGPR and ERT images after more than a month of infiltration. The radar shows that a significant amount of infiltrated water has reached depths of 6 m. Much of the upper portion of the site shows volumetric water contents in excess of 15%. The ERT image shows a similar pattern in terms of an increased region of moisture content; however, the change is less. This possibly indicates that the salinity of the water has decreased, or that the lack of electrodes within this region causes the ERT to be slightly less sensitive to changes in moisture content here. Despite this problem, the ERT still provides a good estimate of the extent of infiltration.

Figure 9 shows the 3-D views of the moisture content that are estimated from ERT

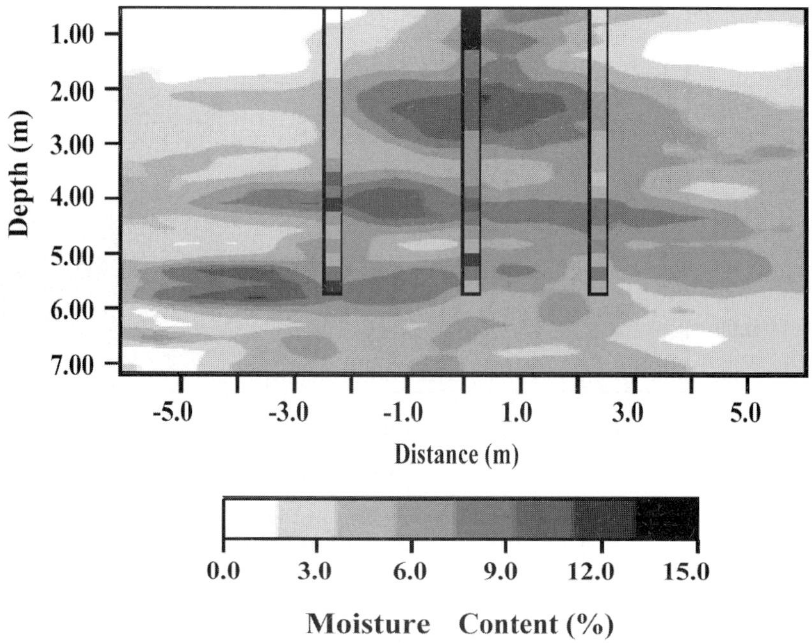

Figure 5. Comparison of moisture content from GPR with neutron data after 8 days of infiltration.

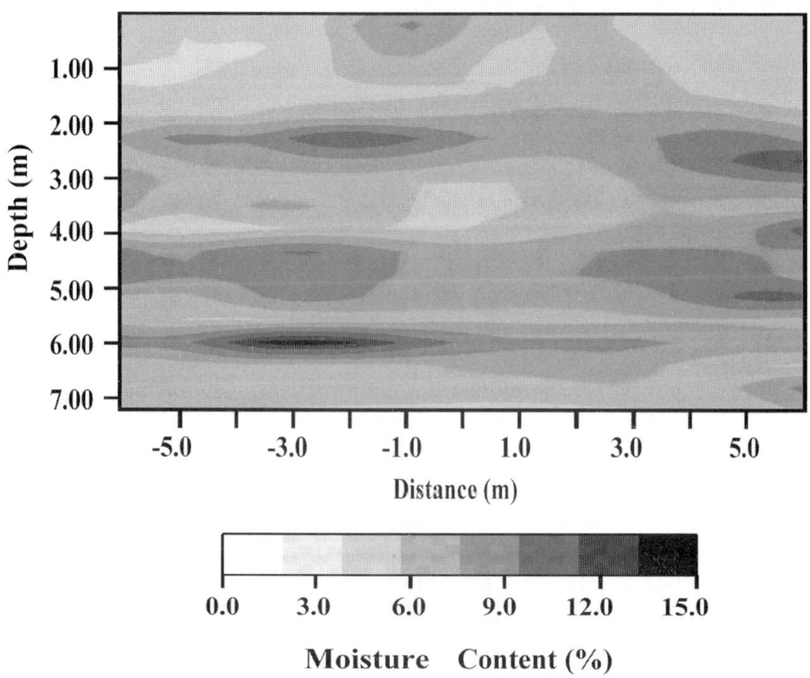

Figure 6. Moisture content estimated from ERT for data after 9 days of infiltration.

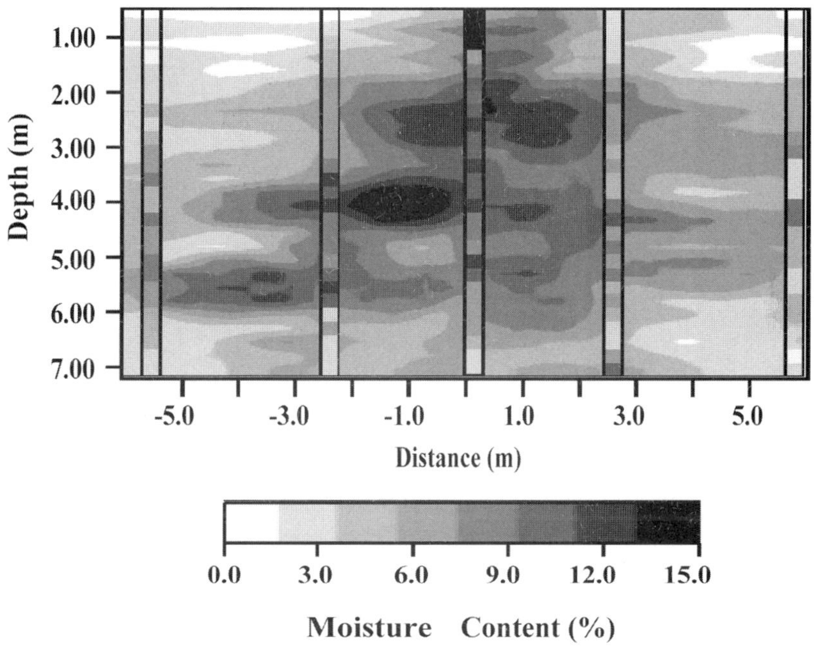

Figure 7. Comparison of moisture content from GPR with neutron data after 35 days of infiltration.

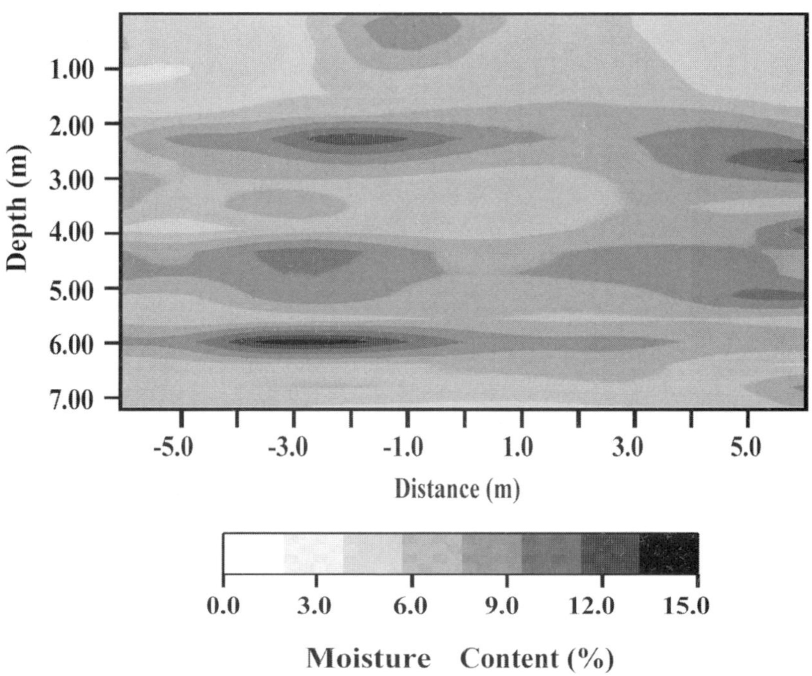

Figure 8. Moisture content estimated from ERT for data after 33 days of infiltration.

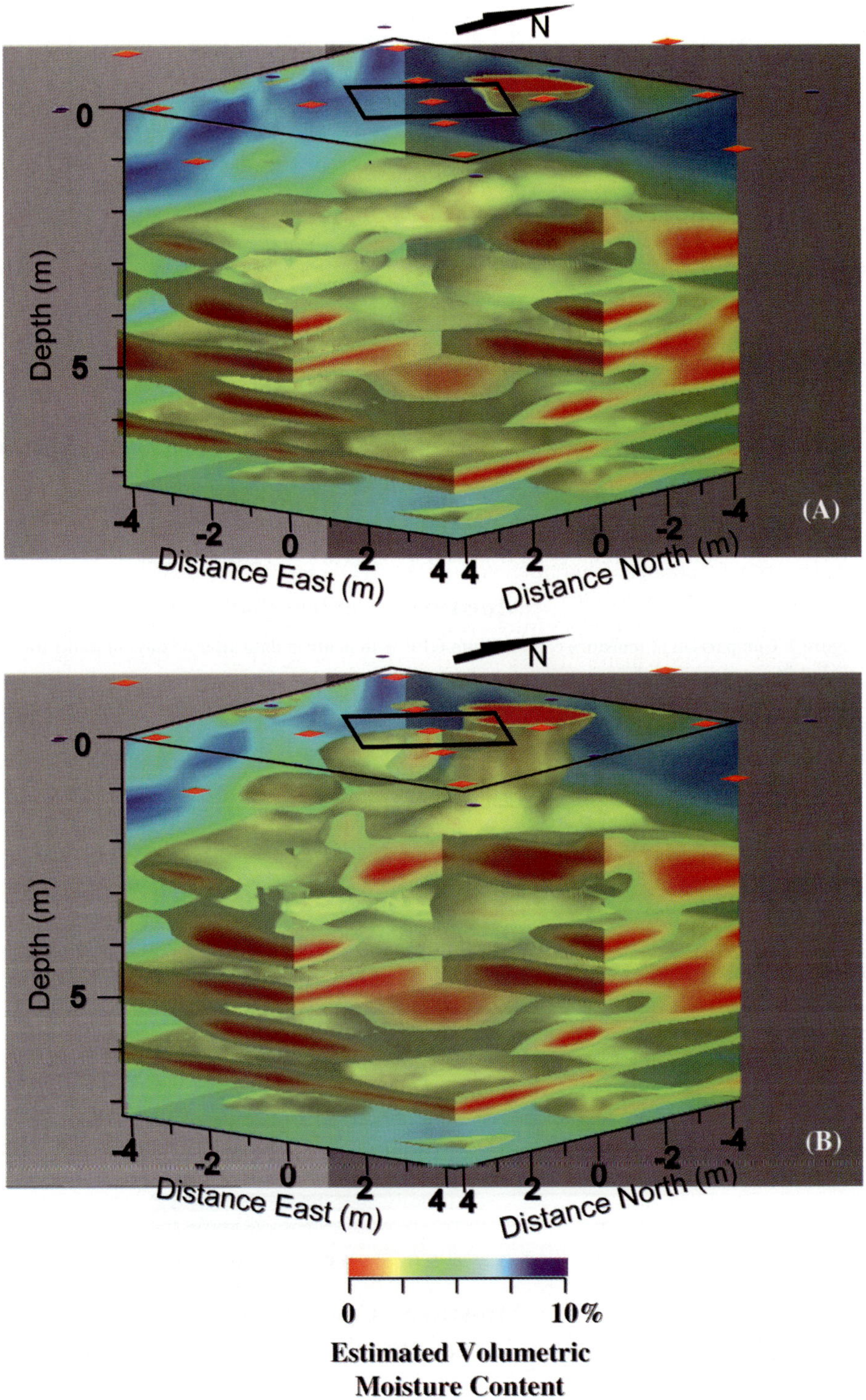
N
Depth (m)
0
5
-4
-2
0
2
4
Distance East (m)
4
2
0
-2
-4
Distance North (m)
(A)
N
Depth (m)
0
5
-4
-2
0
2
4
Distance East (m)
4
2
0
-2
-4
Distance North (m)
(B)
0
10%
Estimated Volumetric
Moisture Content

images. Figure 9A shows the areas of high moisture content (>5%) prior to infiltration that demonstrate the complex, interbedded nature of the site. These results can be compared with the moisture content after about 1 month of infiltration (Figure 9B). Note the vertical zone of high moisture content extending downward from the infiltration pad, and the increase in the moisture content at about 2 m in depth. A more subtle feature is the preferential movement of fluid to the west and out of the image zone.

5. DISCUSSION AND CONCLUSIONS

Results have shown that both XBGPR and ERT provide similar images of in-situ moisture content. Both techniques have their strengths. The XBGPR appears to provide somewhat better moisture content estimates than ERT along the southwest–northeast plane shown. This is in part due to the limited resolution and coarse inverse mesh used for ERT, and in part because the ERT results are excerpted from portions of a 3-D mesh that are not close to VEAs. The resolution of ERT has been shown to be greatest near the VEA. In particular, there are no VEAs near the center of the image plane along which the XBGPR data were collected. The main strength of ERT is that 3-D data sets can be routinely collected and interpreted with existing technology.

Both the ERT and XBGPR results relied on local empirical relations to determine moisture content. For XBGPR, the presence of higher then normal amounts of ferromagnetic minerals made Topp's equation (Topp et al., 1980) inaccurate. For ERT, separate power-law relationships were used to relate background data and post-injection changes to moisture content. These estimates have a disadvantage in that the accuracy of the estimates will decrease with time as the buffering capacity of the shallow sediments is depleted and the salinity of the pore water begins to change. In addition, all of these results demonstrate the importance of collecting samples at a site where this type of conversions from geophysical images to physical properties are going to be attempted, as standard empirical relationships may fail.

ACKNOWLEDGEMENTS

This work was performed at Sandia National Laboratories, the University of Wisconsin–Madison, Multi-Phase Technologies LLC, SteamTech Environmental Services, and New Mexico Tech with funding provided by the U.S. Department of Energy's Office of Energy Research, Environmental Science Management Program. Sandia is a multiprogram laboratory operated by Sandia Corporation, a Lockheed Martin Company, for the United States Department of Energy under Contract DE-AC04-94AL85000.

Figure 9. 3-D views of estimated moisture content from ERT prior to infiltration (A) and 33 days after infiltration began (B). The locations of the neutron/XGPR access tubes, infiltration pad and XGPR profile are shown at the top of the image.

REFERENCES

Aldridge, D.F. and Oldenburg, D.W., 1993. Two dimensional tomographic inversion with finite difference traveltimes. J. Seismic Explor., 2, 257–274.

Archie, G.E., 1942. Electrical resistivity log as an aid in determining some reservoir characteristics. Am. Inst. Min. Metal. (Engr. Transl.), 146, 54–62.

Constable, S., Parker, R.L. and Constable, C.G., 1987. Occam's inversion: a practical algorithm for generating smooth models from electromagnetic sounding data. Geophysics, 52, 289–300.

Davis, J.L. and Annan, A.P., 1989. Ground-penetrating radar for high-resolution mapping of soil and rock stratigraphy. Geophys. Prosp., 37, 531–551.

DeGroot-Hedlin, C. and Constable, S., 1990. Occam's inversion to generate smooth, two-dimensional models from magnetotelluric data. Geophysics, 55, 1613–1624.

LaBrecque, D.J., Morelli, G., Daily, W.D., Ramirez, A. and Lundegard, P., 1995. Occam's inversion of 3-D ERT data. In: Proceedings of the International Symposium on Three-Dimensional Electromagnetics. Schlumberger-Doll Research Center, Ridgefield, CT, pp. 471–482.

Morelli, G. and LaBrecque, D.J., 1996. Robust scheme for ERT inverse modeling. In: Proceedings of the Symposium on the Application of Geophysics to Environmental and Engineering Problems (SAGEEP), Denver, CO, pp. 629–638.

Topp, G.C., Davis, J.L. and Annan, A.P., 1980. Electromagnetic determination of soil water content: Measurements in coaxial transmission lines. Water Resour. Res., 16(3), 574–582.

Waxman, M.H. and Smits, L.J.M., 1968. Electrical conductivities in oil-bearing shaley sands. Soc. Pet. Eng. J., Trans. AIME, 243, 107–122.

Chapter 16

POLARIMETRIC BOREHOLE RADAR APPROACH TO FRACTURE CLASSIFICATION

Motoyuki Sato and Moriyasu Takeshita

Tohoku University, Center for Northeast Asian Studies, Sendai 980-8576, Japan

Abstract: In order to improve the ability of borehole radar, we propose a use of radar polarimetry technology. Scattering from subsurface fractures having a rough surface causes cross-polarized components of the radar field. A 3-D FDTD algorithm was used to simulate the scattering measurement by borehole radar and showed that the degree of depolarization is related to the surface roughness. In order to demonstrate the potential of polarimetric borehole radar, field measurement was carried out at the Mirror Lake fractured-rock research site (NH, USA). We observed many clear reflections from fractures in each polarization state. Even in the raw data, we could find the difference in the radar profile for different polarizations. Polarimetric features of the acquired radar signal were analyzed and it was found that the depolarization effect is independent in each fracture. We propose a technique of classifying fractures by using the energy of the scattering matrix measured by polarimetric borehole radar and show that this information could be used to evaluate water permeability of subsurface fractures.

1. INTRODUCTION

Borehole radar is a developing subsurface sensing technique, which is especially useful for detection of water flow in crystalline rock (Nickel et al., 1983; Wright et al., 1984; Olsson et al., 1992). However, in order to achieve a larger penetration depth, most borehole radar systems operate at the frequencies of 10 MHz–100 MHz, which is relatively low compared to the ground penetrating radar (GPR) used on the ground surface, and this results in a poor radar resolution. In many borehole radar applications, the characterization of subsurface fractures is quite important. Especially, we would like to know the water permeability of each fracture, which is very important in engineering applications. However, it is difficult to determine by conventional borehole radar, because it can only estimate the location and orientation of the fractures.

In order to acquire the information about physical properties of subsurface fractures, we have proposed a polarimetric borehole radar approach for fracture classification (Sato et al., 1995). Radar polarimetry is a new technology which has been intensively tested in airborne and space-borne remote sensing (Ulaby and Elachi, 1990; Mott, 1992). By utilizing the polarization information contained in the radar reflected signal, we can determine more about the radar target. If we apply the radar polarimetry to borehole radar, we believe we may get more detailed information about the subsurface fracture. This paper aims at demonstrating the potential of polarimetric borehole radar by computer simulation and field experiment.

The finite-difference time-domain (FDTD) method is now widely accepted as a numerical simulation technique for electromagnetic scattering. Recently, FDTD has often been used for simulation of ground penetrating radar (GPR) signal (Bourgeois and Smith, 1996). We use a three-dimensional FDTD algorithm for simulation of electromagnetic scattering from a subsurface fracture model having a rough surface. The depolarized component of the scattered wave plays an important role and we propose a technique to evaluate the surface roughness of fractures. To test our technique with real-world data, field measurement was carried out at the Mirror Lake fractured-rock research site. We describe the result from the field measurement and apply the polarimetric borehole radar technique for classification of subsurface fractures.

2. RADAR POLARIMETRY FOR BOREHOLE RADAR

Radar polarimetry is a technique which utilizes the polarization information in the reflected wave from radar targets. Most conventional radar systems, including GPR and borehole radar, have used travel time and reflected power to estimate the position and size of the radar targets. In most cases, these parameters provide good information about the location of the subsurface targets, such as geological formations and subsurface fractures. However, they do not provide us physical parameters of fractures, which are very important in many applications. If we have to know the aperture size, surface shape and mineral content of the fracture, we have to use a radar system, which has a resolving scale higher than the radar target. Borehole radar uses a wavelength of 50 cm–1 m, but its aperture size is sometimes less than 1 mm. In order to increase the radar resolution, we require wideband frequencies, and particular higher frequencies. However, due to the strong attenuation in geological material, the use of high frequency in subsurface targets is limited.

If we use radar polarimetry, we can obtain more information about the radar target, even when radar resolution is poor compared to the size of the radar target. For instance, when we use a linearly polarized wave, the reflection from a linear conductor is polarized in the direction of the conductor. Therefore we can estimate the orientation of the conductor from the polarization angle, where we measure the maximum reflection strength. On the other hand, reflection strength from a flat surface is independent of the polarization angle of the incident wave. Therefore, if we rotate the polarization angle and measure the reflection intensity, we can distinguish a linear conductor from a flat surface. More generally, if we can acquire full information of the polarization state, we can obtain information about the shape and properties of the radar target even if the radar resolution is poor (Ulaby and Elachi, 1990; Mott, 1992).

The authors have proposed in the past to utilize radar polarimetry in borehole radar in order to estimate the subsurface fracture properties (Sato et al., 1995). Subsurface fractures have a rough surface and the surface shape or surface roughness is dependent on the mechanism which created each fracture. Therefore, normally we can classify

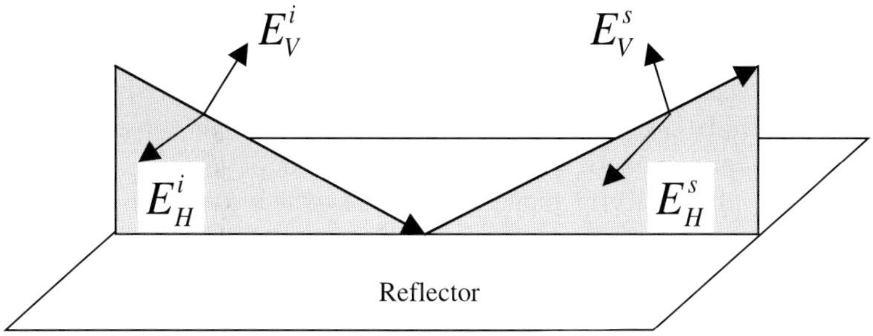

Figure 1. Definition of incident and reflected waves and a reflector for a scattering matrix.

subsurface fractures at a specified site into several clusters. The aperture of each fracture ranges from sub 1 mm to a few meters. Even if the aperture size is less than 1 mm it can be a significant water permeability path, especially for impermeable rock such as granite. When we observe the subsurface fracture in a borehole, for instance by an optical camera or an ultrasonic imaging tool, normally we cannot discriminate a permeable fracture from impermeable fracture. Borehole radar detects the contrast of electrical resistivity and dielectric constant of fractures. Therefore borehole radar can locate fractures, but it does not give us sufficient information to determine the water permeability of the fracture. However, we think, if we measure the shape of the fracture, we can estimate the permeability of the fracture, because an aperture size and a surface roughness of fractures are closely related to the water permeability. It should be noted that the presence of alternation or gouge will still complicate the influence of permeability even if we resolve the roughness.

Theoretically, a flat surface does not cause a cross-polarized component from the incident wave. When the surface of fractures has a roughness, the reflected wave contains the cross-polarized component due to complex mechanisms such as multiple scattering. We think that the strength of the cross-polarized reflected wave is related to the degree of the surface roughness, and inhomogeneity inside the aperture of the subsurface fracture. In other words volume scattering effects will be strong from fractured zones.

Polarimetric status of scattering from a radar target can be generally expressed by using a scattering matrix **S** as follows (Mott, 1992):

$$\begin{pmatrix} E_{\mathrm{H}}^{\mathrm{s}} \\ E_{\mathrm{V}}^{\mathrm{s}} \end{pmatrix} = \mathbf{S} \begin{pmatrix} E_{\mathrm{H}}^{\mathrm{i}} \\ E_{\mathrm{V}}^{\mathrm{i}} \end{pmatrix} = \begin{pmatrix} S_{\mathrm{HH}} & S_{\mathrm{HV}} \\ S_{\mathrm{VH}} & S_{\mathrm{VV}} \end{pmatrix} \begin{pmatrix} E_{\mathrm{H}}^{\mathrm{i}} \\ E_{\mathrm{V}}^{\mathrm{i}} \end{pmatrix} \tag{16.1}$$

where the subscripts 'V' and 'H' denote the vertical and horizontal polarizations, respectively, and the superscripts 'i' and 's' denote the incident and the scattered field, respectively. Figure 1 shows the definition of the incident and reflected waves and the reflector. The scattering matrix can describe polarization scattering properties of a target.

3. NUMERICAL SIMULATION

In this section, we demonstrate simulation examples of a scattered wave from a thin layer having a rough surface. We used a FDTD technique for the 3-dimensional simulation of an electromagnetic wave (Kunz and Luebbers, 1993).

Figure 2 shows the 3-D grid model for FDTD, which we used in the simulation. The transmitter is a short electrical dipole and the receiver position is 1.5 m separated from the transmitter position, which models the actual borehole radar. PML is used for the absorbing boundary condition (Kunz and Luebbers, 1993). In this analysis, all the grid cells have the same size of 15 cm per side, which is fine enough for subsurface fracture modeling in this study. If we need finer modeling of the structure, we would use a sub-grid technique to increase the accuracy of the simulation (Kunz and Luebbers, 1993). Figure 3 shows the pulse form used for the excitation, and its spectrum.

In order to create a numerical model of a rough surface, we use a rough surface having a Gaussian probability density function. This rough surface can be created by assuming a Gaussian power spectrum with a random phase in a 2-D spatial domain

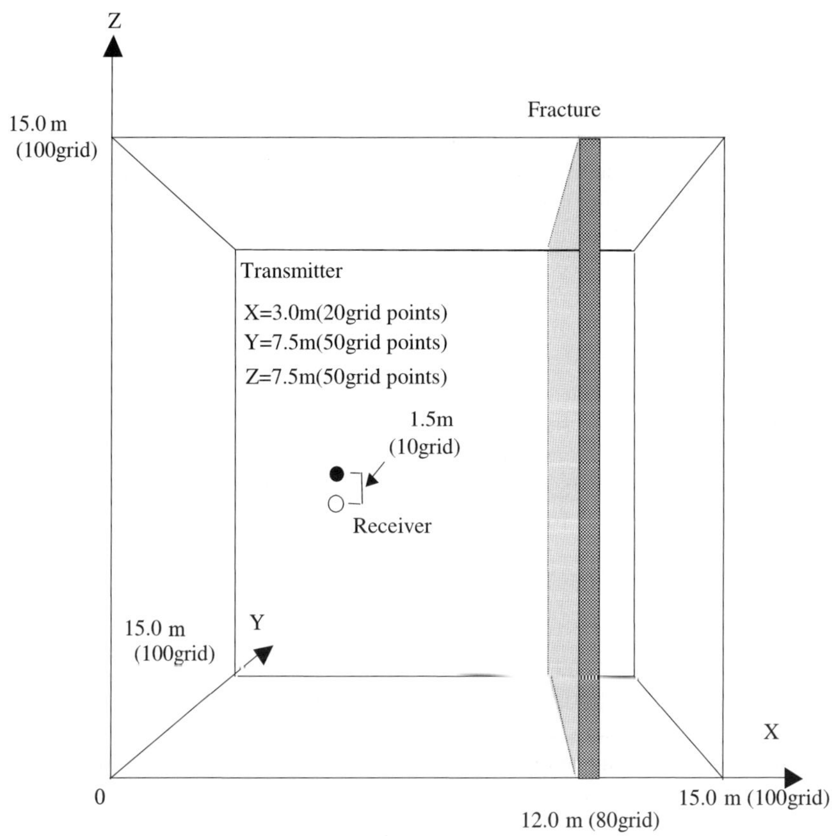

Figure 2. 3-D model space for FDTD simulation. Grid size 100 x 100 x 100 (15 m x 15 m x15 m).

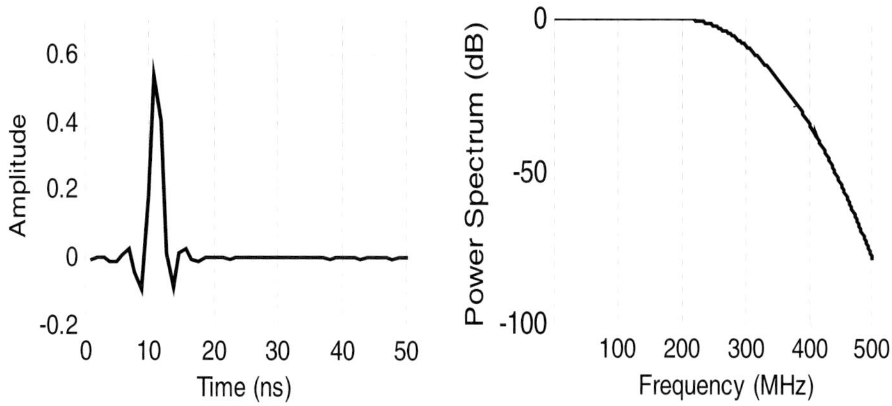

Figure 3. Excitation pulse form used in FDTD simulation: (a) pulse form, (b) spectrum.

and we then obtained the actual shape of the surface by inverse Fourier transformation (Figure 4). The roughness height of each surface position $z(x, y)$ is given by a 2-D FFT of the Gaussian power spectrum as (Phu et al., 1994):

$$z(x,y) = \frac{h}{\sqrt{f(x,y)^2}} f(x,y), \tag{16.2}$$

where

$$f(x,y) = \int_{-\infty}^{\infty}\int_{-\infty}^{\infty} F(k_x,k_y)\exp(jk_x x + jk_y y)\,\mathrm{d}k_x\mathrm{d}k_y \tag{16.3}$$

and

$$F(k_x,k_y) = \exp\left(-\left(\frac{\sqrt{(k_x-k_0)^2+(k_y-k_0)^2}}{\tau}\right)^2\right)\exp(j2\pi N). \tag{16.4}$$

Here, h is the mean roughness height of the subsurface fracture, k_x and k_y are spatial frequencies, and N is a random number used to generate random roughness. We used $k_0 = 0.05$ (1/m) and $\tau = 0.5$ (1/m), in the following models.

Figures 5 and 6 show examples of snap shots of electromagnetic scattering from a subsurface fracture obtained by 3-D FDTD simulation. The ratio of the dielectric constants of the surrounding material and the material inside the fracture is 9. Figure 5 shows scattering from a fracture having a smooth surface and Figure 6 shows that from a rough surface. In Figures 5 and 6, (a) shows the field parallel to the incident field and (b) shows the field orthogonal to the incident field. In the (a) parts of Figures 5 and 6, we can see an incident field to the fracture, which is propagating from the transmitter located near the center, and at the same time, a reflected wave from the fracture surface can be observed. A small part of the wave is transmitted through the fracture. In the (b) parts of Figures 5 and 6, we cannot see the incident field, because it has no signal component in the observing polarization component. No reflection occurs

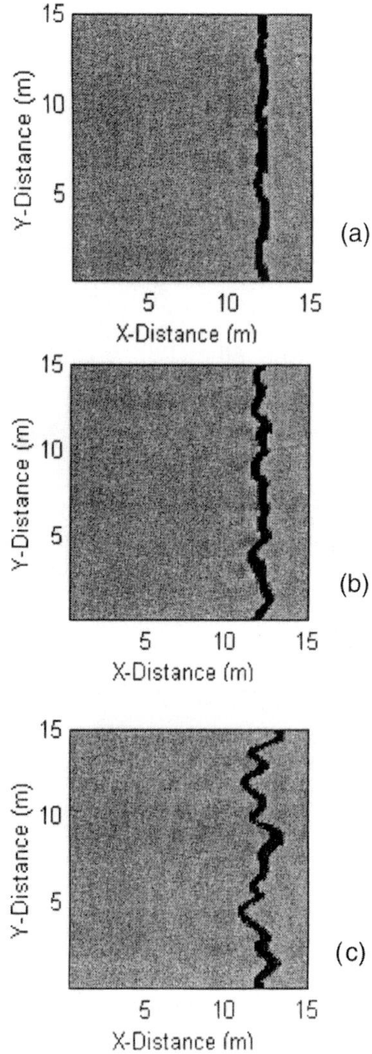

Figure 4. Realized subsurface fracture model. Mean roughness height: (a) 1 grid, (b) 3 grid, (c) 5 grid.

from the fracture in Figure 5b, because the reflected signal has the same polarization to the incident field. However in Figure 6b, we observe that the cross-polarized scattered wave occurs from a rough surface. Even if we measure the polarization states by radar, we cannot improve the radar resolution. However, radar polarization states contain much information about scattering mechanisms. If we can assume a model of fractures having a rough surface, we can determine the parameters such as mean roughness height.

If we assume an electrical dipole having y and z components for the excitation, and measure y and z field components for each excitation, we can numerically determine the four scattering matrix elements. We use this technique to evaluate the polarimetric

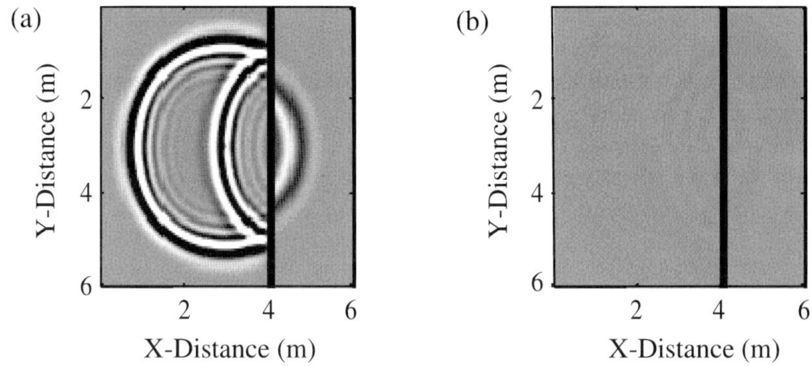

Figure 5. Scattering from a thin layer having a smooth surface: (a) co-polarization, (b) cross-polarization.

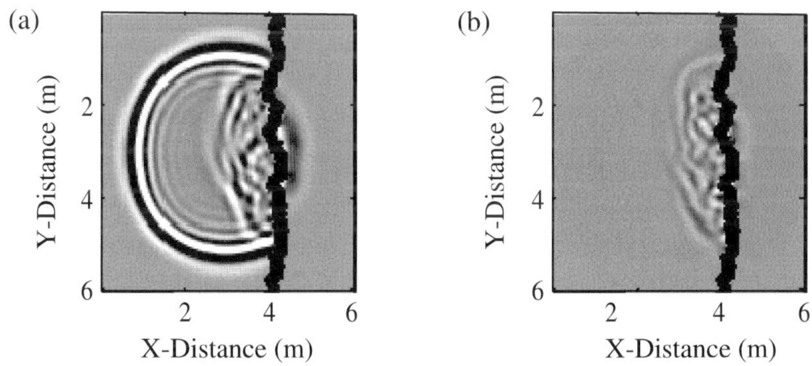

Figure 6. Scattering from a thin layer having a rough surface: (a) co-polarization, (b) cross-polarization.

properties of the scattered wave. Although FDTD gives the full-wave information of the scattered wave, we use the accumulated energy of the scattered wave P_{ij} for polarization evaluation. In the following discussion, we use the definition of the energy as:

$$P_{ij} = \int |S_{ij}|^2 \, dt, \quad \text{where } i, j = \text{H or V}. \tag{16.5}$$

In this subsurface fracture model, the surface roughness is related to a mean roughness height h. Figure 7 shows an example of energy of the scattering matrix element of subsurface fractures having a different surface roughness. Each generated subsurface fracture model has different scattering properties. The energy of each scattering matrix element is normalized by that of VV polarization. Therefore, we took an average of the energy scattering matrixes simulation for nine differently generated fracture models. The relative dielectric constants are 1 and 9 in a surrounding space and a subsurface fracture, respectively. From Figure 7, we can observe a linear relationship between the strength of the cross-polarized component of the scattered wave and the surface roughness of the fracture up to 0.2λ. The subsurface fractures that cause a stronger cross-polarized component can be estimated to have a rougher surface. Also P_{VV} and P_{HH} are almost

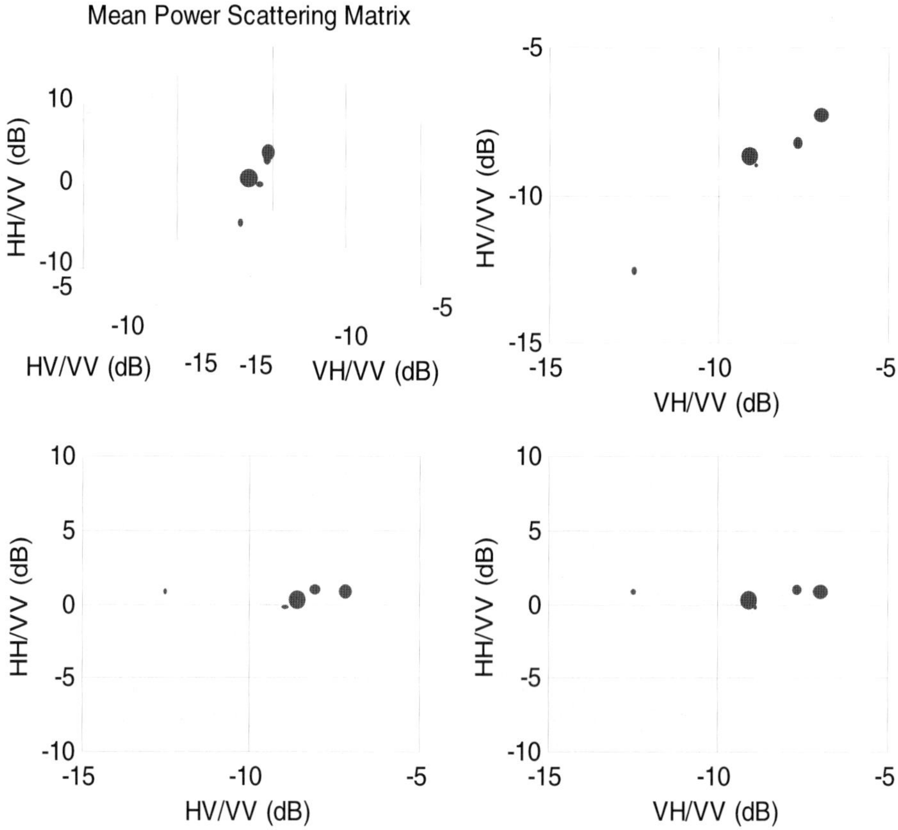

Figure 7. Energy of the scattering matrix elements of the rough fracture surface obtained by FDTD. The size of the plots indicates the mean surface roughness of 1, 2, 3, 4 and 5, respectively.

equal in this simulation, because the subsurface fracture is isotropic. From this result, we suggest that the surface roughness of a subsurface fracture can be evaluated by using the polarimetric borehole radar, if the surface roughness is not very high.

4. POLARIMETRIC BOREHOLE RADAR SYSTEM

The borehole diameter, which we use for radar measurement, is typically 10 cm or less. Most conventional borehole radar systems use dipole antennas for transmitter and receiver, because the thin structure is suitable to fit in a waterproof borehole radar sonde. In order to generate orthogonally polarized waves from a borehole radar sonde, we introduced an axial slot antenna on a conducting cylinder and developed a prototype polarimetric borehole radar system (Sato et al., 1995; Miwa et al., 1999). By combining dipole antennas and cylindrical slot antennas, we achieve full-polarization combinations. We repeat the measurement four times in the same borehole, by changing antenna sets

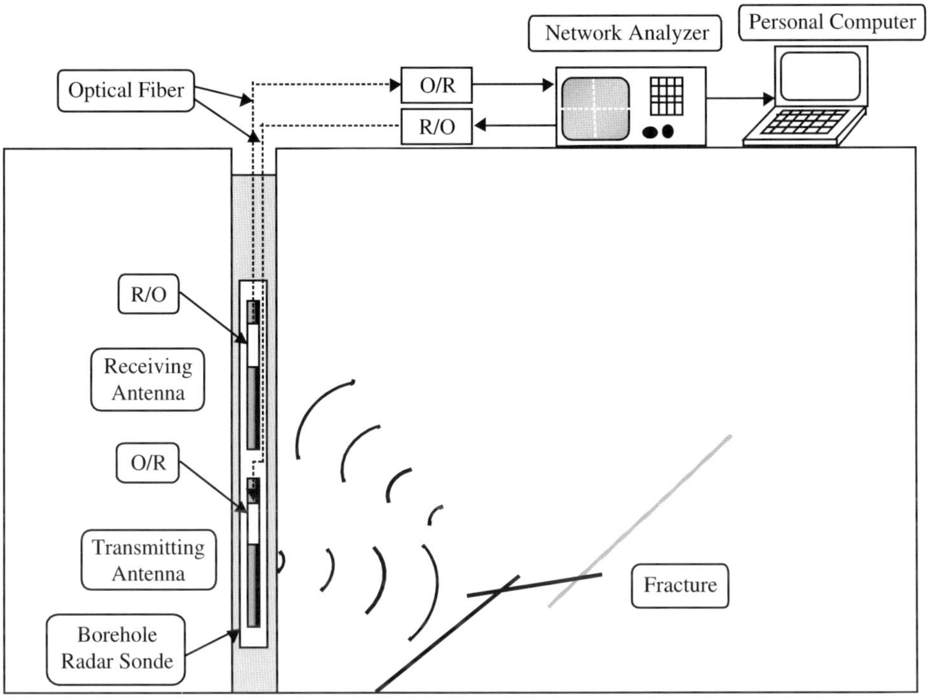

Figure 8. System diagram of the polarimetric borehole radar. R/O and O/R indicate RF and optical signal converters.

in order to acquire full-polarimetric radar signals. One problem of this system is the difference of antenna characteristics. We have developed a signal processing technique, which compensates the antenna characteristics by in situ transmitter–receiver coupling measured in boreholes (Miwa et al., 1999; Sato and Miwa, 2000).

The polarimetric borehole radar system which we have developed is a stepped-frequency radar system based on a network analyzer. The system diagram of the radar is shown in Figure 8. The network analyzer on the ground surface is connected to the antennas in a downhole sonde by an analog optical link. A nonmetallic optical fiber cable suspends the downhole radar sonde. The frequency bandwidth of the analog optical link is 1 MHz–500 MHz. Currently 100 m is the maximum depth of measurement by this system because of the optical fiber cable length. We have established that the optical fiber cable can be extended to 1000 m by laboratory test.

5. FIELD MEASUREMENT

We have tested our polarimetric borehole radar system in several test field sites. Here we show one of the field experiments. A joint research group of Tohoku University and U.S. Geological Survey carried out field measurements at the Mirror Lake fractured-rock research site (NH, USA) (Lane et al., 1996, 1998), during 14–19 October 1998.

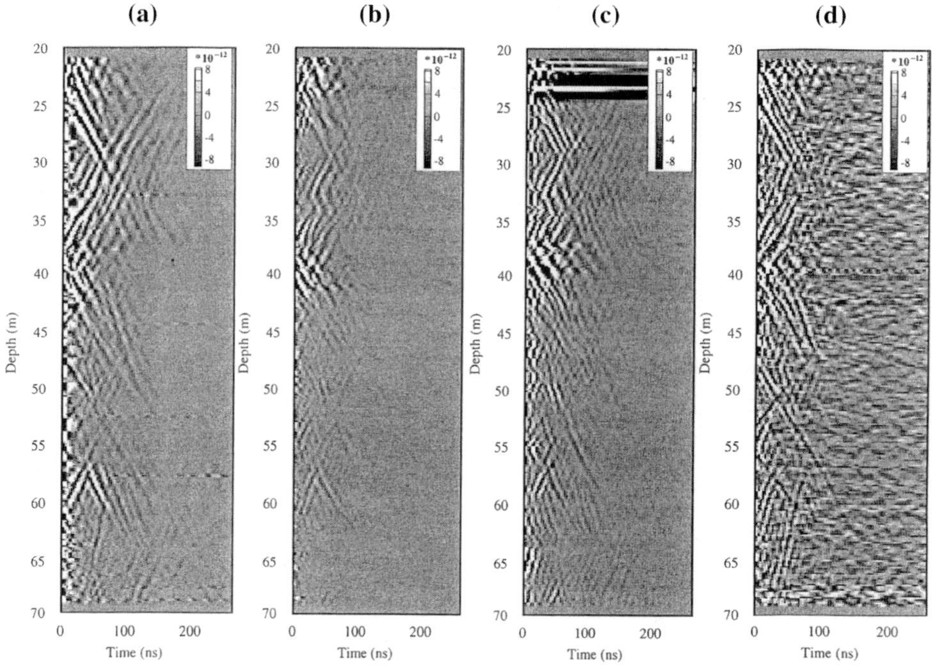

Figure 9. Polarimetric borehole radar profile: (a) VV, (b) VH, (c) HV, (d) HH.

Single-hole reflection measurements were conducted at boreholes FSE-3 and FSE-1, with an antenna separation of 1.6 m in all the antenna arrangements. The cross-hole arrangement was carried out between FSE-1 and FSE-3 for all the combinations of antennas. The separation of the FSE-1 and FSE-3 is 13 m. In all the measurements, frequency domain data were acquired between 2 MHz and 402 MHz at 2 MHz frequency intervals.

We can select the frequency bandwidth for the time-domain signal from the raw frequency-domain signal. We optimize the frequency bandwidth by observing the time-domain radar profile. In this study, the raw frequency spectrum between 50 MHz and 110 MHz was used and the time-domain radar signal was obtained by inverse Fourier transformation. The original time-domain radar signal contains a strong direct wave component and it obscures the reflected signal. An averaged signal at adjacent depths was removed from the original signal to enhance the reflection. After the reflection enhancement, the antenna compensation (Sato et al., 1998; Sato and Miwa, 2000) was applied to the four radar profiles by using the single-hole calibration technique.

Figure 9 shows the processed polarimetric borehole radar profile of FSE-1. Typical radar reflection from subsurface fractures is observed. The radar reflection from a subsurface fracture, which extends more than several meters can be approximated as a V-shape in the radar signature having an apex at the depth of cross point to the borehole. The borehole radar has an omni-directional antenna pattern and we cannot estimate the azimuth orientation of each fracture. The dielectric constant of the host rock is about

8 and the conductivity is less than 0.001 S/m. The maximum detectable range of this measurement was about 10 m.

6. DISCUSSION

From the radar profiles in Figure 9, we can observe that the reflection occurs at the same depth in all of the polarization states. The radar detectability is not strongly dependent on the polarization. Also, we can see that the radar resolution is almost the same in all the figures. However, we can see that the intensity of the reflection is different in each polarization state.

When the radar resolution is high enough, we can evaluate the characteristics of the fracture by using the intensity ratio of the reflection in each polarization, i.e., by using a scattering matrix. However, it is not easy to determine the scattering matrix from the measured polarimetric borehole radar profile because some reflections are always superimposed and separation of each reflection cannot be exactly achieved. Here, we tried to evaluate them by comparing the energy ratio of a reflection wave train from each fracture. In order to evaluate the energy from each fracture, we applied a gating window on the radar profile and picked up signal trains and evaluated the energy contained in the train. Figure 10 shows the energy ratio obtained by this technique. These values are

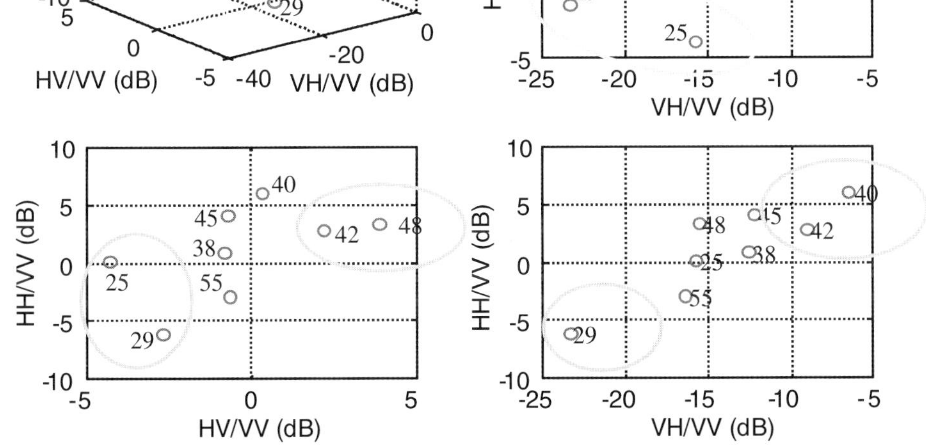

Figure 10. Energy of the scattering matrix elements of rough fracture surface obtained by polarimetric borehole radar measurement. Mirror Lake FSE-1.

equivalent to the P_{ij} defined by (16.5). In this figure, the numbers by each dot indicate the depth of the fracture. We think we can classify the fractures in this borehole into two clusters. The fractures in the same cluster are distributed close to each other. From the hydraulic tracer test in this borehole, we know that there is a permeable zone around 45 m. Therefore we estimate that one of the clusters that includes fractures at 40, 42 and 48 m will be water-permeable (Lane et al., 1996, 1998). This cluster has higher cross-polarization energy to the co-polarization, compared to other clusters. We interpret that these fractures have a rough surface and the aperture size is larger than the other fractures, which caused the high water permeability.

7. CONCLUSIONS

In this paper, we demonstrate the potential of the polarimetric borehole radar for subsurface fracture evaluation. Inverse scattering approaches will be very difficult for precise subsurface fracture estimation, because the size and shapes which have to be determined are much smaller than the wavelength of radar. However, modeling of polarimetric scattering to estimate the surface roughness is possible, as shown by FDTD simulation. We found a clear linearity between the depolarized components of the reflected wave to the surface roughness of the fracture. We believe that studying more examples of polarimetric scattering from geological structures will provide more efficient parameters to be determined in radar measurement for classification of fractures. FDTD and other numerical techniques along with actual radar measurement will improve these techniques.

Field measurement at Mirror Lake site shows that polarimetric borehole radar gives us more information about fracture characteristics than the conventional borehole radar by polarimetry. Classification techniques have been investigated, and show that the fractures which are water-permeable could be identified by the polarimetric borehole radar information.

ACKNOWLEDGEMENTS

We acknowledge John W. Lane Jr., Marc L. Buursink, Christopher J. Powers and Eric A. White for their support for this field measurement. This research was supported by the Japan Society for the Promotion of Science, Grant-in-Aid for International Scientific Research (A), 10044122 and for Scientific Research (B) 10450389, 1998–1999.

REFERENCES

Bourgeois, J. and Smith, G., 1996. A fully three-dimensional simulation of a ground-penetrating radar. FDTD theory compared with experiment. IEEE Trans. Geosci. Remote Sensing, 34(1), 36–44.

Kunz, K. and Luebbers, R., 1993. The Finite Difference Time Domain for Electromagnetics. CRC Press, Boca Raton, FL.

Lane, J.W., Haeni, F. and Placzek, G., 1996. Use of borehole-radar methods to detect a saline tracer in fractured crystalline bedrock at Mirror Lake, Grafton county, New Hampshire, USA. Proc. 6th Int. Conf. Ground Penetrating Radar, Sendai, pp. 185–190.

Lane, J.W., Haeni, F., Placzek, G. and Day-Lewis, F., 1998. Use of time-lapse attenuation-difference radar tomography methods to monitor saline tracer transport in fractured crystalline bedrock. Proc. 7th Int. Conf. Ground Penetrating Radar, Lawrence, KS, pp. 533–538.

Miwa, T., Sato, M. and Niitsuma, H., 1999. Subsurface fracture measurement with polarimetric borehole radar. IEEE Trans. Geosci. Remote Sensing, 37(2), 828–837.

Mott, H., 1992. Antennas for Radar and Communications. Wiley, New York.

Nickel, H., Sender, F., Thierbach, R. and Weichart, H., 1983. Exploring the interior of salt domes from boreholes. Geophys. Prospect., 31, 131–148.

Olsson, L., Falk, O., Forslund, L., Lundmark and Sandberg, F., 1992. Borehole radar applied to the characterization of hydraulically conductive fracture zones in crystalline rock. Geophys. Prospect., 40, 109–142.

Phu, P., Ishinaru, A. and Kuga, Y., 1994. Copolarized and cross-polarized enhanced backscattering from two-dimensional very rough surface at millimeter wave frequencies. Radio Sci., 29(5) 1275–1291.

Sato, M. and Miwa, T., 2000. Polarimetric borehole radar system for fracture measurement. Subsurf. Sensing Technol. Appl., 1(1), 161–175.

Sato, M., Ohkubo, T. and Niitsuma, H., 1995. Cross-polarization borehole radar measurements with a slot antenna. Appl. Geophys., 33(1–3), 53–61.

Sato, M., Takeshita, M., Miwa, T. and Niitsuma, H., 1998. Polarimetric borehole radar applied to geological exploration. Proc. 7th Int. Conf. Ground Penetrating Radar, Lawrence, KS, pp. 7–12.

Ulaby, F. and Elachi, C., 1990. Radar Polarimetry for Geoscience Applications. Artech House, Norwood, MA.

Wright, D.L., Watts, R.D. and Bramsoe, E., 1984. A short-pulse electromagnetic transponder for hole-to-hole use. IEEE Trans. Geosci. Remote Sensing, GE-22(6), 720–725.

AUTHOR INDEX

KEYWORD INDEX

The subjects in the index are mentioned in the chapter starting at the indicated page.